MEMS and
Microstructures
in Aerospace
Applications

Edited by

Robert Osiander

M. Ann Garrison Darrin

John L. Champion

Taylor & Francis
Taylor & Francis Group

Boca Raton   London   New York

A CRC title, part of the Taylor & Francis imprint, a member of the
Taylor & Francis Group, the academic division of T&F Informa plc.

Published in 2006 by
CRC Press
Taylor & Francis Group
6000 Broken Sound Parkway NW, Suite 300
Boca Raton, FL 33487-2742

International Standard Book Number-10: 0-8247-2637-5 (Hardcover)
International Standard Book Number-13: 978-0-8247-2637-9 (Hardcover)
Library of Congress Card Number 2005010800

Library of Congress Cataloging-in-Publication Data

Osiander, Robert.
    MEMS and microstructures in aerospace applications / Robert Osiander, M. Ann Garrison Darrin, John Champion.
        p. cm.
    ISBN 0-8247-2637-5
    1. Aeronautical instruments. 2. Aerospace engineering--Equipment and supplies. 3. Microelectromechanical systems. I. Darrin, M. Ann Garrison. II. Champion, John. III. Title.

TL589.O85 2005
629.135--dc22   2005010800

Taylor & Francis Group
is the Academic Division of T&F Informa plc.

Visit the Taylor & Francis Web site at
http://www.taylorandfrancis.com

and the CRC Press Web site at
http://www.crcpress.com

Preface

MEMS and Microstructures in Aerospace Applications is written from a programmatic requirements perspective. MEMS is an interdisciplinary field requiring knowledge in electronics, micromechanisms, processing, physics, fluidics, packaging, and materials, just to name a few of the skills. As a corollary, space missions require an even broader range of disciplines. It is for this broad group and especially for the system engineer that this book is written. The material is designed for the systems engineer, flight assurance manager, project lead, technologist, program management, subsystem leads and others, including the scientist searching for new instrumentation capabilities, as a practical guide to MEMS in aerospace applications. The objective of this book is to provide the reader with enough background and specific information to envision and support the insertion of MEMS in future flight missions. In order to nurture the vision of using MEMS in microspacecraft — or even in spacecraft — we try to give an overview of some of the applications of MEMS in space to date, as well as the different applications which have been developed so far to support space missions. Most of these applications are at low-technology readiness levels, and the expected next step is to develop space qualified hardware. However, the field is still lacking a heritage database to solicit prescriptive requirements for the next generation of MEMS demonstrations. (Some may argue that that is a benefit.) The second objective of this book is to provide guidelines and materials for the end user to draw upon to integrate and qualify MEMS devices and instruments for future space missions.

Editors

Robert Osiander received his Ph.D. at the Technical University in Munich, Germany, in 1991. Since then he has worked at JHU/APL's Research and Technology Development Center, where he became assistant supervisor for the sensor science group in 2003, and a member of the principal professional staff in 2004. Dr. Osiander's current research interests include microelectromechanical systems (MEMS), nanotechnology, and Terahertz imaging and technology for applications in sensors, communications, thermal control, and space. He is the principal investigator on "MEMS Shutters for Spacecraft Thermal Control," which is one of NASA's New Millenium Space Technology Missions, to be launched in 2005. Dr. Osiander has also developed a research program to develop carbon nanotube (CNT)-based thermal control coatings.

M. Ann Garrison Darrin is a member of the principal professional staff and is a program manager for the Research and Technology Development Center at The Johns Hopkins University Applied Physics Laboratory. She has over 20 years experience in both government (NASA, DoD) and private industry in particular with technology development, application, transfer, and insertion into space flight missions. She holds an M.S. in technology management and has authored several papers on technology insertion along with coauthoring several patents. Ms. Darrin was the division chief at NASA's GSFC for Electronic Parts, Packaging and Material Sciences from 1993 to 1998. She has extensive background in aerospace engineering management, microelectronics and semiconductors, packaging, and advanced miniaturization. Ms. Darrin co-chairs the MEMS Alliance of the Mid Atlantic.

John L. Champion is a program manager at The Johns Hopkins University Applied Physics Laboratory (JHU/APL) in the Research and Technology Development Center (RTDC). He received his Ph.D. from The Johns Hopkins University, Department of Materials Science, in 1996. Dr. Champion's research interests include design, fabrication, and characterization of MEMS systems for defense and space applications. He was involved in the development of the JHU/APL Lorentz force xylophone bar magnetometer and the design of the MEMS-based variable reflectivity concept for spacecraft thermal control. This collaboration with NASA–GSFC was selected as a demonstration technique on one of the three nanosatellites for the New Millennium Program's Space Technology-5 (ST5) mission. Dr. Champion's graduate research investigated thermally induced deformations in layered structures. He has published and presented numerous papers in his field.

Contributors

James J. Allen
Sandia National Laboratory
Albuquerque, New Mexico

Bradley G. Boone
The Johns Hopkins University Applied
  Physics Laboratory
Laurel, Maryland

Stephen P. Buchner
NASA Goddard Space Flight Center
Greenbelt, Maryland

Philip T. Chen
NASA Goddard Space Flight Center
Greenbelt, Maryland

M. Ann Garrison Darrin
The Johns Hopkins University Applied
  Physics Laboratory
Laurel, Maryland

Cornelius J. Dennehy
NASA Goddard Space Flight Center
Greenbelt, Maryland

Dawnielle Farrar
The Johns Hopkins University Applied
  Physics Laboratory
Laurel, Maryland

Samara L. Firebaugh
United States Naval Academy
Annapolis, Maryland

Thomas George
Jet Propulsion Laboratory
Pasadena, California

R. David Gerke
Jet Propulsion Laboratory
Pasadena, California

Brian Jamieson
NASA Goddard Space Flight Center
Greenbelt, Maryland

Robert Osiander
The Johns Hopkins University Applied
  Physics Laboratory
Laurel, Maryland

Robert Powers
Jet Propulsion Laboratory
Pasadena, California

Keith J. Rebello
The Johns Hopkins University Applied
  Physics Laboratory
Laurel, Maryland

Jochen Schein
Lawrence Livermore National
  Laboratory
Livermore, California

Theodore D. Swanson
NASA Goddard Space Flight Center
Greenbelt, Maryland

Danielle M. Wesolek
The Johns Hopkins University Applied
  Physics Laboratory
Laurel, Maryland

Acknowledgments

Without technology champions, the hurdles of uncertainty and risk vie with certainty and programmatic pressure to prevent new technology insertions in spacecraft. A key role for these champions is to prevent obstacles from bringing development and innovation to a sheer halt.

The editors have been fortunate to work with the New Millennium Program (NMP) Team for Space Technology 5 (ST5) at the NASA Goddard Space Flight Center (GSFC). In particular, Ted Swanson, as technology champion, and Donya Douglas, as technology leader, created an environment that balanced certainty, uncertainties, risks and pressures for ST5, micron-scale machines open and close to vary the emissivity on the surface of a microsatellite radiator. These "VARI-E" microelectromechanical systems (MEMS) are a result of collaboration between NASA, Sandia National Laboratories, and The Johns Hopkins University Applied Physics Laboratory (JHU/APL). Special thanks also to other NASA "tech champions" Matt Moran (Glenn Research Center) and Fred Herrera (GSFC) to name a few! Working with technology champions inspired us to realize the vast potential of "small" in space applications.

A debt of gratitude goes to our management team Dick Benson, Bill D'Amico, John Sommerer, and Joe Suter and to the Johns Hopkins University Applied Physics Laboratory for its support through the Janney Program. Our thanks are due to all the authors and reviewers, especially Phil Chen, NASA, in residency for a year at the laboratory. Thanks for sharing in the pain.

There is one person for whom we are indentured servants for life, Patricia M. Prettyman, whose skills and abilities were and are invaluable.

Contents

1 Overview of Microelectromechanical Systems and Microstructures in Aerospace Applications

Robert Osiander and M. Ann Garrison Darrin

CONTENTS

The machine does not isolate man from the great problems of nature but plunges him more deeply into them.
Saint-Exupéry, *Wind, Sand, and Stars,* **1939**

1.1 INTRODUCTION

To piece together a book on microelectromechanical systems (MEMS) and microstructures for aerospace applications is perhaps foolhardy as we are still in the infancy of micron-scale machines in space flight. To move from the infancy of a technology to maturity takes years and many awkward periods. For example, we did not truly attain the age of flight until the late 1940s, when flying became accessible to many individuals. The insertion or adoption period, from the infancy of flight, began with the Wright Brothers in 1903 and took more than 50 years until it was popularized. Similarly, the birth of MEMS began in 1969 with a resonant gate field-effect transistor designed by Westinghouse. During the next decade, manufacturers began using bulk-etched silicon wafers to produce pressure sensors, and experimentation continued into the early 1980s to create surface-micromachined polysilicon actuators that were used in disc drive heads. By the late 1980s, the potential of MEMS devices was embraced, and widespread design and implementation grew in the microelectronics and biomedical industries. In 25 years, MEMS moved from the technical curiosity realm to the commercial potential world. In the 1990s, the U.S. Government and relevant agencies had large-scale MEMS support and projects underway. The Air Force Office of Scientific Research (AFOSR) was supporting basic research in materials while the Defense Advanced Research Projects Agency (DARPA) initiated its foundry service in 1993. Additionally, the National Institute of Standards and Technology (NIST) began supporting commercial foundries.

In the late 1990s, early demonstrations of MEMS in aerospace applications began to be presented. Insertions have included Mighty Sat 1, Shuttle Orbiter STS-93, the DARPA-led consortium of the flight of OPAL, and the suborbital ride on Scorpius® (Microcosm). These early entry points will be discussed as a foundation for the next generation of MEMS in space. Several early applications emerged in the academic and amateur satellite fields. In less than a 10-year time frame, MEMS advanced to a full, regimented, space-grade technology. Quick insertion into aerospace systems from this point can be predicted to become widespread in the next 10 years.

This book is presented to assist in ushering in the next generation of MEMS that will be fully integrated into critical space-flight systems. It is designed to be used by the systems engineer presented with the ever-daunting task of assuring the mitigation of risk when inserting new technologies into space systems.

To return to the quote above from Saint Exupéry, the application of MEMS and microsystems to space travel takes us deeper into the realm of interactions with environments. Three environments to be specific: on Earth, at launch, and in orbit. Understanding the impacts of these environments on micron-scale devices is essential, and this topic is covered at length in order to present a springboard for future generations.

1.2   IMPLICATIONS OF MEMS AND MICROSYSTEMS IN AEROSPACE

The starting point for microengineering could be set, depending on the standards, sometime in the 15th century, when the first watchmakers started to make pocket watches, devices micromachined after their macroscopic counterparts. With the introduction of quartz for timekeeping purposes around 1960, watches became the first true MEMS device.

When we think of MEMS or micromachining, wrist and pocket watches do not necessarily come to our mind. While these devices often are a watchmaker's piece of art, they are a piece of their own, handcrafted in single numbers, none like the other. Today, one of the major aspects of MEMS and micromachining is batch processing, producing large numbers of devices with identical properties, at the same time assembled parallel in automatic processes. The introduction of microelectronics into watches has resulted in better watches costing a few dollars instead of a few thousand dollars, and similarly the introduction of silicon surface micromachining on the wafer level has reduced, for example, the price of an accelerometer, the integral part of any car's airbag, to a few dimes.

Spacecraft application of micromachined systems is different in the sense that batch production is not a requirement in the first place — many spacecraft and the applications are unique and only produced in a small number. Also, the price tag is often not based on the product, but more or less determined by the space qualification and integration into the spacecraft. Reliability is the main issue; there is typically only one spacecraft and it is supposed to work for an extended time without failure.

In addition, another aspect in technology development has changed over time. The race into space drove miniaturization, electronics, and other technologies. Many enabling technologies for space, similar to the development of small chronometers in the 15th and 16th centuries, allowed longitude determination, brought accurate navigation, and enabled exploration. MEMS (and we will use MEMS to refer to any micromachining technique) have had their success in the commercial industries — automotive and entertainment. There, the driver as in space is cost, and the only solution is mass production. Initially pressure sensors and later accelerometers for the airbag were the big successes for MEMS in the automotive industry which reduced cost to only a few dimes. In the entertainment industry, Texas Instruments' mirror array has about a 50% market share (the other devices used are liquid crystal-based electronic devices), and after an intense but short development has helped to make data projectors available for below $1000 now. One other MEMS application which revolutionized a field is uncooled IR detectors. Without sensitivity losses, MEMS technology has also reduced the price of this equipment by an order of magnitude, and allowed firefighters, police cars, and luxury cars to be equipped with previously unaffordable night vision. So the question is, what does micromachining and MEMS bring to space?

Key drivers of miniaturization of microelectronics are the reduced cost and mass production. These drivers combine with the current significant trend to integrate more and more components and subsystems into fewer and fewer chips, enabling increased functionality in ever-smaller packages. MEMS and other sensors and actuator technologies allow for the possibility of miniaturizing and integrating entire systems and platforms. This combination of reduced size, weight, and cost per unit with increased functionality has significant implications for Air Force missions, from global reach to situational awareness and to corollary civilian scientific and commercial based missions. Examples include the rapid low-cost global deployment of sensors, launch-on-demand tactical satellites, distributed sensor networks, and affordable unmanned aerial vehicles (UAVs). Collective arrays of satellites that function in a synchronized fashion promise significant new opportunities in capabilities and robustness of satellite systems. For example, the weight and size reduction in inertial measurement units (IMUs) composed of MEMS accelerometers and rate gyros, global positioning system (GPS) receivers for navigation and attitude determination, and MEMS-based microthruster systems are enablers for small spacecraft, probes, space robotics, nanosatellites, and small planetary landers.

The benefits include decreased parts count per spacecraft, increased functionality per unit spacecraft mass, and the ability to mass produce micro-, nano-, and picosatellites for launch-on-demand tactical applications (e.g., inspector spacecraft) and distributed space systems. Microlaunch vehicles enabled by micromachined subsystems and components such as MEMS liquid rocket engines, valves, gyros, and accelerometers could deliver 1 or 2 kg to low-Earth orbit. Thus, it will be possible to place a payload (albeit a small one) as well as fully functional micro-satellites into orbit for $10,000 to $50,000, rather than the $10 million to $50 million required today.[1]

In fact, researchers at the SouthWest Research Institute have performed extensive tests and determined that the vacuum of space produces an ideal environment for some applications using MEMS devices. MEMS devices processed in a vacuum for 10^{10} cycles had improved motion with decreased voltage.[2] MEMS devices for space applications will be developed and ultimately flown in optimized MEMS-based scientific instruments and spacecraft systems on future space missions.

1.3 MEMS IN SPACE

While many of the MEMS devices developed within the last decade could have applications for space systems, they were typically developed for the civilian or military market. Only a few devices such as micropropulsion and scientific instrumentation have had space application as a driving force from the beginning. In both directions, there have been early attempts in the 1990s to apply these devices to the space program and investigate their applicability. A sample of these demonstrations are listed herein and acknowledged for their important pathfinding roles.

> He who would travel happily must travel light.
> **Antoine de Saint-Exupéry**

1.3.1 DIGITAL MICRO-PROPULSION PROGRAM STS-93

The first flight recorded for a MEMS device was on July 23, 1999, on the NASA flight STS-93 with the Space Shuttle *Columbia*. It was launched at 12:31 a.m. with a duration of 4 days and carried a MEMS microthruster array into space for the first time. DARPA funded the TRW/Aerospace/Caltech MEMS Digital Micro-Propulsion Program which had two major goals: to demonstrate several types of MEMS microthrusters and characterize their performance, and to fly MEMS microthrusters in space and verify their performance during launch, flight, and landing.

1.3.2 PICOSATELLITE MISSION

Six picosatellites, part of the payload on OPAL, were launched on January 26, 2000 at Vandenberg Air Force Base. The picosatellites were deployed on February 4, 2000 and performed for 6 days until February 10, 2000, when the batteries were drained. Rockwell Science Center (RSC) designed and implemented a MEMS-based radio frequency switch experiment, which was integrated into the miniature satellite (picosat) as an initial demonstration of MEMS for space applications. This effort was supported by DARPA Microsystems Technology Office (MTO), and the mission was conducted with Aerospace Corporation and Stanford University as partners. MEMS surface-micromachined metal contacting switches were manufactured and used in a simple experiment aboard the miniature satellites to study the device behavior in space, and its feasibility for space applications in general. During the entire orbiting period, information was collected on both the communications and networking protocols and MEMS RF switch experiments. The performance of RF switches has been identical to their performance before the launch.[3]

1.3.3 SCORPIUS SUB-ORBITAL DEMONSTRATION

A microthruster array measuring one fourth the size of a penny, designed by a TRW-led team for use on micro-, nano- and picosatellites, has successfully demonstrated its functionality in a live fire test aboard a Scorpius® sub-orbital sounding rocket built by Microcosm on March 9, 2000. Individual MEMS thrusters, each a poppy seed-sized cell fueled with lead styphnate propellant, fired more than 20 times at 1-sec intervals during the test staged at the White Sands Missile Range. Each thruster delivered 10^{-4} newton sec of impulse.[4]

1.3.4 MEPSI

The series of MEMS-based Pico Sat Inspector (MEPSI) space flight experiments demonstrated the capability to store a miniature (less than 1 kg) inspector (PICOSAT) agent that could be released upon command to conduct surveillance of the host spacecraft and share collected data with a dedicated ground station. The DoD has approved a series of spiral development flights (preflights) leading up to a final flight that will perform the full MEPSI mission. The first iteration of the MEPSI PICOSAT was built and flown on STS-113 mission in December 2002.

All MEPSI PICOSATs are 4 × 4 × 5 in. cube-shaped satellites launched in tethered pairs from a special PICOSAT launcher that is installed on the Space Shuttle, an expandable launch vehicle (ELV) or a host satellite. The launcher that will be used for STS/PICO2 was qualified for shuttle flight during the STS-113 mission and will not need to be requalified.[5]

1.3.5   MISSILES AND MUNITIONS — INERTIAL MEASUREMENT UNITS

On June 17, 2002, the success of the first MEMS-based inertial measurement units (IMU) guided flight test for the Army's NetFires Precision Attack Missile (PAM) program served as a significant milestone reached in the joint ManTech program's efforts to produce a smaller, lower cost, higher accuracy, tactical grade MEMS-based IMU. During the 75 sec flight, the PAM flew to an altitude of approximately 20,000 ft and successfully executed a number of test maneuvers using the navigation unit that consisted of the HG-1900 (MEMS-based) IMU integrated with a GPS receiver. The demonstration also succeeded in updating the missile's guidance point in midflight, resulting in a successful intercept.[6]

1.3.6   OPAL, SAPPHIRE, AND Emerald

Satellite Quick Research Testbed (SQUIRT) satellite projects at Stanford University demonstrate micro- and nanotechnologies for space applications. SAPPHIRE is a testbed for MEMS tunneling infrared horizon detectors. The second microsatellite, OPAL, is named after its primary mission as an Orbiting Picosatellite Launcher. OPAL explores the possibilities of the mothership–daughtership mission architecture using the SQUIRT bus to eject palm-sized, fully functional picosatellites. OPAL also provides a testbed for on-orbit characterization of MEMS accelerometers, while one of the picosatellites is a testbed for MEMS RF switches. Emerald is the upcoming SQUIRT project involving two microsatellites, which will demonstrate a virtual bus technology that can benefit directly from MEMS technology. Its payloads will also include a testbed dedicated to comprehensive electronic and small-scale component testing in the space environment. Emerald will also fly a colloid microthruster prototype, a first step into the miniaturization of thruster subsystems that will eventually include MEMS technology. The thruster is being developed jointly with the Plasma Dynamic Laboratory at Stanford University.[7–9]

1.3.7   INTERNATIONAL EXAMPLES

It would truly be unfair after listing a series of United States originated demonstrations to imply that this activity was limited to the U.S. On the international field, there is significant interest, effort, and expertise. The European Space Agency (ESA)[10,11] and Centre National d'Etudes Spatiales (CNES)[12] have significant activity. Efforts in Canada at the University of Victoria[13] include MEMS adaptive optics for telescopes. In China, it is being experimented with "Yam-Sat" and on silicon satellites,[14] while work in Japan includes micropropulsion[15] and other activities too numerous to include herein. Many of these efforts cross national boundaries and are large collaborations.

1.4   MICROELECTROMECHANICAL SYSTEMS AND MICROSTRUCTURES IN AEROSPACE APPLICATIONS

MEMS and Microstructures in Aerospace Applications is loosely divided into the following four sections:

1.4.1 An Understanding of MEMS and the MEMS Vision

It is exciting to contemplate the various space mission applications that MEMS technology could possibly enable in the next 10–20 years. The two primary objectives of Chapter 2 are to both stimulate ideas for MEMS technology infusion on future NASA space missions and to spur adoption of the MEMS technology in the minds of mission designers. This chapter is also intended to inform non-space-oriented MEMS technologists, researchers, and decision makers about the rich potential application set that future NASA Science and Exploration missions will provide. The motivation for this chapter is therefore to lead the reader to identify and consider potential long-term, perhaps disruptive or revolutionary, impacts that MEMS technology may have for future civilian space applications. A general discussion of the potential of MEMS in space applications is followed by a brief showcasing of a few selected examples of recent MEMS technology developments for future space missions. Using these recent developments as a point of departure, a vision is then presented of several areas where MEMS technology might eventually be exploited in future science and exploration mission applications. Lastly, as a stimulus for future research and development, this chapter summarizes a set of barriers to progress, design challenges, and key issues that must be overcome for the community to move on from the current nascent phase of developing and infusing MEMS technology into space missions, in order to achieve its full potential.

Chapter 3 discusses the fundamentals of the three categories of MEMS fabrication processes. Bulk micromachining, sacrificial surface micromachining, and LIGA have differing capabilities that include the achievable device aspect ratio, materials, complexity, and the ability to integrate with microelectronics. These differing capabilities enable their application to a range of devices. Commercially successful MEMS devices include pressure sensors, accelerometers, gyroscopes, and ink-jet nozzles. Two notable commercial successes include the Texas Instruments Digital Mirror Device (DMD®) and the Analog Devices ADXL® accelerometers and gyroscopes. The paths for the integration of MEMS as well as some of the advanced materials that are being developed for MEMS applications are discussed.

Chapter 4 discusses the space environment and its effects upon the design, including material selection and manufacturing controls for MEMS. It provides a cursory overview of the thermal, mechanical, and chemical effects that may impact the long-term reliability of the MEMS devices, and reviews the storage and application conditions that the devices will encounter. Space-mission environmental influences, radiation, zero gravity, zero pressure, plasma, and atomic oxygen and their potential concerns for MEMS designs and materials selection are discussed. Long-life requirements are included as well. Finally, with an understanding of the concerns unique to hardware for space environment operation, materials selection is included. The user is cautioned that this chapter is barely an introduction, and should be used in conjunction with the sections of this book covering reliability, packaging, contamination, and handling concerns.

An entire chapter, Chapter 5, deals with radiation-induced performance degradation of MEMS. It begins with a discussion on the space radiation environment encountered in any space mission. The radiation environment relevant to MEMS consists primarily of energetic particles that originate in either the sun (solar particles) or in deep space (cosmic rays). Spatial and temporal variations in the particle densities are described, together with the spectral distribution. This is followed by a detailed discussion on the mechanisms responsible for radiation damage that give rise to total ionizing dose, displacement damage dose, and single event effects. The background information serves as a basis for understanding the radiation degradation of specific MEMS, including accelerometers, microengines, digital mirror devices, and RF relays. The chapter concludes by suggesting some approaches for mitigating the effects of radiation damage.

1.4.2 MEMS IN SPACE SYSTEMS AND INSTRUMENTATION

Over the past two decades, micro- or nanoelectromechanical systems (MEMS and NEMS) and other micronanotechnologies (MNT) have become the subjects of active research and development in a broad spectrum of academic and industrial settings. From a space systems perspective, these technologies promise exactly what space applications need, that is, high-capability devices and systems with low mass and low power consumption. Yet, very few of these technologies have been flown or are currently in the process of development for flight. Chapter 6 examines some of the underlying reasons for the relatively limited infusion of these exciting technologies in space applications. A few case studies of the "success stories" are considered. Finally, mechanisms for rapidly and cost-effectively overcoming the barriers to infusion of new technologies are suggested. As evidenced by the numerous MNT-based devices and systems described in this and other chapters of this book, one is essentially limited only by one's imagination in terms of the diversity of space applications, and consequently, the types of MNT-based components and systems that could be developed for these applications. Although most MNT concepts have had their birthplace in silicon-integrated circuit technology, the field has very rapidly expanded into a multidisciplinary arena, exploiting novel physical, chemical, and biological phenomena, and utilizing a broad and diverse range of materials systems.

Chapter 7 discusses science instrumentation applications for microtechnologies. The size and weight reduction offered by micromachining approaches has multiple insertion points in the development of spacecraft science instrumentation. The use of MEMS technology is particularly attractive where it provides avenues for the reduction of mission cost without the sacrifice of mission capability. Smaller instruments, such as nuclear magnetic resonance MEMS probes to investigate environmental conditions, can essentially reduce the weight and size of planetary landers, and thereby reduce launch costs. MEMS technology can generate new capabilities such as the multiple object spectrometers developed for the James Webb Space Telescope, which is based on MEMS shutter arrays. New missions can be envisioned that use a large number of small satellites with micromachined instruments, magnetometers or plasma spectrometers to map, for example, the spatial and temporal magnetic field distribution (MagConn). A number of science instruments will be discussed, where the application of MEMS technologies will provide new capabilities, performance improvement, or a reduction in size and weight without performance sacrifice.

1.4.3 MEMS in Satellite Subsystems

The topic area of MEMS in satellite subsystems covers communication, guidance, navigation and control, and thermal and micropropulsion. Chapter 8 reviews MEMS devices and their applicability in spacecraft communication. One of the most exciting applications of MEMS for microwave communications in spacecraft concerns the implementation of "active aperture phase array antennas." These systems consist of groups of antennas phase-shifted from each other to take advantage of constructive and destructive interference in order to achieve high directionality. Such systems allow for electronically steered, radiated, and received beams which have greater agility and will not interfere with the satellite's attitude. Such phase array antennas have been implemented with solid-state components; however, these systems are power-hungry and have large insertion losses and problems with linearity. In contrast, phase shifters implemented with microelectromechanical switches have lower insertion loss and require less power. This makes MEMS an enabling technology for lightweight, low-power, electronically steerable antennas for small satellites. A very different application is the use of microoptoelectromechanical systems (MOEMS) such as steerable micromirror arrays for space applications. Suddenly, high transfer rates in optical systems can be combined with the agility of such systems and allow optical communications with full pointing control capabilities. While this technology has been developed during the telecom boom in the early 2000s, it is in its infancy in space application. The chapter discusses a number of performance tests and applications.

Thermal control systems are an integral part of all spacecraft and instrumentation, and they maintain the spacecraft temperature within operational temperature boundaries. For small satellite systems with reduced thermal mass, reduced surface and limited power, new approaches are required to enable active thermal control using thermal switches and actively controlled thermal louvers. MEMS promises to offer a solution with low power consumption, low size, and weight as required for small satellites. Examples discussed in Chapter 9 are the thermal control shutters on NASA's ST5 New Millennium Program, thermal switch approaches, and applications of MEMS in heat exchangers. Active thermal control systems give the thermal engineer the flexibility required when multiple identical satellites are developed for different mission profiles with a reduced development time.

Chapter 10 discusses the use of MEMS-based microsystems to the problems and challenges of future spacecraft guidance, navigation, and control (GN&C) mission applications. Potential ways in which MEMS technology can be exploited to perform GN&C attitude sensing and control functions are highlighted, in particular, for microsatellite missions where volume, mass, and power requirements cannot be satisfied with conventional spacecraft component technology. A general discussion on the potential of MEMS-based microsystems for GN&C space applications is presented, including the use of embedded MEMS gyroscopes and accelerometers in modular multifunction GN&C systems that are highly integrated, compact, and at low power and mass. Further, MEMS technology applied to attitude sensing and control actuation functions is discussed with brief descriptions of several selected examples of specific recent MEMS technology developments for GN&C applications. The chapter concludes with an overview of future insertion points of MEMS GN&C applications in space systems.

The different micropropulsion systems, which are divided into the two major groups of electric and chemical propulsion, are discussed in Chapter 11. Each propulsion system is discussed with respect to its principle of operation, its current state-of-the-art, and its MEMS or micromachined realization or potential thereof. It is shown that the number of pure MEMS propulsion devices is limited, and that there are still significant challenges ahead for other technologies to make the leap. The major challenge to produce a MEMS-based propulsion system including control, propellant, and thruster is in the miniaturization of all components combined.

1.4.4 TECHNICAL INSERTION OF MEMS IN AEROSPACE APPLICATIONS

The last section of the book is in one aspect different from the previous sections; it cannot be based on historical data. Even with the number of MEMS devices flown on the shuttle in some experiments, there has not been a sincere attempt to develop requirements for the space qualification of MEMS devices. Most of the authors in this section have been involved in the development of the MEMS thermal control shutters for the ST5 space mission, and have tried to convey this experience in these chapters, hoping to create a basic understanding of the complexities while dealing with MEMS devices and the difference to well understood integration of microelectronics.

At some point, every element is a packaging issue. In order to achieve high performance or reliability of MEMS for space applications, the importance of MEMS packaging must be recognized. Packaging is introduced in Chapter 12 as a vital part of the design of the device and the system that must be considered early in the product design, and not as an afterthought. Since the evolution of MEMS packaging stems from the integrated circuit industry, it is not surprising that some of these factors are shared between the two. However, many are specific to the application, as will be shown later. A notable difference between a MEMS package and an electronics package in the microelectronics industry is that a MEMS package provides a window to the outside world to allow for interaction with its environment. Furthermore, MEMS packaging must account for a more complex set of parameters than what is typically considered in the microelectronics industry, especially given the harsh nature of the space and launch environments.

Chapter 13 is entirely devoted to handling and contamination controls for MEMS in space applications due to the importance of the topic area to final mission success. Handling and contamination control is discussed relative to the full life cycle from the very basic wafer level processing phase to the orbit deployment phase. MEMS packaging will drive the need to tailor the handling and contamination control plans in order to assure adequacy of the overall program on a program-by-program basis. Plan elements are discussed at length to assist the user in preparing and implementing effective plans for both handling and contamination control to prevent deleterious effects.

The space environment provides for a number of material challenges for MEMS devices, which will be discussed in Chapter 14. This chapter addresses both the known failure mechanisms such as stiction, creep, fatigue, fracture, and material incompatibility induced in the space environment. Environmentally induced stresses such as shock and vibration, humidity (primarily terrestrial), radiation, electrical stresses and thermal are reviewed along with the potential for combinations of stress factors. The chapter provides an overview on design and material precautions to overcome some of these concerns.

Chapter 15 begins with a discussion on several approaches for assessing the reliability of MEMS for space flight applications. Reliability for MEMS is a developing field and the lack of a historical database is truly a barrier to the insertion of MEMS in aerospace applications. The use of traditional statistically derived reliability approaches from the microelectronic military specification arena and the use of physics of failure techniques, are introduced.

Chapter 16 on "Quality Assurance Requirements, Manufacturing and Test" addresses the concerns of the lack of historical data and well-defined test methodologies to be applied for assuring final performance for the emerging MEMS in space. The well-defined military and aerospace microcircuit world forms the basis for assurance requirements for microelectromechanical devices. This microcircuit base, with its well-defined specifications and standards, is supplemented with MEMS-specific testing along with the end item application testing as close to a relevant environment as possible. The objective of this chapter is to provide a guideline for the user rather than a prescription; that is, each individual application will need tailored assurance requirements to meet the needs associated with each unique situation.

1.5 CONCLUSION

Within the next few years, there will be numerous demonstrations of MEMS and microstructures in space applications. MEMS developments tend to look more like the growth of the Internet rather than the functionality growth seen in microcircuits and quantified by Moore's law. Custom devices in new applications will be found and will be placed in orbit. As shown in this overview, many of the journeys of MEMS into space, to date, have been of university or academic grade, and have yet to find their way into critical embedded systems. This book may be premature as it is not written on a vast basis of knowledge gleaned from the heritage flights for MEMS and microstructures. However, it is hoped that this work will help prepare the way for the next generation of MEMS and microsystems in space.

As for the future, your task is not to foresee it, but to enable it.
Antoine de Saint-Exupéry, *The Wisdom of the Sands*

REFERENCES

1. *Implications of Emerging Micro- and Nanotechnologies Committee on Implications of Emerging Micro- and Nanotechnologies.* Air Force Science and Technology Board Division on Engineering and Physical Sciences, 2002.
2. McComas, D.J., et al., Space applications of microelectromechanical systems: Southwest Research Institute® vacuum microprobe facility and initial vacuum test results. *Review of Scientific Instruments*, 74, (8), 3874–3878, 2003.
3. Yao, J.J., et al., Microelectromechanical system radio frequency switches in a picosatellite mission. *Smart Materials and Structures*, 10, (6), 1196–1203, 2001.
4. Micro Thrusters built by TRW Team targets future microsatellites. *Small Times* "Business Wire," May 16, 2001.
5. http://www.darpa.mil/mto/mems/summaries/2004_summaries/afrl.html
6. http://www.ml.afrl.af.mil/stories/mlm_asc_03_1429.html
7. Twiggs, R., Space system developments at Stanford University — from launch experience of microsatellites to the proposed future use of picosatellites. *Proceedings of SPIE* 4136, 79–86, 2000.
8. Kitts, C.A. and Twiggs, R.J., Initial developments in the Stanford SQUIRT program. *Proceedings of SPIE* 2317, 178–185, 1995.
9. Kitts, C., et al., Emerald: A low-cost spacecraft mission for validating formation flying technologies. *Proceedings of the 1999 IEEE Aerospace Conference, Mar 6–Mar 13 1999*, 2, 217226, 1999.
10. Sekler, J., et al., COPS — a novel pressure gauge using MEMS devices for space, European Space Agency, (Special Publication) ESA SP, 439–443, 2003.
11. Sekler, J. and Wobmann, L., Development of an European QCM — outgassing detector with miniaturised interfaces, European Space Agency, (Special Publication) ESA SP, 515–519, 2003.
12. Lafontan, X., et al., The advent of MEMS in space. *Microelectronics Reliability*, 43, (7), 1061–1083, 2003.
13. Hampton, P., et al., Adaptive optics control system development. *Proceedings of SPIE* 5169, 321–330, 2003.
14. Liang, X., et al., Silicon solid-state small satellite design based on IC and MEMS. *Proceedings of the 1998 5th International Conference on Solid-State and Integrated Circuit Technology*, 932–935, 1998.
15. Tanaka, S., et al., MEMS-based solid propellant rocket array thruster with electrical feedthroughs. *Transactions of the Japan Society for Aeronautical and Space Sciences*, 46, (151), 47–51, 2003.

2 Vision for Microtechnology Space Missions

Cornelius J. Dennehy

CONTENTS

2.1   INTRODUCTION

We live in an age when technology developments combined with the innate human urge to imagine and innovate are yielding astounding inventions at an unprecedented rate. In particular, the past 20 years have seen a disruptive technology called microelectromechanical systems (MEMS) emerge and blossom in multiple ways. The commercial appeal of MEMS technologies lies in their low cost in high-volume production, their inherent miniature-form factor, their ultralow mass and power, their ruggedness, all with attendant complex functionality, precision, and accuracy. We are extremely interested in utilizing MEMS technology for future space mission for some of the very same reasons.

Recently dramatic progress has been occurring in the development of ultraminiature, ultralow power, and highly integrated MEMS-based microsystems that can sense their environment, process incoming information, and respond in a precisely controlled manner. The capability to communicate with other microscale devices and, depending on the application, with the macroscale platforms they are hosted on, will permit integrated and collaborative system-level behaviors. These attributes, combined with the potential to generate power on the MEMS scale, provide a potential for MEMS-based microsystems not only to enhance, or even replace, today's existing macroscale systems but also to enable entirely new classes of microscale systems.

As described in detail in subsequent chapters of this book, the roots of the MEMS technology revolution can be found in the substantial surface (planar) micromachining technology investments made over the last 30 years by integrated circuit (IC) semiconductor production houses worldwide. Broadly speaking, it is also a revolution that exploits the integration of multidisciplinary engineering processes and techniques at the submillimeter (hundreds of microns) device size level. The design and development of MEMS devices leverages heavily off of well-established, and now standard, techniques and processes for 2-D and 3-D semiconductor fabrication and packaging. MEMS technology will allow us to field new generations of sensors and devices in which the functions of detecting, sensing, computing, actuating, controlling, communicating, and powering are all colocated in assemblies or structures with dimensions of the order of 100–200 μm or less.

Over the past several years, industry analysts and business research organizations have pointed to the multibillion dollar-sized global commercial marketplace for MEMS-based devices and microsystems in such areas as the automotive industry, communications, biomedical, chemical, and consumer products. The MEMS-enabled ink jet printer head and the digital micromirror projection displays are often cited examples of commercially successful products enabled by MEMS technology. Both the MEMS airbag microaccelerometer and the tire air-pressure sensors are excellent examples of commercial applications of MEMS in the automotive industry sector. Implantable blood pressure sensors and fluidic micropumps for *in situ* drug delivery are examples of MEMS application in the biomedical arena.

Given the tremendous rapid rate of technology development and adoption over the past 100 years, one can confidently speculate that MEMS technology, especially when coupled with the emerging developments in nanoelectromechanical systems (NEMS) technology, has the potential to change society as did the introduction of the telephone in 1876, the tunable radio receiver in 1916, the electronic transistor in 1947, and the desktop personal computer (PC) in the 1970s. In the not too distant future, once designers and manufacturers become increasingly aware of the possibilities that arise from this technology, it may well be that MEMS-based devices and microsystems become as ubiquitous and as deeply integrated in our society's day-to-day existence as the phone, the radio, and the PC are today.

Perhaps it is somewhat premature to draw MEMS technology parallels to the technological revolutions initiated by such — now commonplace — household electronics. It is, however, very probable that as more specific commercial applications are identified where MEMS is clearly the competitively superior alternative, and the low-cost fabrication methods improve in device quality and reliability, and industry standard packaging and integration solutions are formulated, more companies focusing solely on commercializing MEMS technology will emerge and rapidly grow to meet the market demand. What impact this will have on society is unknown, but it is quite likely that MEMS (along with NEMS), will have an increasing presence in our home and our workplace as well as in many points in between. One MEMS industry group has gone so far as to predict that before 2010 there will be at least five MEMS devices per person in use in the United States.

It is not the intention of this chapter to comprehensively describe the far-reaching impact of MEMS-based microsystems on humans in general. This is well beyond the scope of this entire book, in fact. The emphasis of this chapter is on how the space community might leverage and exploit the billion-dollar worldwide investments being made in the commercial (terrestrial) MEMS industry for future space applications. Two related points are relevant in this context. First, it is unlikely that without this significant investment in commercial MEMS, the space community would even consider MEMS technology. Second, the fact that each year companies around the world are moving MEMS devices out of their research laboratories into commercial applications — in fields such as biomedicine, optical communications, and information technology — at an increasing rate can only be viewed as a very positive influence on transitioning MEMS technology toward space applications. The global commercial investments in MEMS have created the foundational physical infrastructure, the highly trained technical workforce, and most importantly, a deep scientific and engineering knowledge base that will continue to serve, as the strong intellectual springboard for the development of MEMS devices and microsystems for future space applications.

Two observations can be made concerning the differences between MEMS in the commercial world and the infusion of MEMS into space missions. First, unlike the commercial marketplace where very high-volume production and consumption is the norm, the niche market demand for space-qualified MEMS devices will be orders of magnitude less. Second, it is obvious that transitioning commercial MEMS designs to the harsh space environment will not be necessarily trivial. Their inherent mechanical robustness will clearly be a distinct advantage in surviving the dynamic shock and vibration exposures of launch, orbital maneuvering, and lunar or planetary landing. However, it is likely that significant modeling, simulation, ground test, and flight test will be needed before space-qualified MEMS devices, which satisfy the stringent reliability requirements traditionally imposed upon space platform components, can routinely be produced in reasonable volumes. For example, unlike their commercial counterparts, space MEMS devices will need to simultaneously provide radiation hardness (or at least radiation tolerance), have the capability to operate over wide thermal extremes, and be insensitive to significant electrical or magnetic fields.

In the remainder of this chapter, recent examples of MEMS technologies being developed for space mission applications are discussed. The purpose of providing this sampling of developments is to provide the reader with insight into the current state of the practice as an aid to predicting where this technology might eventually take us. A vision will then be presented, from a NASA perspective, of application areas where MEMS technology can possibly be exploited for science and exploration-mission applications.

2.2 RECENT MEMS TECHNOLOGY DEVELOPMENTS FOR SPACE MISSIONS

It is widely recognized that MEMS technology should and will have many useful applications in space. A considerable amount of the literature has been written describing in general terms the ways in which MEMS technology might enable constellations of cost-effective microsatellites[1] for various types of missions and highly miniaturized science instruments[2] as well as such advancements as "Lab on a Chip" microsensors for remote chemical detection and analysis.[3]

Recently, several of the conceptual ideas for applying MEMS in future space missions have grown into very focused technology development and maturation projects. The activities discussed in this section have been selected to expose the reader to some highly focused and specific applications of MEMS in the areas of spacecraft thermal control, science sensors, mechanisms, avionics, and propulsion. The intent here is not to provide design or fabrication details, as each of these areas will be addressed more deeply in the following chapters of this book, but rather to showcase the wide range of space applications in which MEMS can contribute.

While there is clearly a MEMS-driven stimulus at work today in our community to study ways to re-engineer spacecraft of the future using MEMS technology, one must also acknowledge the reality that the space community collectively is only in the nascent phase of applying MEMS technology to space missions. In fact, our community probably does not yet entirely understand the full potential that MEMS technology may have in the space arena. True understanding and the knowledge it creates will only come with a commitment to continue to create innovative designs, demonstrate functionality, and rigorously flight-validate MEMS technology in the actual space environment.

2.2.1 NMP ST5 THERMAL LOUVERS

The Space Technology-5 (ST5) project, performed under the sponsorship of NASA's New Millennium Program (NMP), has an overall focus on the flight validation of advanced microsat technologies that have not yet flown in space in order to reduce the risk of their infusion in future NASA missions. The NMP ST5 Project is designing and building three miniaturized satellites, shown in Figure 2.1, that are approximately 54 cm in diameter, 28 cm in height, and with a mass less than 25 kg per vehicle. As part of the ST5 mission these three microsats will perform some of the same functions as their larger counterparts.

One specific technology to be flight validated on ST5 is MEMS shutters for "smart" thermal control conceptualized and tested by NASA's Goddard Space Flight

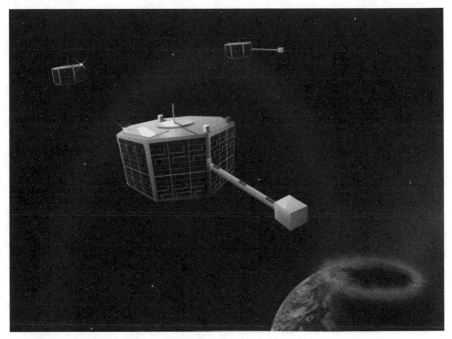

FIGURE 2.1 The NMP ST5 Project is designing and building three miniature satellites that are approximately 54 cm in diameter and 28 cm in height with a mass less than 25 kg per vehicle. (*Source*: NASA.)

Center (GSFC), developed by the Johns Hopkins University Applied Physics Laboratory (JHU/APL) and fabricated at the Sandia National Laboratory. In JHU/APL's rendition, the radiator is coated with arrays of micro-machined shutters, which can be independently controlled with electrostatic actuators, and which controls the apparent emittance of the radiator.[1] The latest prototype devices are 1.8 mm × 0.88 mm arrays of 150 × 6 mm shutters that are actuated by electrostatic comb drives to expose either the gold coating or the high-emittance substrate itself to space. Figure 2.2 shows an actuator block with the arrays. Prototype arrays designed by JHU/APL have been fabricated at the Sandia National Laboratories using their SUMMiT V® process. For the flight units, about 38 dies with 72 shutter arrays each will be combined on a radiator and independently controlled.

The underlying motivation for this particular technology can be summarized as follows: Most spacecraft rely on radiative surfaces (radiators) to dissipate waste heat. These radiators have special coatings that are intended to optimize performance under the expected heat load and thermal sink environment. Typically, such radiators will have a low absorptivity and a high infrared emissivity. Given the variable dynamics of the heat loads and thermal environment, it is often a challenge to properly size the radiator. For the same reasons, it is often necessary to have some means of regulating the heat-rejection rate in order to achieve proper thermal

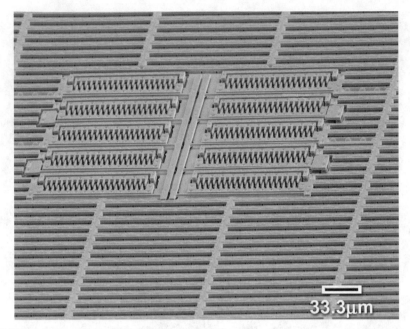

FIGURE 2.2 The NMP ST5 MEMS thermal louver actuator block with shutter array. (*Source*: JHU/APL.)

balance. One potential solution to this design problem is to employ the MEMS micromachined shutters to create, in essence, a variable emittance coating (VEC). Such a VEC yields changes in the emissivity of a thermal control surface to allow the radiative heat transfer rate to be modulated as needed for various spacecraft operational scenarios. In the case of the ST5 flight experiment, the JHU/APL MEMS thermal shutters will be exercised to perform adaptive thermal control of the spacecraft by varying the effective emissivity of the radiator surface.

2.2.2 JWST MICROSHUTTER ARRAY

NASA's James Webb Space Telescope (JWST) is a large (6.5-m primary mirror diameter) infrared-optimized space telescope scheduled for launch in 2011. JWST is designed to study the earliest galaxies and some of the first stars formed after the Big Bang. When operational, this infrared observatory will take the place of the Hubble Space Telescope and will be used to study the universe at the important but previously unobserved epoch of galaxy formation. Over the past several years, scientists and technologists at NASA GSFC have developed a large format MEMS-based microshutter array that is ultimately intended for use in the JWST near infrared spectrometer (NIRSpec) instrument. It will serve as a programmable field selector for the spectrometer and the complete microshutter system will be composed of four 175 by 384 pixel modules. This device significantly enhances the capability of the JWST since the microshutters can be selectively configured to make highly efficient use of nearly the entire NIRSpec detector, obtaining hundreds of object spectra simultaneously.

Micromachined out of a silicon nitride membrane, this device, as shown in Figure 2.3 and Figure 2.4, consists of a 2-D array of closely packed and independently selectable shutter elements. This array functions as an adaptive input mask for the multiobject NIRSpec, providing very high contrast between its open and closed states. It provides high-transmission efficiency in regions where shutters are commanded open and where there is sufficient photon blocking in closed areas. Operationally, the desired configuration of the array will be established via ground command, then simultaneous observations of multiple celestial targets can be obtained.

Some of the key design challenges for the microshutter array include obtaining the required optical (contrast) performance, individual shutter addressing, actuation, latching, mechanical interfaces, electronics, reliability, and environment requirements. For this particular NIRSpec application, the MEMS microshutter developers also had to ensure the device would function at the 37 K operating temperature of the spectrometer as well as meet the demanding low-power dissipation requirement.

Figure 2.5 shows the ability to address or actuate and provide the required contrast demonstrated on a fully functional 128 by 64 pixel module in 2003 and the development proceeding the 175 by 384 pixel flight-ready microshutter module that will be used in the JWST NIRSpec application. This is an outstanding example of applying MEMS technology to significantly enhance the science return from a space-based observatory.

FIGURE 2.3 JWST microshutters for the NIRSpec detector. (*Source*: NASA.)

├────────────────────┤ 50 μm

FIGURE 2.4 Individual shuttle element of the JWST shuttle array. (*Source*: NASA).

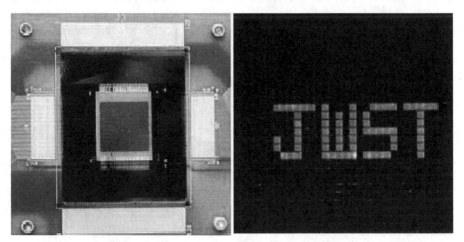

FIGURE 2.5 Ability to address or actuate and provide the required contrast demonstrated on a fully functional 128 by 64 pixel module of the MEMS microshutter array. (*Source*: NASA.)

2.2.3 INCHWORM MICROACTUATORS

The NASA Jet Propulsion Laboratory (JPL) is currently developing an innovative inchworm microactuator[4] for the purpose of ultraprecision positioning of the mirror segments of a proposed Advanced Segmented Silicon Space Telescope (ASSiST). This particular activity is one of many diverse MEMS or NEMS technology developments for space mission applications being pursued at NASA/JPL.[5]

2.2.4 NMP ST6 INERTIAL STELLAR CAMERA

NASA's NMP is sponsoring the development of the inertial stellar compass (ISC) space avionics technology that combines MEMS inertial sensors (gyroscopes) with a wide field-of-view active pixel sensor (APS) star camera in a compact, multifunctional package.[6] This technology development and maturation activity is being performed by the Charles Stark Draper Laboratory (CSDL) for a Space Technology-6 (ST6) flight validation experiment now scheduled to fly in 2005. The ISC technology is one of several MEMS technology development activities being pursued at CSDL[7] and, in particular, is an outgrowth of earlier CSDL research focused in the areas of MEMS-based guidance, navigation, and control (GN&C) sensors or actuators[8] and low-power MEMS-based space avionic systems for space.[9]

The ISC, shown in Figure 2.6, is a miniature, low-power, stellar inertial attitude determination system that provides an accuracy of better than 0.1° (1-Sigma) in three axes while consuming only 3.5 W and is packaged in a 2.5-kg housing.[10]

The ISC MEMS gyro assembly, as shown in Figure 2.7, incorporates CSDL's tuning fork gyro (TFG) sensors and mixed signal application specific integrated

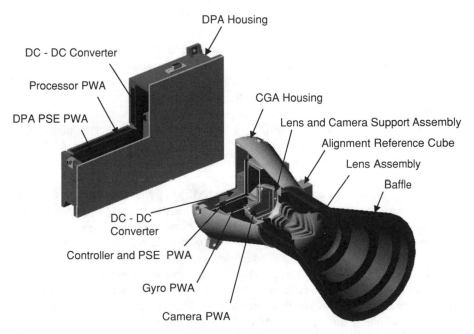

FIGURE 2.6 The NMP ST6 inertial stellar camera. (*Source*: NASA JPL/CALTECH ST6.)

FIGURE 2.7 NMP ST6 ISC MEMS 3-axis gyro assembly. (*Source*: Charles Stark Draper Laboratory.)

circuit (ASIC) electronics designs. Inertial systems fabricated from similar MEMS gyro components have been used in precision-guided munitions (PGMs), autonomous vehicles, and other space-related mission applications. The silicon MEMS gyros sense angular rate by detecting the Coriolis effect on a sense mass that is driven into oscillation by electrostatic motors. Coriolis forces proportioned to the rotational rate of the body cause the sense mass to oscillate out of plane. This change is measured by capacitive plates. A more detailed discussion of MEMS inertial sensors, both gyros and accelerometers, is presented in Chapter 10 of this book.

The ISC technology, enabled by embedded MEMS gyroscopes, is a precursor of things to come in the spacecraft avionics arena as the push toward much more highly integrated, GN&C systems grows in the future. There is a wide range of science and exploration mission applications that would benefit from the infusion of the compact, low-power ISC technology. Some envisioned applications include using the ISC as a "single sensor" solution for attitude determination on medium-performance spacecraft, as a "Bolt On" — independent safehold sensor for any spacecraft, or as an acquisition sensor for rendezvous applications. It has been estimated that approximately 1.5 kg of mass and 26 W of power can be saved by employing a single MEMS-based attitude sensor such as the ISC to replace the separate and distinct star tracker and inertial reference units typically used on spacecraft.[10] So in this case, MEMS is an enhancing technology that serves to free up precious spacecraft resources. For example, the mass savings afforded by using the MEMS-based ISC could be allocated for additional propellant or, likewise, the power savings could potentially be directly applied to the mission payload. These are some of the advantages afforded by using MEMS technology.

2.2.5 MICROTHRUSTERS

Over the past several years MEMS catalytic monopropellant microthruster research and development has been conducted at NASA's GSFC.[11] MEMS-based propulsion systems have the potential to enable missions that require micropropulsive maneuvers for formation flying and precision pointing of micro-, nano-, or pico-sized satellites. Current propulsion technology cannot meet the minimum thrust requirements (10–1000 μN) or impulse-bit requirements (1–1000 μN·sec), or satisfy the severely limited system mass (<0.1 kg), volume (<1 cm^3), and power constraints (<1 W). When compared to other proposed micropropulsion concepts, MEMS catalytic monopropellant thrusters show the promise of the combined advantages of high specific density, low system power and volume, large range of thrust levels, repeatable thrust vectors, and simplicity of integration. Overall, this approach offers an attractive technology solution to provide scalable micro-Newton level microthrusters. This particular MEMS microthruster design utilizes hydrogen peroxide as the propellant and the targeted thrust level range is between 10 and 500 μN with impulse bits between 1 and 1000 μN·sec and a specific impulse (I_{sp}) greater than 110 sec.

A prototype MEMS microthruster hardware has been fabricated as seen in Figure 2.8, using GSFC's detector development laboratory (DDL) facilities and equipment. Individual MEMS fabricated reaction chambers are approximately 3.0 × 2.5 × 2.0 mm. Thrust chambers are etched in a 0.5 mm silicon substrate and the vapor is deposited with silver using a catalyst mask.

2.2.6 OTHER EXAMPLES OF SPACE MEMS DEVELOPMENTS

The small sampling of space MEMS developments given earlier can be categorized as some very significant technological steps toward the ultimate goal of routine and systematic infusion of this technology in future space platforms. Clearly NASA researchers have identified several areas where MEMS technology will substantially improve the performance and functionality of the future spacecraft. NASA is currently investing at an increasing rate in a number of different MEMS technology areas. A review of the NASA Technology Inventory shows that in fiscal year 2003 there were a total of 111 distinct MEMS-based technology development tasks being funded by NASA. Relative to GFY02 where 77 MEMS-based technology tasks were cataloged in the NASA Technology Inventory, this is over a 40% increase in MEMS tasks. It is almost a 90% increase relative to GFY01 where 59 MEMS R&D tasks were identified. The MEMS technologies included in the NASA inventory are:

FIGURE 2.8 A prototype MEMS microthruster hardware fabricated in GSFC's detector development laboratory (DDL). (*Source*: NASA.)

- MEMS Stirling coolers
- MEMS liquid–metal microswitches
- MEMS inertial sensors
- MEMS microwave RF switches and phase shifters
- MEMS thrusters
- MEMS deformable mirrors
- MEMS pressure or temperature sensors
- MEMS power supplies.

To sum up this section, it should be stressed that the few selected developments highlighted above are not intended to represent a comprehensive list[12,13] of recent or ongoing space MEMS technology developments. In fact, there are a number of other very noteworthy space MEMS technology projects in various stages of developments. Among these are:

- Flat plasma spectrometer[14] for space plasma and ionospheric–thermospheric scientific investigations
- Miniature mass spectrometer[3,14] for planetary surface chemistry investigations
- Switch-reconfigurable antenna array element[15] for space-based radar applications

- Microheat-sinks for microsat thermal control applications
- Tunable Fabry–Perot etalon optical filters for remote sensing applications[5]
- Two-axis fine-pointing micromirrors for intersatellite optical communications applications.[16]

2.3 POTENTIAL SPACE APPLICATIONS FOR MEMS TECHNOLOGY

It should be apparent that the near-term benefit of MEMS technology is that it allows developers to rescale existing macrosystems down to the microsystem level. However, beyond simply shrinking today's devices, the true beauty of MEMS technology derives from the system redefinition freedom it provides to designers, leading to the invention of entirely new classes of highly integrated microsystems.

It is envisioned that MEMS technology will serve as both an "enhancing" and an "enabling" technology for many future science and exploration missions. Enabling technologies are those that provide the presently unavailable capabilities necessary for a mission's implementation and are vital to both intermediate and long-term missions. Enhancing technologies typically provide significant mission performance improvements, mitigations of critical mission risks, and significant increases in mission critical resources (e.g., cost, power, and mass).

MEMS technology should have a profound and far-reaching impact on many of NASA's future space platforms. Satellites in low-Earth orbit, deep-space interplanetary probes, planetary rovers, advanced space telescopes, lunar orbiters, and lunar landers could all likely benefit in some way from the infusion of versatile MEMS technology. Many see the future potential for highly integrated spacecraft architectures where boundaries between traditional, individual bus and payload subsystems are at a minimum blurred, or in some extreme applications, nonexistent with the infusion of multifunctional MEMS-based microsystems.

NASA's GSFC has pursued several efforts not only to increase the general awareness of MEMS within the space community but also to spur along specific mission-unique infusions of MEMS technology where appropriate. Over the past several years the space mission architects at the GSFC's Integrated Mission Design Center (IMDC), where collaborative end-to-end mission conceptual design studies are performed, have evaluated the feasibility of using MEMS technology in a number of mission applications. As part of this MEMS technology "push" effort, many MEMS-based devices emerging from research laboratories have been added to the IMDC's component database used by the mission conceptual design team. The IMDC is also a rich source of future mission requirements and constraints data that can be used to derive functional and performance specifications to guide MEMS technology developments. Careful analysis of these data will help to identify those missions where infusing a specific MEMS technology will have a significant impact, or conversely, identifying where an investment in a broadly applicable "crosscutting" MEMS technology will yield benefits to multiple missions.

The remainder of this section covers some high-priority space mission application areas where MEMS technology infusion would appear to be beneficial.

2.3.1 INVENTORY OF MEMS-BASED SPACECRAFT COMPONENTS

It is expected that MEMS technology will offer NASA mission designers very attractive alternatives for challenging applications where power, mass, and volume constraints preclude the use of the traditional components. MEMS technologies will enable miniaturized, low-mass, low-power, modular versions of many of the current inventory of traditional spacecraft components.

2.3.2 AFFORDABLE MICROSATELLITES

A strong driver for MEMS technology infusion comes from the desire of some space mission architects to implement affordable constellations of multiple micro-satellites. These constellations, of perhaps as many as 30–100 satellites, could be deployed either in loosely controlled formations to perform spatial or temporal space environment measurements, or in tightly controlled formations to synthesize distributed sparse aperture arrays for planet finding.

A critical aspect to implementing these multisatellite constellations in today's cost-capped fiscal environment will be the application of new technologies that reduce the per unit spacecraft cost while maintaining the necessary functional performance. The influence of technology in reducing spacecraft costs evaluated by NASA[17] through analysis of historical trend data leads us to the conclusion that, on average, the use of technologies that reduce spacecraft power will reduce spacecraft mass and cost. Clearly a large part of solving the affordable microsatellite problem will involve economies of scale. Identifying exactly those technologies that have the highest likelihood of lowering spacecraft cost is still in progress. However, a case can be made that employing MEMS technology, perhaps in tandem with the ultra-low power electronics[18] technology being developed by NASA and its partners, will be a significant step toward producing multiple microsatellite units in a more affordable way.

It should also be pointed out that another equally important aspect to lowering spacecraft costs will be developing architectures that call for the use of standard-off-the-shelf and modular MEMS-based microsystems. Also, there will be a need to fundamentally shift away from the current "hands on" labor-intensive limited-production spacecraft manufacturing paradigm toward a high-volume, more "hands off" production model. This would most likely require implementing new cost-effective manufacturing methodologies where such things as parts screening, subsystem testing, spacecraft-level integration and testing, and documentation costs are reduced.

One can anticipate the "Factory of the Future," which produces microsatellites that are highly integrated with MEMS-based microsubsystems, composed of miniaturized electronics, devices and mechanisms, for communications, power, and attitude control, extendable booms and antennas, microthrusters, and a broad range of microsensor instrumentation. The multimission utility of having a broadly capable nano- or microspacecraft has not been overlooked by NASA's mission architects. New capabilities such as this will generate new concepts of space operations to perform existing missions and, of greater import, to enable entirely new types of missions.

Furthermore, because the per unit spacecraft cost has been made low enough through the infusion of MEMS technology, the concept of flying "replaceable" microsatellites is both technically and economically feasible. In such a mission concept, the requirements for redundancy or reliability will be satisfied at the spacecraft level, not at the subsystem level where it typically occurs in today's design paradigm. In other words, MEMS-based technology, together with appropriate new approaches to lower spacecraft-level integration, test and launch costs, could conceivably make it economical to simply perform an on-orbit spacecraft replacement of a failed spacecraft. This capability opens the door to create new operational concepts and mission scenarios.

2.3.3 SCIENCE SENSORS AND INSTRUMENTATION

As described in Chapter 7 of this book, the research topic of MEMS-based science sensors and instruments is an incredibly rich one. Scientists and MEMS technologists are collaborating to first envision and then rapidly develop highly integrated, miniaturized, low-mass and power-efficient sensors for both science and exploration missions. The extreme reductions in sensor mass and power attainable via MEMS technology will make it possible to fly multiple high performance instrumentation suites on microsatellites, nanosatellites, planetary landers, and autonomous rovers, entry probes, and interplanetary platforms. The ability to integrate miniaturized sensors into lunar or planetary *In Situ* Resource Utilization (ISRU) systems and/or robotic arms, manipulators, and tools (i.e., a drill bit) will have high payoff on future exploration missions. Detectors for sensing electromagnetic fields and particles critical to several future science investigations of solar terrestrial interactions are being developed in a MEMS format. Sensor technologies using micromachined optical components, such as microshutters and micromirrors for advanced space telescopes and spectrometers, are also coming of age. One exciting research area is the design and development of adaptive optics devices made up of either very dense arrays of MEMS micromirrors or membrane mirrors to perform wavefront aberration correction functions in future space observatories. These technologies have the potential to replace the very expensive and massive high-precision optical mirrors traditionally employed in large space telescopes. Several other MEMS-based sensing systems are either being actively developed or are in the early stages of innovative design. Examples of these include, but are not limited to, micromachined mass spectrometers (including MEMS microvalves) for chemical analysis, microbolometers for infrared spectrometry, and entire laboratory-on-a-chip device concepts. One can also envision MEMS-based environmental and state-of-health monitoring sensors being embedded into the structures of future space transportation vehicles and habitats on the lunar (or eventually on a planetary) surface as described in the following section on exploration applications for MEMS.

2.3.4 EXPLORATION APPLICATIONS

There are a vast number of potential application areas for MEMS technology within the context of the U.S. Vision for Space Exploration (VSE). We explore some of those here.

In the integrated vehicle health management (IVHM) arena, emphasis will be placed upon developing fault detection, diagnosis, prognostics, information fusion, degradation management capabilities for a variety of space exploration vehicles and platforms. Embedded MEMS technology could certainly play a significant role in implementing automated spacecraft IVHM systems and the associated crew emergency response advisory systems.

Developing future ISRU systems will dictate the need for automated systems to collect lunar regolith for use in the production of consumables. Innovative ISRU systems that minimize mass, power, and volume will be part of future power system and vehicle refueling stations on the lunar surface and planetary surfaces. These stations will require new techniques to produce oxygen and hydrogen from lunar regolith, and further, new systems to produce propellants and other consumables from the Mars atmosphere will need to be developed.

MEMS technology should also play a role in the development of the space and surface environmental monitoring systems that will support exploration. Clearly the observation, knowledge, and prediction of the space, lunar, and planetary environments will be important for exploration. MEMS could also be exploited in the development of environmental monitoring systems for lunar and planetary habitats. This too would be a very suitable area for MEMS technology infusion.

2.3.5 SPACE PARTICLES OR MORPHING ENTITIES

Significant technological changes will blossom in the next few years as the multiple developments of MEMS, NEMS, micromachining, and biochemical technologies create a powerful confluence. If the space community at large is properly prepared and equipped, the opportunity to design, develop, and fly revolutionary, ultra-integrated mechanical, thermal, chemical, fluidic, and biologic microsystems can be captured. Building these type of systems is not feasible using conventional space platform engineering approaches and methods.

Some space visionaries are so enthused by this huge "blue sky" potential as to blaze completely new design paths over the next 15–25 years. They envision the creation of such fundamentally new mission ideas as MEMS-based "spaceborne sensor particles" or autonomously morphing space entities that would resemble today's state-of-the-art space platforms as closely as the currently ubiquitous PCs resemble the slide rules used by an earlier generation of scientists and engineers. These MEMS-enabled "spaceborne sensor particles" could be used to make very dense *in situ* science observations and measurements. One can even envision these "spaceborne sensor particles" breaking the access-to-space bottleneck — which significantly limits the scope of what we can do in space — by being able to take advantage of novel space launch systems innovations such as electromagnetic or light-gas cannon launchers where perhaps thousands of these devices could be dispensed at once.

2.4 CHALLENGES AND FUTURE NEEDS

In this section, it will be stressed that while some significant advancements are being made to develop and infuse MEMS technology into space mission applications, there is much more progress to be made. There are still many challenges, barriers, and issues (not all technical or technological) yet to be dealt with to fully exploit the potential of MEMS in space. The following is a brief summary of some of the key considerations and hurdles to be faced.

2.4.1 CHALLENGES

History tells us that the infusion of new technological capabilities into space missions will significantly lag behind that of the commercial or the industrial sector. Space program managers and other decision makers are typically very cautious about when and where new technology can be infused into their missions. New technologies are often perceived to add unnecessary mission risk.

Consequently, MEMS technology developers must acknowledge this barrier to infusion and strive to overcome it by fostering a two-way understanding and interest in MEMS capabilities with the mission applications community. This motivates the need, in addition to continually maturing the Technology Readiness Level (TRL) of their device or system, to proactively initiate and maintain continuing outreach with the potential space mission customers to ensure a clear mutual understanding of MEMS technology benefits, mission requirements and constraints (in particular the "Mission Assurance" space qualification requirements), risk metrics, and potential infusion opportunities.

2.4.2 FUTURE NEEDS

It is unlikely that the envisioned proliferation of MEMS into future science and exploration missions will take place without significant future technological and engineering investments focused on the unique and demanding space applications arena. Several specific areas where such investments are needed are suggested here.

Transitioning MEMS microsystems and devices out of the laboratory and into operational space systems will not necessarily be straightforward. The overwhelming majority of current MEMS technology developments have been targeted at terrestrial, nonspace applications. Consequently, many MEMS researchers have never had to consider the design implications of having to survive and operate in the space environment. An understanding of the space environment will be a prerequisite for developing "flyable" MEMS hardware. Those laboratory researchers who are investigating MEMS technology for space applications must first take the time to study and understand the unique challenges and demanding requirements imposed by the need to first survive the rigors of the short-term dynamic space launch environment as well as the long-term on-orbit operating environments found in various mission regimes. Chapter 4 of this book is intended to provide just such a broad general background on the space environment and will be a valuable reference for MEMS technologists. In a complementary effort, the space system professionals in industry and in government, to whom the demanding space environmental requirements are routine, must do a much better job of guiding the MEMS technology community through the hurdles of designing, building, and qualifying space hardware.

The establishment of much closer working relationships between MEMS technologists and their counterparts in industry is certainly called for. Significantly more industry–university collaborations, focused on transitioning MEMS microsystems and devices out of the university laboratories, will be needed to spur the infusion of MEMS technology into future space missions. It is envisioned that these collaborative teams would target specific space mission applications for MEMS. Appropriate mission assurance product reliability specifications, large-scale manufacturing considerations, together with industry standard mechanical or electrical interface requirements, would be combined very early in the innovative design process. In this type of collaboration, university-level pilot production would be used to evaluate and path find viable approaches for the eventual large volume industrial production process yielding space-qualified commercial-off-the-shelf (COTS) MEMS flight hardware.

On a more foundational level, continued investment in expanding and refining the general MEMS knowledge base will be needed. The focus here should be on improving our understanding the mechanical and electrical behaviors of existing MEMS materials (especially in the cryogenic temperature regimes favored by many space-sensing applications) as well as the development of new exotic MEMS materials. New techniques for testing materials and methods for performing standardized reliability assessments will be required. The latter need will certainly drive the development of improved high-fidelity, and test-validated, analytical software models. Exploiting the significant recent advances in high-performance computing and visualization would be a logical first step here.

Another critical need will be the development of new techniques and processes for precision manufacturing, assembly and integration of silicon-based MEMS devices with macroscale nonplanar components made from metals, ceramics, plastics, and perhaps more exotic materials. The need for improved tools, methods, and processes for the design and development of the supporting miniature, low-power mixed-signal (analog and digital) electronics, which are integral elements of the MEMS devices, must also be addressed.

The investigation of innovative methods for packing and tightly integrating the electrical drive signal, data readout, and signal conditioning elements of the MEMS devices with the mechanical elements should be aggressively pursued. In most applications, significant device performance improvements, along with dramatic reductions in corrupting electrical signal noise, can be accomplished by moving the electronics as physically close as possible to the mechanical elements of the MEMS device. This particular area, focused on finding new and better ways to more closely couple the MEMS electronics and mechanical subelements, can potentially have high payoffs and should not be overlooked as an important research topic.

Lastly it is important to acknowledge that a unified "big picture" systems approach to exploiting and infusing MEMS technology in future space missions is currently lacking and, perhaps worse, nonexistent. While there are clearly many localized centers of excellence in MEMS microsystem and device technology development within academia, industry, nonprofit laboratories, and federal government facilities, there are few, if any, comparable MEMS systems engineering and integration centers of excellence. Large numbers of varied MEMS "standalone" devices are being designed and developed, but there is not enough work being done currently on approaches, methods, tools, and processed to integrate heterogenous MEMS elements together in a "system of systems" fashion. For example, in the case of the affordable microsatellite discussed earlier, it is not at all clear how one would go about effectively and efficiently integrating a MEMS microthruster or a MEMS microgyro with other MEMS-based satellite elements such as a command or telemetry system, a power system, or on-board flight processor. We certainly should not expect to be building future space systems extensively composed of MEMS microsystems and devices using the integration and interconnection approaches currently employed. These are typically labor-intensive processes using interconnection technologies that are both physically cumbersome and resource (power or mass) consuming. The cost economies and resource benefits of using miniature mass-produced MEMS-based devices may very well be lost if a significant level of "hands-on" manual labor is required to integrate the desired final payload or platform system. Furthermore, it is quite reasonable to expect that future space systems will have requirements for MEMS-based payloads and platforms that are both modular and easily reconfigurable in some "plug and play" fashion. The work to date on such innovative technology as MEMS harnesses and MEMS switches begins to address this interconnection or integration need, but significant work remains to be done in the MEMS flight system engineering arena. In the near future, to aid in solving the dual scale (macro-to-MEMS) integration problem, researchers could pursue ways to better exploit newly emerging low power or radiation hard microelectronics packaging and high-density interconnect technologies as well as Internet-based wireless command or telemetry interface technology. Researchers should also evaluate methods to achieve a zero integration time (ZIT) goal for MEMS flight systems using aspects of today's plug and play component technology, which utilizes standard data bus interfaces. Later on, we most likely will need to identify entirely new architectures and approaches to accomplish the goal of simply and efficiently interconnecting MEMS microsystems and devices composed of various types of metals, ceramics, plastics, and exotic materials.

Balancing our collective technological investments between the intellectually stimulating goal of developing the next best MEMS standalone device in the laboratory and the real world problem that will be faced by industry of effectively integrating MEMS-based future space systems is a recommended strategy for ultimate success. Significant investments are required to develop new space system engineering approaches to develop adaptive and flexible MEMS flight system architectures and the supporting new MEMS-scale interconnection hardware or software building blocks. Likewise the closely associated need to test and validate these highlyintegrated MEMS "system of systems" configurations prior to launch will drive the need for adopting (and adapting) the comprehensive, highly autonomous built-in test (BIT) functions commonly employed in contemporary nonaerospace commercial production lines.

Research in this arena could well lead to the establishment of a new MEMS microsystems engineering discipline. This would be a very positive step in taking the community down the technological path toward the ultimate goal of routine, systematic, and straightforward infusion of MEMS technology in future space missions.

There are several important interrelated common needs that span all the emerging MEMS technology areas. Advanced tools, techniques, and methods for high-fidelity dynamic modeling and simulation of MEMS microsystems will certainly be needed, as will be multiple MEMS technology ground testbeds, where system functionality can be demonstrated and exercised. These testbed environments will permit the integration of MEMS devices in a flight configuration like hardware-in-the-loop (HITL) fashion. The findings and the test results generated by the testbeds will be used to update the MEMS dynamic models. The last common need is for multiple and frequent opportunities for the on-orbit demonstration and validation of emerging MEMS-based technologies for space. Much has been accomplished in the way of technology flight validation under the guidance and sponsorship of such programs as NASA's NMP, but many more such opportunities will be required to propel the process of validating the broad family of MEMS technologies needed to build new and innovative space systems. The tightly interrelated areas of dynamic models and simulations, ground testbeds, and on-orbit technology validation missions will all be essential to fully understand and to safely and effectively infuse the MEMS into future missions.

2.5 CONCLUSIONS

The success of future science and exploration missions quite possibly will be dependent on the development, validation, and infusion of MEMS-based microsystems that are not only highly integrated, power efficient, and minimally packaged but also flexible and versatile enough to satisfy multimission requirements. Several MEMS technology developments are already underway for future space applications. The feasibility of many other MEMS innovations for space is currently being studied and investigated.

The widespread availability and increasing proliferation of MEMS technology specifically targeted for space applications will lead future mission architects to evaluate entirely new design trades and options where MEMS can be effectively infused to enhance current practices or perhaps enable completely new mission opportunities. The space community should vigorously embrace the potential disruptive technological impact of MEMS on how space systems are designed, built, and operated. One option is to adopt a technology infusion approach similar to the one the Defense Advanced Research Projects Agency (DARPA) has pursued for the development and widespread integration of MEMS-based microsystems to revolutionize our military's capabilities on future battlefields. Technologists, researchers, and decision makers interested in developing truly innovative and enabling MEMS-based microsystems that will support the VSE goals of affordability, reliability, effectiveness, and flexibility would do well to study the DARPA approach, where multiple high-risk or high-payoff military MEMS technologies are being pursued to dramatically improve the agility, accuracy, lethality, robustness, and reliability of warfighter systems.

Transitioning MEMS microsystems and devices out of the laboratory and into operational space systems will present many challenges. Clearly much has been accomplished but several critical issues remain to be resolved in order to produce MEMS microsystems that will satisfy the demanding performance and environmental requirements of space missions. In the spirit of Rear Admiral Grace Murray Hopper (who is quoted as saying "If it's a good idea, go ahead and do it. It's much easier to apologize than it is to get permission") the community must continue to innovate with open minds for if we constrain our vision for MEMS in space, an opportunity may be missed to bend (or even break) current space platform design and production paradigms.

REFERENCES

1. Osiander, R., S.L. Firebaugh, J.L. Champion, et al., Microelectromechanical devices for satellite thermal control, *IEEE Sensors Journal Microsensors and Microacuators: Technology and Applications* **4**(4), pp. 525 (2004).
2. Wesolek, D.M., J.L. Champion, F.A. Hererro, et al., A micro-machined flat plasma spectrometer (FlaPS), *Proceedings of SPIE — The International Society for Optical Engineering* 5344, pp. 89 (2004).
3. Sillon, N. and R. Baptist, Sensors and actuators B (chemical), *Proceedings of 11th International Conference on Solid State Sensors and Actuators Transducers '01/Eurosensors XV*, Elsevier, Switzerland, Vol. B83, pp. 129 (2002).
4. Mott, D.B., R. Barclay, A. Bier, et al., Micromachined tunable Fabry–Perot filters for infrared astronomy, *Proceedings of SPIE — The International Society for Optical Engineering* 4841, pp. 578 (2002).
5. George, T., Overview of MEMS/NEMS technology development for space applications at NASA/JPL, *Smart Sensors, Actuators, and MEMS, May 19–21 2003*, The International Society for Optical Engineering, Maspalonas, Gran Canaria, Spain (2003).
6. Brady, T., et al., The inertial stellar compass: a new direction in spacecraft attitude determination, *16th Annual AIAA/USU Conference on Small Satellites*, Logan, UT (2002).
7. Duwel, A. and N. Barbour, MEMS development at Draper lab, *Society for Experimental Mechanics (SEM) Annual Conference* (2003).
8. Connelly, J.A., et al., Alignment and performance of the infrared multi-object spectrometer, *Cryogenic Optical Systems and Instruments X, Aug 6 2003*, The International Society for Optical Engineering, San Diego, CA (2003).

9. Johnson, W.M. and R.E. Phillips, Space avionics stellar-inertial subsystem, *20th Digital Avionics Systems Conference Proceedings, Oct 14–18 2001*, Institute of Electrical and Electronics Engineers, Inc., Daytona Beach, FL (2001).

10. Brady, T., et al., A multifunction, low-power attitude determination technology breakthrough, *AAS G&C Conference, AAS 03–003* (2003).

11. Hitt, D.L., C.M. Zakrzwski, and M.A. Thomas, MEMS-based şatellite micropropulsion via catalyzed hydrogen peroxide decomposition, *Smart Materials and Structures* **10**(6), pp. 1163–1175 (2001).

12. Caffey, J.R. and P.E. Kladitis, The effects of ionizing radiation on microelectromechanical systems (MEMS) actuators: electrostatic, electrothermal, and bimorph, *17th IEEE International Conference on Micro Electro Mechanical Systems (MEMS): Maastricht MEMS 2004 Technical Digest, Jan 25–29 2004*, Maastricht, Netherlands, Institute of Electrical and Electronics Engineers Inc., Piscataway, United States (2004).

13. Hewagama, T., et al., Spectral contrast enhancement techniques for extrasolar planet imaging, *High-Contrast Imaging for Exo-Planet Detection, Aug 23–26 2002*. The International Society for Optical Engineering, Waikoloa, Hawaii (2002).

14. Siebert, P. G., Petzold, , and J. Muller, Processing of complex microsystems: a micro mass spectrometer, *Proceedings of the SPIE — The International Society for Optical Engineering* 3680, pp. 562 (1999).

15. Bernhard, J.T., et al., Stacked reconfigurable antenna elements for space-based radar applications, *2001 IEEE Antennas and Propagation Society International Symposium — Historical Overview of Development of Wireless, Jul 8–13 2001*, Institute of Electrical and Electronics Engineers, Inc., Boston (2001).

16. Graeffe, J., et al., Scanning micromechanical mirror for fine-pointing units of intrasatellite optical links, *Design, Test, Integration, and Packaging of MEMS/MOEMS, May 9–11 2000*, Paris, Fr, Society of Photo-Optical Instrumentation Engineers, Bellingham, Washington (2000).

17. Buehler, M.G., et al., Technologies for affordable SEC missions, *IEEE Big Sky Conference* (2004).

18. Gambles, J., et al., An ultra-low-power, radiation-tolerant Reed–Solomon Encoder for space applications. *Proceedings of the Custom Integrated Circuits Conference*, pp. 631–634 (2003).

3 MEMS Fabrication

James J. Allen

CONTENTS

3.1 INTRODUCTION

Making devices small has long had engineering, scientific, and esthetic motivations. John Harrison's quest[1] to make a small (e.g., hand-sized) chronometer in the 1700s for nautical navigation was motivated by the desire to have an accurate time-keeping instrument that was insensitive to temperature, humidity, and motion. A small chronometer could meet these objectives and allow for multiple instruments on a ship for redundancy and error averaging. The drive toward miniaturization of various mechanical and electrical devices advanced over the years, but in the 1950s several key events occurred that would motivate development at an increased pace.

The development of the transistor[2] in 1952, and a manufacturing method for a planar silicon transistor[3,4] in 1960 set the stage for development of fabrication processes to achieve small feature sizes. The drive for microelectronic devices with smaller and smaller features continues to the present day.

Dr. Richard Feynman presented a seminal talk "There's Plenty of Room at the Bottom" on December 29, 1959 at the annual meeting of the American Physical

Society at the California Institute of Technology (Caltech), which was first published in the 1960 issue of Caltech's *Engineering and Science* magazine,[5] and it has since been reprinted several times.[6,7] In the talk, Feynman conceptually presented, motivated, and challenged people with the desire, and advantages of exploring engineered devices at a small scale. This talk is frequently cited as the conceptual beginnings of the fields of microelectromechanical systems (MEMS), and nanotechnology. Feynman offered two US \$1000 prizes for the following achievements:

- Building a working electric motor no larger than a 1/64th-in. (400-μm) cube
- Printing text at a scale that the *Encyclopedia Britannica* could fit on the head of a pin.

In less than a year, a Caltech engineer, William McLellan, constructed a 250-μm, 2000-rpm electric motor using 13 separate parts to collect his prize.[8] This illustrated that technology was constantly moving toward miniaturization and that some aspects of the technology already existed. However, the second prize was not awarded until 1985, when T. Newman, and R.F.W. Pease used e-beam lithography to print the first page of *A Tale of Two Cities* within a 5.9-μm square.[9] The achievement of the second prize was enabled by the developments of the microelectronics industry in the ensuing 25 years. The first indication that MEMS can be realized came in the early days of microelectronic development. The fabrication processes that were being developed for microelectronics would eventually be utilized and further developed for MEMS fabrication at the micron scale. Sense mechanisms such as the piezoresistive properties of the microelectronic materials (silicon, germanium)[10] and the mechanical motion of a silicon transistor gate[11] provided the indication that MEMS sensors could be developed. During the 1960s and 1970s, initial devices (strain gauges, pressure sensors, accelerometers) that utilized piezoresistive sensing were developed. However, it was not until the early 1980s that the field of MEMS was launched in earnest, stimulated in part by the consideration of silicon as a mechanical material.[12] In the ensuing years, the development of fabrication processes for MEMS, the demonstration of MEMS devices, and the commercialization of MEMS devices have occurred.

3.2 MEMS FABRICATION TECHNOLOGIES

MEMS fabrication technologies are a part of a spectrum of fabrication technologies that also include traditional precision machining processes. Traditional machining processes can utilize a large variety of materials, fabricate complex three-dimensional devices, and produce precise devices. MEMS fabrication processes are generally more limited in the materials utilized, but they can produce functional devices with micron-scale dimensions. Table 3.1 is a comparison of the MEMS fabrication processes and conventional machining processes.

TABLE 3.1
Comparison of the Capabilities of MEMS Fabrication Technologies and Conventional Machining

Capability	LIGA	Bulk Micromachining	Surface Micromachining	Conventional Machining
Feature size	~3 to 5 μm	~3 to 5 μm	1 μm	~10 to 25 μm
Device thickness	>1 mm	>1 mm	13 μm	Very large
Lateral dimension	>2 mm	>2 mm	2 mm	>10 m
Relative tolerance	~10^{-2}	~10^{-2}	~10^{-1}	>10^{-3}
Materials	Electroplated metals or injection molded plastics	Very limited material suite	Very limited material suite	Extremely large material suite
Assembly requirements	Assembly required	Assembly required	Assembled as fabricated	Assembly required
Scalability	Limited	Limited	Yes	Yes
MicroElectronic integratability	No	Yes for SOI bulk processes	Yes	No
Device geometry	Two-dimensional high aspect ratio	Two-dimensional high aspect ratio	Multi-layer Two-dimensional	Very flexible Three-dimensional
Processing	Parallel processing at the wafer level	Parallel processing at the wafer level	Parallel processing at the wafer level	Serial processing

The evaluation of a fabrication process for an application requires the assessment of a number of factors:

- The process-critical dimension (i.e., the smallest dimension that can be fabricated)
- The process precision (i.e., dimensional accuracy or nominal device dimension)
- Materials available for fabrication
- Assembly requirements to produce a functioning device
- Process scalability (i.e., can large quantities of devices be produced?)
- Integrability with other fabrication processes (e.g., microelectronics)

A large assortment of MEMS fabrication processes have been developed, but they may be grouped into three broad categories, which are discussed in further detail in subsequent sections.

- Lithographie, Galvanoformung, Abformung (LIGA)
- Bulk micromachining
- Sacrificial surface micromachining

Figure 3.1 shows the basic concepts of each fabrication category. Bulk micromachining and sacrificial surface micromachining are frequently silicon based and are generally very synergistic to the microelectronics industry since they tend to use common tool sets.

Bulk micromachining utilizes wet- or dry-etch processes to produce an isotropic or anisotropic etch profile in a material. Bulk micromachining can create large MEMS structures (tens of microns to millimeters thick) that can be used for applications such as inertial sensing or fluid flow channels. Commercial applications of bulk micromachining have been available since the 1970s. These applications include pressure sensors, inertial sensors, and ink-jet nozzles.

Sacrificial surface micromachining (SSM) is a direct outgrowth of the processes of the microelectronic industry and the materials used are largely silicon based. This technology has had several commercial successes in the last decade, including in optical mirror arrays and inertial sensors. Both these applications include integrated microelectronics for sensing and control functions. This technology is generally limited to film thicknesses of 2–6 μm; however, the resulting devices are assembled as fabricated. This gives SSM technology a significant advantage for applications involving large arrays of devices. Also, SSM technology has a path toward integration of electronics with the MEMS structures that will allow for control or sensing applications.

LIGA technology was demonstrated in the 1980s. This technology can fabricate devices with small critical dimension and high aspect ratio (i.e., thickness or width) from metallic materials that can be electroplated. This provides advantages in applications requiring a broad set of materials. However, assembly of large numbers or arrays of devices is an issue.

3.3 LIGA

The LIGA process[13] is capable of making complex structures of electroplatable metals with very high aspect ratios and thicknesses of several hundred microns. The LIGA process utilizes x-ray lithography, thick resist layers, and electroplated metals to form complex structures. Since x-ray synchrotron radiation is used as the exposure source for LIGA, the mask substrate is made of materials transparent to x-rays (e.g., silicon nitride, polysilicon). An appropriate mask-patterned layer would be a high atomic weight material (e.g., gold).

The LIGA fabrication sequence shown schematically in Figure 3.2 starts with the deposition of a sacrificial material such as polyimide, which is used for separating the LIGA part from the substrate after fabrication. The sacrificial material should have good adhesion to the substrate yet be readily removed when desired. A thin seed layer of material is then deposited, which

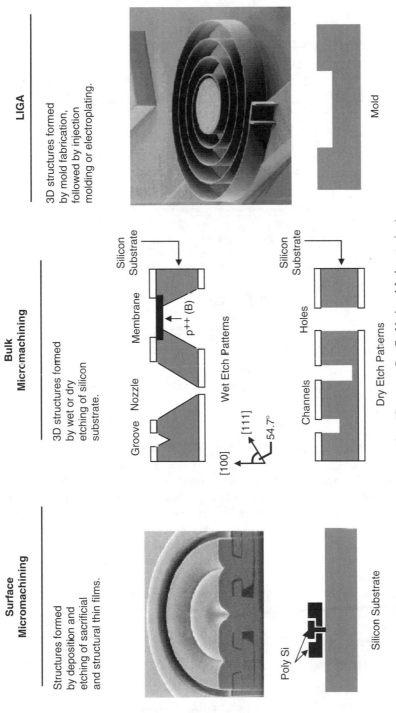

FIGURE 3.1 MEMS fabrication technology categories. (Courtesy: Sandia National Laboratories.)

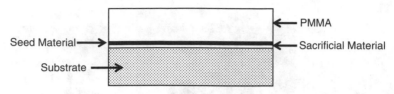

a. Substrate with sacrificial material, seed material, and PMMA applied.

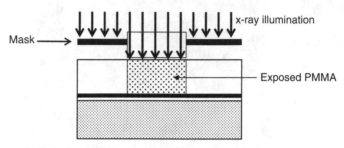

b. Exposing PMMA with x-ray synchrotron radiation.

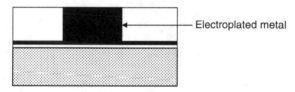

c. Electroplated metal in the developed PMMA mold.

FIGURE 3.2 LIGA fabrication sequence.

will enable the electroplating of the LIGA base material. A frequently used seed material would be a sputter-deposited alloy of titanium and nickel. Then a thick layer of the resist material, polymethylmethacrylate (PMMA), is applied. The synchrotron provides a source of high-energy collimated x-ray radiation needed to expose the thick layer of resist material. The exposure system of the mask and x-ray synchrotron radiation can produce vertical sidewalls in the developed PMMA layer. The next step is the electroplating of the base material (e.g., nickel) and polishing of the top layer of the deposited base material. Then the PMMA and sacrificial material are removed to produce a complete LIGA part.

Since LIGA can produce metal parts, magnetic actuation is feasible. Figure 3.3 shows an assembled LIGA mechanism. Assembly of LIGA devices for large-scale manufacturing is a challenging issue.

3.4   BULK MICROMACHINING

Bulk micromachining uses wet-[14] and dry etching methods[15,16] to achieve isotropic and anisotropic etches of features in materials. In order to manufacture items of

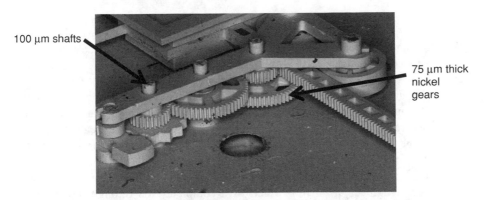

100 μm shafts

75 μm thick nickel gears

FIGURE 3.3 Assembled LIGA fabricated mechanism. (Courtesy: Sandia National Laboratories.)

practical interest, a number of different aspects of the etch processes need to be considered:

- Masking
- Etch selectivity due to crystallographic orientation or materials
- Etch stop and endpoint detection

3.4.1 WET ETCHING

Wet etching is purely a chemical process that can be isotropic in amorphous materials such as silicon dioxide and directional in crystalline materials such as silicon. Contaminants and particulates in this type of process are purely a function of the chemical purity or of chemical system cleanliness. Agitation of the wet chemical bath is frequently used to aid the movement of reactants and by-products to and from the surface. Agitation will also aid the uniformity of etch, since the by-products may be in the form of solids or gases that must be removed. A modern, wet-chemical bench will usually have agitation, temperature, and time controls as well as filtration to remove particulates.

The etching of silicon dioxide (SiO_2) is a common wet-etch process employed in a surface micromachining release etch or etch of isotropic features. This may be done with water to HF mixture in the ratio of a 6:1 by volume.

$$SiO_2 + 6HF \rightarrow H_2 + SiF_6 + 2H_2O \qquad (3.1)$$

Since HF is consumed in this reaction, the concentration will decrease as the etch proceeds, which would require that more HF be added to maintain concentration. Alternatively, a buffering agent could be used to help maintain the concentration and pH in this reaction. Equation (3.2) shows the chemical reaction that would enable NH_4F to be used as a buffering agent in the HF etches.

$$NH_4F \leftrightarrow NH_3 + HF \qquad\qquad (3.2)$$

Wet-etching methods can also be used on crystalline materials to achieve aniso-tropic directional etches. For example, a common directional wet etchant for crystalline silicon is potassium hydroxide (KOH). KOH etches 100 times faster in the (1 0 0) direction than the (1 1 1) direction. Patterned silicon dioxide can be used as an etch mask for these types of etches. Very directional etches can be achieved with these techniques as illustrated in Figure 3.4. Note the angular features (54.7°) that can be etched in silicon. Table 3.2 lists some of the common etchants for crystalline silicon and their selectivity.

If there are no etch stops in a wet-etching process the two options available to the process engineer are a timed etch or a complete etch through the material. A timed etch is difficult to control accurately due to the many other variables in the process such as temperature, chemical agitation, purity, and concentration. If this is not satisfactory, etch stops can be used to define a boundary for the etch to stop on. There are several etch-stop methods that can be utilized in wet etching:

- p^+ (boron diffusion or implant) etch stop
- Material-selective etch stop
- Electrochemical etch stop

Boron-doped silicon has a greatly reduced etch rate in KOH. The use of born-doped regions, which are either diffused or implanted, has been used either to form features or as an etch stop as seen in Figure 3.5. Also, a thin layer of a material such as silicon nitride, which has a greatly reduced etch rate, can be deposited on a material to form a membrane on which etching will stop.

An electrochemical etch stop can also be used as shown in Figure 3.6. Silicon is a material that readily forms a silicon oxide layer, which will impede etching of the bulk material. The formation of the oxide layer is a reduction–oxidation reaction that can be impeded by a reverse-biased p–n junction, which prevents the current

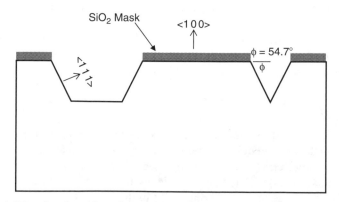

FIGURE 3.4 Directional etching of crystalline silicon.

TABLE 3.2
Common Crystalline Silicon Etchants' Selectivity and Etch Rates

Etchant	Etch Rate
$18HF + 4HNO_3 + 3Si \rightarrow 2H_2SiF_6 + 4NO + 8H_2O$	Nonselective
$Si + H_2O + 2KOH \rightarrow K_2SiO_3 + 2H_2$	{1 0 0} 0.14 μ/min
	{1 1 1} 0.0035 μ/min
	SiO_2 0.0014 μ/min
	SiN_4 not etched
Ethylene diamine pyrocatechol (EDP)	{1 0 0} 0.75 μ/min
	{1 1 1} 0.021 μ/min
	SiO_2 0.0002 μ/min
	SiN_4 0.0001 μ/min
Tetramethylammonium hydroxide (TMAH)	{1 0 0} 1.0 μ/min
	{1 1 1} 0.029 μ/min
	SiO_2 0.0002 μ/min
	SiN_4 0.0001 μ/min

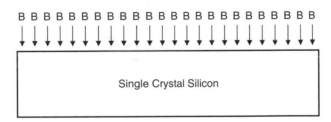

(a) Implant Boron in Single Crystal Silicon wafer

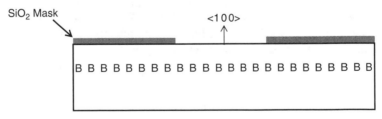

(b) Deposit and Pattern Silicon Dioxide Etch Mask

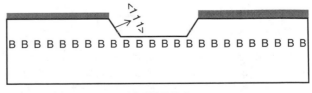

(c) KOH Etch

FIGURE 3.5 Boron-doped silicon used to form features or an etch stop.

Container

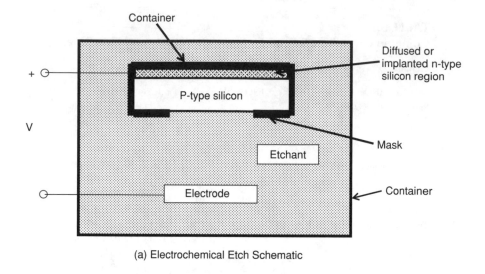

Diffused or
implanted n-type
silicon region

+ ⊙

P-type silicon

V

Etchant

Mask

⊙

Electrode

Container

(a) Electrochemical Etch Schematic

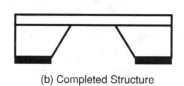

(b) Completed Structure

FIGURE 3.6 Electrochemical etch stop process schematic.

flow necessary for the reaction to occur. The p–n junction can be formed on a p-type silicon wafer with an n-type region diffused or implanted with an n-type dopant (e.g., phosphorus, arsenic) to a prescribed depth. With the p–n junction reverse biased, the p-type silicon will be etched because a protective oxide layer cannot be formed and the etch will stop on the n-type material.

3.4.2 PLASMA ETCHING

Plasma etching offers a number of advantages compared to wet etching:

- Easy to start and stop the etch process
- Repeatable etch process
- Anisotropic etches
- Few particulates

Plasma etching includes a large variety of etch processes and associated chemistries that involve varying amounts of physical and chemical attack. The plasma provides a flux of ions, radicals, electrons, and neutral particles to the surface to be etched. Ions produce both physical and chemical attack of the surface, and the radicals contribute to chemical attack.

The details and types of etch chemistries involved in plasma etching are varied and quite complex. This topic is too voluminous to be discussed in detail here, but there exist a number of excellent references on this subject.[15] The proper choice of these chemistries produces various etch rates and selectivity of material etch rates, which is essential to the integration of processes to produce microelectronics or MEMS devices. Fluoride etch chemistries is one of the most widely studied for silicon etches. Equations (3.3), (3.4), and (3.5) illustrate some of the fluoride reactions involved in the etching of silicon, silicon dioxide, and silicon nitride, respectively. There are a number of feed gases that can produce the free radicals involved in these reactions:

$$Si + 4F^* \rightarrow SiF_4 \tag{3.3}$$

$$3SiO_2 + 4CF_3^+ \rightarrow 2CO + 2CO_2 + 3SiF_4 \tag{3.4}$$

$$Si_3N_4 + 12F^* \rightarrow 3SiF_4 + 2N_2 \tag{3.5}$$

The anisotropy of the plasma etch can be increased by the formation of nonvolatile fluorocarbons that deposit on the sidewalls as seen in Figure 3.7. This process is

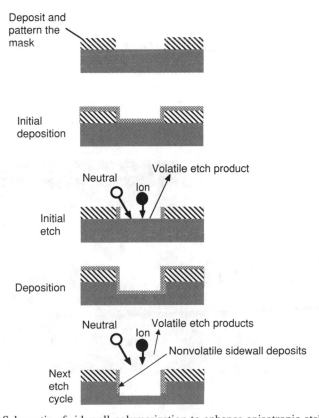

FIGURE 3.7 Schematic of sidewall polymerization to enhance anisotropic etching.

called *polymerization* and is controlled by the ratio of fluoride to carbon in the reactants. The sidewall deposits produced by polymerization can only be removed by physical ion collisions. Etch products from the resist masking are also involved in the polymerization.

Etch *endpoint detection* is important in controlling the etch depth or minimizing the damage to underlying films. Endpoint detection is accomplished by analysis of the etch effluents or spectral analysis of the plasma glow discharge to detect.

The type of plasma etches include reactive ion etching (RIE), high-density plasma etching (HDP), and deep reactive ion etching (DRIE). RIE utilizes low-pressure plasma. Chlorine (Cl)-based plasmas are commonly used to etch silicon, GaAs, and Al. RIE may damage the material due to the impacts of the ions. The damage can be mitigated by annealing at high temperatures. HDP etches utilize magnetic and electric fields to dramatically increase the distance that free electrons can travel in the plasma. HDP etches have good selectivity of Si to SiO_2 and resist. The DRIE etch cycles between the etch chemistry and deposition of the sidewall polymer, which enables the high aspect ratio and vertical sidewalls attainable with this process.[16]

Figure 3.8 shows two sample applications of bulk micromachining utilizing DRIE to produce deep channels and an electrostatic resonator.

3.5   SACRIFICIAL SURFACE MICROMACHINING

The basic concept of surface micromachining fabrication process has had its roots as far back as in the 1950s and 1960s with electrostatic shutter arrays[17] and a resonant gate transistor.[11] However, it was not until the 1980s that surface micro-machining utilizing the microelectronics toolset received significant attention.

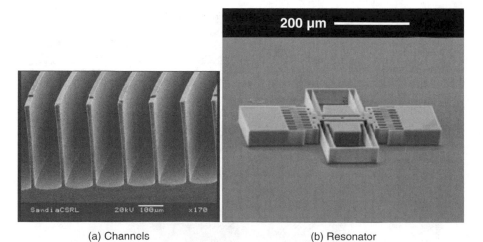

(a) Channels   (b) Resonator

FIGURE 3.8 Bulk micromachined channels and resonator. (Courtesy: Sandia National Laboratories.)

Howe and Muller[18] provided a basic definition of polycrystalline silicon surface micromachining, and Fan et al.[19] illustrated an array of mechanical elements such as fixed-axle pin joints, self-constraining pin joints, and sliding elements. Pister et al.[20] demonstrated the design for microfabricated hinges, which enable the erection of optical mirror elements.

Surface micromachining is a fabrication technology based upon the deposition, patterning, and etching of a stack of materials upon a substrate. The materials consist of alternating layers of a *structural* material and a *sacrificial* material. The sacrificial material is removed at the end of the fabrication process via a *release etch*, which yields an assembled mechanical structure or mechanism. Figure 3.9 illustrates the fabrication sequence for a cantilever beam fabrication in a surface micromachine process that has two structural layers and one sacrificial layer.

Surface micromachining uses the planar fabrication methods common to the microelectronics industry. The tools for depositing alternating layers of structural and sacrificial materials, photolithographical patterning, and etching the layers have their roots in the microelectronics industry. Etches of the structural layers define the shape of the mechanical structure, while the etching of the sacrificial layers define the anchors of the structure to the substrate and between structural layers. Deposition of a low-stress structural layer is a key goal in a surface micromachine process. From a device-design standpoint, it is preferable to have a slightly tensile average residual stress with minimal or zero residual stress gradient, which eliminates the design consideration of structural buckling. The stress in a thin film is a function of the deposition conditions such as temperature. A postdeposition anneal is frequently used to reduce the layer stress levels. For polysilicon the anneal step can require several hours at 1100°C.

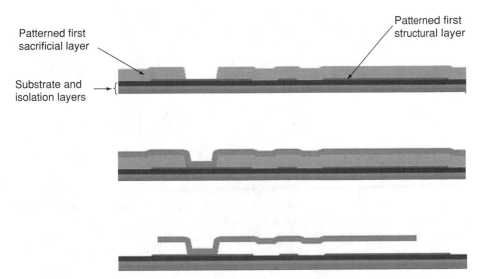

FIGURE 3.9 Surface micromachined cantilever beam with underlying electrodes showing the effect of topography induced by conformal layers.

TABLE 3.3
Example Surface Micromachining Technologies
Material Systems

Structural	Sacrificial	Release	Application
PolySi	SiO_2	HF	SUMMiT V™
SiN	polySi	XeF_2	GLV™
Al	Resist	Plasma etch	TI DMD™
SiC	PolySi	XeF_2	MUSIC™

Note: SUMMiT™ — Sandia Ultra-planar, Multi-level MEMS Technology
GLV™ — Grating Light Valve (Silicon Light Machines)
TI DMD™ — Digital Mirror Device (Texas Instruments)
MUSIC™ — Multi User Silicon Carbide (FLX micro)

Polycrystalline silicon (polysilicon) and silicon dioxide are a common set of structural and sacrificial materials, respectively, used in surface micromachining. The release etch for this situation is HF, which readily etches silicon dioxide but minimally attacks the polysilicion layers. A number of different combinations of structural, sacrificial materials and release etches have been utilized in surface micromachining processes. Table 3.3 summarizes a sample of surface micromachining material systems that have been utilized in commercial and foundry processes. Material system selection depends on several issues such as the structural layer mechanical properties (e.g., residual stress, Young's modulus, hardness, etc.) or the thermal budget required in the surface micromachining processes, which may affect additional processing necessary to develop a product.

Even though surface micromachining leverages the fabrication processes and tool set of the microelectronics industry, there are several distinct differences and challenges shown in Table 3.4. The surface micromachine MEMS devices are generally larger and they are composed of much thicker films than microelectronic devices. The repeated deposition and patterning of the thick films used in surface micromachining will produce a topography of increasing complexity as more layers are added to the process. Figure 3.9 shows the topography induced on an upper structural layer by the patterning of lower levels caused by the conformal films deposited by processes such as chemical vapor deposition (CVD). Figure 3.10 shows a scanning electron microscopic image of this effect in an inertial sensor made in a two-level surface micromachine process.

In addition to the topography induced in the higher structural levels by the patterning of lower structural and sacrificial layers, there are two significant process difficulties encountered. The first difficulty results from the anisotropic plasma etch used for the definition of the layer features to attain vertical sidewalls. The topography in the layer will inhibit the removal of material in the steps of the topographical features. This is illustrated in Figure 3.11, which shows there is an increased vertical layer height at the topographical steps that prevents removal of

TABLE 3.4
A Comparison of MEMS and Microelectronics

Criteria	Microelectronics	MEMS
Feature size	Sub-micron	1–3 μm
Device size	Sub-micron	~50 μm to 1mm
Materials	Silicon-based	Varied (silicon, metals, plastics)
Fundamental devices	Limited Set:	Widely Varied:
	Transistor, capacitor, resistor	Fluidic, mechanical, optical, electrical elements (sensors, actuators, switches, mirrors, etc.)
Fabrication process	Standardized:	Varied:
	Planar silicon process	Three main categories of MEMS fabrication processes plus variants.
		• Bulk micromachining
		• Surface micromachining
		• LIGA

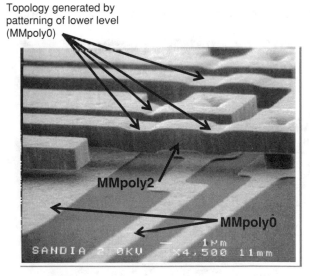

Topology generated by
patterning of lower level
(MMpoly0)

MMpoly2

MMpoly0

FIGURE 3.10 Scanning electron microscope image of topography in a two-level surface micromachine process. (Courtesy: Sandia National Laboratories.)

material at these discontinuities. This will give rise to the generation of small particles of material, *stringers*, that can either be attached to the underlying layers or float away during the release etch as shown Figure 3.12. Stringers can hamper a MEMS device from functioning properly due to mechanical interference or electrical shorting.

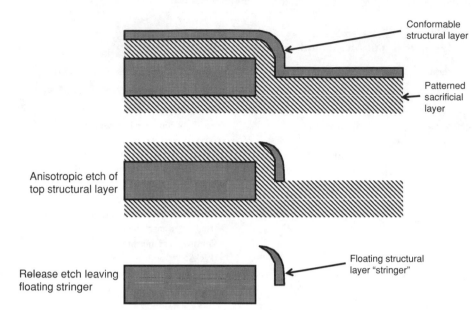

FIGURE 3.11 Illustration of stringer formation at a topographical discontinuity.

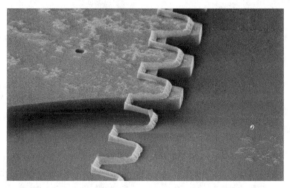

FIGURE 3.12 Scanning electron microscope image of a stringer that was formed and floated to another location on the die during the release etch. (Courtesy: Sandia National Laboratories.)

The second process difficulty is the challenge of photolithographic definition of layers with severe topography. The photoresist coating is difficult to apply and the depth of focus will lead to a decreased resolution of patterned features.

The application of chemical mechanical polishing (CMP) to surface microma-chine MEMS processes directly addresses the issues of topography as shown in Figure 3.13. CMP was originally utilized in the microelectronics industry for global planarization,[21] which is needed as the levels of electrical interconnect increase. CMP planarization was first reported in the MEMS by Nasby et al.[22,23] Figure 3.14 shows a linkage that has been fabricated in a surface micromachined process,

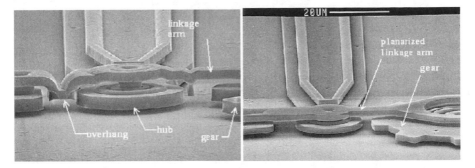

(a) Example of a conformable layer   (b) Example of topography removed
                                                by Chemical Mechanical Polishing

FIGURE 3.13 Example of a linkage fabricated in SUMMiT™ with and without CMP. (Courtesy: Sandia National Laboratories.)

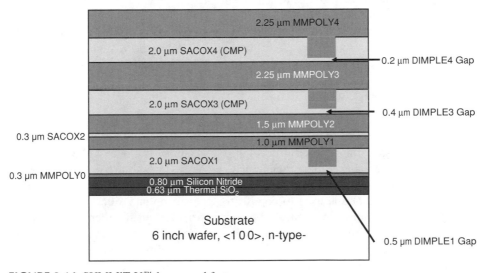

FIGURE 3.14 SUMMiT V™ layers and features.

SUMMiT™ (Sandia National Laboratories, Albuquerque, New Mexico), before and after CMP, was included in the process. In addition to solving the fabrication issues of topography, the use of CMP also aids in realizing designs without range of motion and interference constraints imposed by topography issues. CMP will also aid in the development of MEMS optical devices by enhancing the optical quality of surface micromachined MEMS mirrors.[24]

The release etch is the last step in the surface micromachine fabrication sequence. For a polysilicon surface micromachine process, the release etch involves a wet etch in HF to remove the silicon dioxide sacrificial layers. The removal of the sacrificial layers will yield a mechanically free device capable of motion. For very long or wide structures, etch-release holes are frequently incorporated into the structural layers to provide access for HF to the underlying sacrificial silicon dioxide. This will reduce the etch-release process time. Since the MEMS device is immersed in a liquid during the release etch, an issue is the adhesion and stiction of the MEMS layers upon removal from the liquid release etchant.[25] Since poly-silicon surfaces are hydrophilic the removal of liquids from the MEMS device can be problematic. Surface tension of the liquid between the MEMS layers will produce large forces, pulling the layers together. Stiction of the MEMS layers after the release etch can be addressed in several ways:

- Making the MEMS device very stiff to resist the surface tension forces
- Fabricating a bump (i.e., *dimple*) on the MEMS surfaces, which will prevent the layers from coming into large area contact
- Using a fusible link to hold the MEMS device in place during the release etch, which can be mechanically or electrically removed subsequently[26]
- Using a release process, which avoids the liquid meniscus during drying, such as supercritical carbon dioxide drying[27] or freeze sublimation[28]
- Use a release process that will make the surface hydrophobic, by using self-assembled monolayer (SAM) coatings.[29] It has been reported that SAM coatings also have the affect of reducing friction and wear

3.5.1 SUMMiT V™

An example of a surface micromachined MEMS fabrication process is SUMMiT (Sandia Ultra-planar, Multi-level MEMS Technology), a state-of-the-art five-level surface micromachine process developed by Sandia National Laboratories.[30,31] SUMMiT processing utilizes standard IC processes, which are optimized for the thicker films required in MEMS applications. Low-pressure chemical vapor deposition (LPCVD) is used to deposit the polysilicon and silicon dioxide films. Optical photolithography is utilized to transfer the designed patterns on the mask to the photosensitive material that is applied to the wafer (e.g., photoresist or resist). Reactive ion etches are used to etch the defined patterns into the thin films of the various layers. A wet chemical etch is also used to define a hub feature, as well as the final release etch of the SUMMiT process. Figure 3.14 schematically shows the layers and features in the SUMMiT V process. The SUMMiT V process uses 14 photolithography steps and masks to define the required features. Table 3.5 lists the layer and mask names and a summary of their use.

The SUMMiT fabrication process begins with a bare n-type, <100> silicon wafer. A 0.63 μm layer of SiO_2 is thermally grown on the bare wafer. This layer of oxide acts as an electrical insulator between the single-crystal silicon substrate and the first polycrystalline silicon layer (MMPOLY0). A 0.8 μm thick layer of low-stress silicon nitride (SiN_x) is deposited on top of the oxide layer. The nitride layer is an electrical insulator, but it also acts as an etch stop protecting the underlying oxide from wet etchants during processing. The nitride layer can be patterned with the NITRIDE_CUT mask to establish electrical contact with the

TABLE 3.5
SUMMiT V™ Layer Names, Mask Names, and Purposes

SUMMiT V™ Layer	Mask	Purpose
NITRIDE	NITRIDE_CUT	Electrical contact to the substrate
MMPOLY0	MMPOLY0	Electrical interconnect
SACOX1	DIMPLE1_CUT	Dimple
	SACOX1_CUT	Anchors
MMPOLY1	MMPOLY1	Structural layer definition
	PIN_JOINT	Hub formation
SACOX2	SACOX2	Hub formation
MMPOLY2	MMPOLY2	Structural layer
SACOX3	DIMPLE3_CUT	Anchors
	SACOX3_CUT	
SACOX3		Dimple
MMPOLY3	MMPOLY3	Structural layer definition
SACOX4	DIMPLE4_CUT	Dimple sacrificial layer definition
	SACOX4_CUT	
MMPOLY4	MMPOLY4	Structural layer definition

substrate. A 0.3-μm thick layer of doped polycrystalline silicon known as MMPOLY0 is deposited on top of the nitride layer. MMPOLY0 is not a structural layer, but is usually patterned for use as a mechanical anchor, electrical ground, or electrical wiring layer. Following MMPOLY0 deposition, the first sacrificial layer of 2 μm of silicon dioxide (SACOX1) is deposited. SACOX1 is a *conformable* layer that will reflect any patterning of the underlying MMPOLY0 layer. Upon deposition of the SACOX1 layer, dimples are patterned and etched into the oxide. The dimples (primarily used for antistiction purposes) are formed in the MMPOLY1 (the next polysilicon deposition) by filling the holes etched into the SACOX1 layer. The dimple depth is controlled via timed 1.5 μm deep etch.

Following the dimple etches, the SACOX1 layer is patterned again with the SACOX1_cut mask and etched to form anchor sites through the depth of SACOX1 to the MMPoly0 layer. With the anchor sites defined, a 1-μm thick layer of doped polysilicon (MMPOLY1) is deposited. MMPOLY1 deposited over the SACOX1 layer will be anchored or bonded to MMPOLY0 at the SACOX1 cuts and will also act as an electrical connection between MMPOLY0 and MMPOLY1.

The MMPOLY1 layer can be patterned with the MMPOLY1 mask to define a pattern in the polysilicon layer, or the PIN_JOINT_CUT mask to define a feature used in the formation of a rotational hub or pin-joint structure. The hub or pin-joint is defined at the PIN_JOINT_CUT site by the combination of an anisotropic reactive ion etch and a wet etch to undercut the MMPOLY1 layer. This feature will be used to form a *captured rivet head* for the hub or pin-joint.

A 0.3-μm layer of silicon dioxide, SACOX2 is then deposited and patterned with the SACOX2 mask. The SACOX2 is deposited by an LPCVD process, which is conformable and will deposit on the inside wall of the hub structure. The thickness of SACOX2 defines the clearance of the hub structure. SACOX2 can also be used as a hard mask to define MMPOLY1 using the subsequent etch that also defines MMPOLY2.

Upon completion of the SACOX2 deposition, pattern, and etch, a 1.5-μm thick layer of doped polysilicon, MMPOLY2 is deposited. Any MMPOLY2 layer material that is deposited directly upon MMPOLY1 (i.e., not separated by SACOX2) will be bonded together. Following the MMPOLY2 deposition, an anisotropic reactive ion etch is performed to etch MMPoly2 and composite layers of MMPOLY1 and MMPOLY2 (laminated together to form a single layer 2.5-μm thick). The MMPOLY2 etch will stop on silicon dioxide, hence MMPOLY1 will be protected by any SACOX2 on top of MMPOLY1 and the SACOX2 layer can be used as a hard mask to define a pattern in MMPOLY1.

At this point in the SUMMiT V process all the layers have been *conformable* (i.e., assume the shape of the underlying patterned layers). To enable the addition of subsequent structural and sacrificial levels without the fabrication and design constraints of the conformable layers, CMP is used to planarize the sacrificial oxide layers. With the MMPOLY2 etch complete, approximately 6 μm of TEOS (tetraethoxysilane) silicon dioxide (SACOX3) is deposited. CMP is used to planarize the oxide to a thickness of about 2 μm above the highest point of MMPOLY2. Following planarization, SacOx3 is patterned and etched to provide dimples and anchors to the MMPOLY2 layer using the DIMPLE3_CUT and SACOX3_CUT masks, respectively. The DIMPLE3_CUT etch is performed by etching all the way through the SACOX3 layer, stopping on MMPOLY2. Then 0.4 μm of silicon is deposited to backfill the dimple hole to provide the 0.4 μm standoff distance. The processing of the DIMPLE3 feature will provide a repeatable standoff distance.

A 2-μm thick layer of doped poly (MMPoly3) is deposited on the CMP planarized SACOX3 layer. The MMPOLY3 layer will be flat and not have the topography due to the patterning of the underlying layers. This will ease design constraint on the higher levels and enhance the use of MMPOLY3 and MMPOLY4 layers as mirror surfaces in optical applications. The MMPOLY3 layer is patterned and etched using the MMPOLY3 mask.

The processing for the SACOX4 and MMPOLY4 layers proceeds using the SACOX4_CUT, DIMPLE4_CUT, and MMPOLY4 mask in an analogous fashion to the SACOX3 and MMPOLY3 layers, except that the DIMPLE4 standoff distance is 0.2 μm.

Release and drying of the SUMMiT V die are the final fabrication steps. The device is released by etching all the exposed silicon dioxide away with a 100:1 HF:HCl wet etch. Following the wet release etch, a drying process can be employed using either simple air evaporation, supercritical CO_2 drying,[27] or CO_2 freeze sublimation.[28] The choice of the drying process will depend upon the design of the particular devices. Structures that are very stiff will be less sensitive to the surface tension forces, and they can be processed by simple air drying. Supercritical CO_2 drying processing for large devices is a better option.

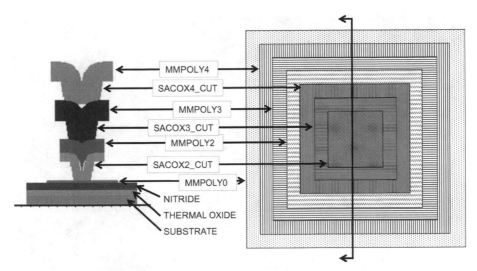

FIGURE 3.15 Masks and cross-section of a post composed of anchored layers.

Figure 3.15 illustrates the SUMMiT V masks and layers to fabricate a post containing all the structural layers. For this particular structure the dimple and the hub capabilities of SUMMiT are not utilized.

The SUMMiT V sacrificial surface micromachine fabrication process is capable of fabricating complex mechanisms and actuators. The ability to fabricate a low-clearance hub enables the rotary mechanisms and gear reduction systems shown in Figure 3.16. Figure 3.17 shows a vertically erected mirror that is held in place by elastic snap hinges. The vertical mirror is mounted upon a rotationally indexable table driven by an electrostatic comb drive actuator. SUMMiT V has also been used to fabricate large arrays of devices that are enabled by the fact that surface micro-machined devices are assembled when they are fabricated.

3.6 INTEGRATION OF ELECTRONICS AND MEMS TECHNOLOGY

The integration of electronic circuitry with MEMS technology becomes essential for sensing applications, which require increased sensitivity (e.g., Analog Devices ADXL accelerometers[32]), or actuation applications, which require the control of large arrays of MEMS devices (e.g., Texas Instruments Digital Mirror Device [DMD®][33]). For sensor applications the packaging integration of a MEMS device and an electronic ASIC becomes unacceptable when the parasitic capacitances and wiring resistances impact sensor performance (i.e., RC time constants of the integrated MEMS system are significant). For actuation applications such as a large array of optical devices that require individual actuation and control circuitry, a packaging solution becomes untenable with large device count.

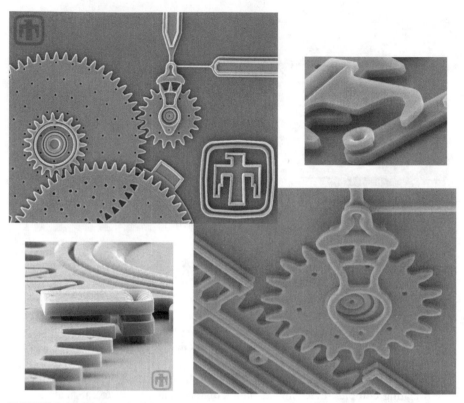

FIGURE 3.16 Rack and pinion drive, gear reduction system fabricated in SUMMiT V™. (Courtesy: Sandia National Laboratories.)

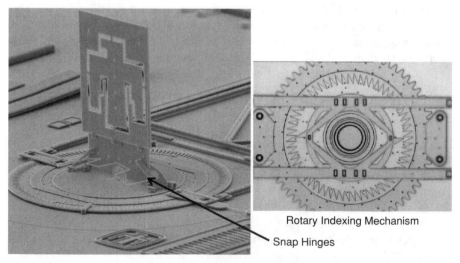

Rotary Indexing Mechanism

Snap Hinges

FIGURE 3.17 Rotary indexing device and vertically erected mirror with snap hinges fabricated in SUMMiT V™. (Courtesy: Sandia National Laboratories.)

Of the three MEMS fabrication technologies previously discussed, surface micromachining is the most amenable to integration with electronics to form an integration of electonics and MEMS technology (IMEMS) process. There are several challenges to the development of an IMEMS process:

- *Large vertical topologies:* Microelectronic fabrication requires planar substrates due to the use of precision photolithographic processes. Surface micromachine topologies can exceed 10 μm due to the thickness of the various layers.
- *High-temperature anneals:* The mitigation of the residual stress of the surface micromachine structural layers can require extended period time at high temperatures (such as several hours at 1100°C for polysilicon). This would have adverse effects due to the thermal budget of microelectronics that is limited due to dopant diffusion and metallization.

There are three strategies for the development of an IMEMS process.[34]

- *Microelectronics first:* This approach overcomes the planarity constraint imposed by the photolithographic processes by building the microelectronics before the nonplanar micromechanical devices. The need for extended high temperature anneals is mitigated by the selection of MEMS materials (e.g., aluminum, amorphous diamond[35]), and selection of the microelectronic metallization (e.g., tungsten instead of aluminum), which make the MEMS and microelectronic processing compatible. Examples of this IMEMS approach include an all-tungsten CMOS process that was developed by researchers at the Berkeley Sensor and Actuator Center[36] seen in Figure 3.18. The TI DMD (Texas Instruments Incorporated, Dallas, TX)[33] uses the microelectronics first approach and utilizes an aluminum structural layer MEMS and photoresist sacrificial layer MEMS, which enables low-temperature processing.
- *Interleave the microelectronics and MEMS fabrication:* This approach may be the most economical for large-scale manufacturing since it optimizes and combines the manufacturing processes for MEMS and microelectronics. However, this requires extensive changes to the overall manufacturing flow in order to accommodate the changes in the microelectronic device or the MEMS device. Analog devices has developed and marketed an accelerometer and gyroscope that illustrates the viability and commercial potential of the interleaving integration approach.[32]
- *MEMS fabrication first:* This approach fabricates, anneals, and planarizes the micromechanical device area before the microelectronic devices are fabricated, which eliminates the topology and thermal processing constraints. The MEMS devices are built in a trench, which is then refilled with oxide, planarized, and sealed to form the starting wafer for the CMOS processing as seen in Figure 3.19. This technology was targeted for inertial sensor

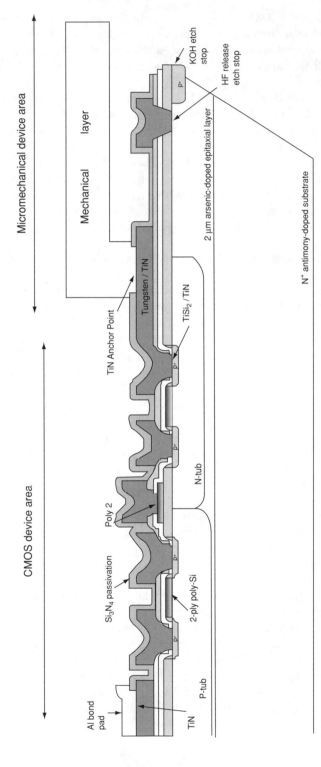

FIGURE 3.18 Microelectronics first approach to MEMS-microelectronics process integration.

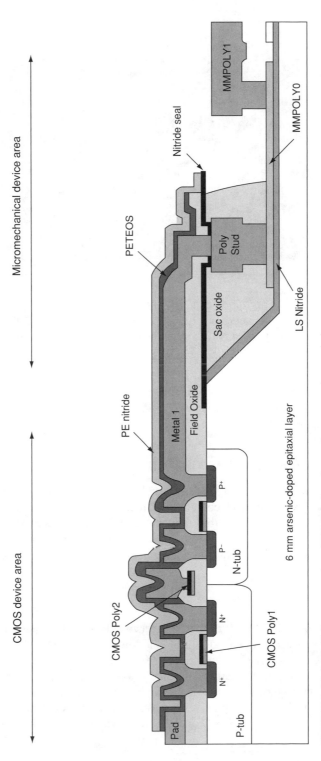

FIGURE 3.19 MEMS first approach to MEMS-microelectronics process integration.

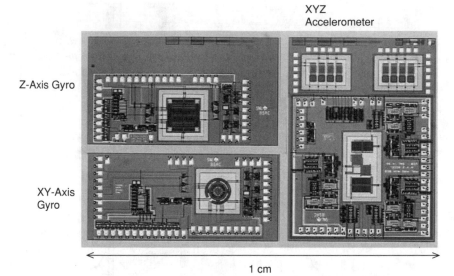

FIGURE 3.20 Inertial measurement unit fabricated in the MEMS first approach to MEMS-microelectronics process integration method. Designed at University of California, Berkeley, Berkeley Sensor Actuator Center. Fabricated by Sandia National Laboratories.

applications. Prototypes were designed by the Berkeley Sensor and Actuator Center (BSAC), University of California, and fabricated by Sandia National Laboratories shown in Figure 3.20.

3.7 ADDITIONAL MEMS MATERIALS

In addition to silicon-based materials and electroplated metals that have been discussed for use in MEMS technologies, a number of other materials are available, which may have unique properties that enable particular applications. For example, the high-temperature properties of silicon carbide, the hardness of diamond and silicon carbide, or the low deposition temperatures of silicon–germanium alloys and diamond.

3.7.1 SILICON CARBIDE

Silicon carbide (SiC) has outstanding mechanical properties, particularly at high temperatures. Silicon is generally limited to lower temperatures due to a reduction in the mechanical elastic modulus above 600°C and a degradation of the electrical pn-junctions above 150°C. Silicon carbide is a wide bandgap semiconductor (2.3–3.4 eV), which suggests the promise of high-temperature electronics.[37] It also has outstanding mechanical properties of hardness, elastic modulus, and wear resistance,[38] as seen in Table 3.6. SiC does not melt but sublimes above 1800°C, and it

TABLE 3.6
Comparative Properties of Silicon, Silicon Carbide, and Diamond

Property	3C-SiC	Diamond	Si
Young's modulus E (GPa)	448	800	160
Melting point (°C)	2830 (*sublimation*)	1400 (*phase change*)	1415
Hardness (kg/mm^2)	2840	7000	850
Wear resistance	9.15	10.0	$\ll 1$

has excellent chemical properties as well. Therefore, SiC is an outstanding material for harsh environments.[39]

SiC has a large number (>250) crystal variations,[40] polytopes. Of these polytopes, 6H-SiC and 4H-SiC are common for microelectronics and 3C-SiC are attractive for MEMS applications. Technology exists for the growth of high-quality 6H-SiC and 4H-SiC 50 mm wafers. Single-crystal 3C-SiC wafers have not been produced but 3C-SiC can be grown on (100–150 mm) Si wafers. However, polycrystalline 3C-SiC wafers are available.

The chemical inertness of SiC or polycrystalline SiC presents challenges for the micromachining of these materials. Uses of conventional RIE techniques for SiC result in relatively low etch rates compared to polysilicon surface micromachining, and the etch selectivity of SiC to either Si or SiO$_2$ is poor, which makes them inadequate etch stop materials.

An approach for patterning SiC is a micromolding technique.[41] The micromold process consists of forming mold upon a substrate and depositing the material, which fills the mold, and covering the surface. The surface is then polished such that only the material within the mold remains. Therefore, the micromolding process is able to bypass the RIE etch rate and selectivity issues to yield a planarized wafer that is amenable to multilayer processing.

SiC micromachining technologies have been used to fabricate prototype devices[42] that are required to operate under extreme conditions of temperature, wear, and chemical environments. However, control of the in-plane stress and stress gradients of SiC is still under development.

3.7.2 SILICON–GERMANIUM

Polycrystalline silicon–germanium alloys (poly-Si$_{1-x}$Ge$_x$) have been extensively investigated for electronic devices, but they also present some attractive features as a MEMS material.[43] Poly-Si$_{1-x}$Ge$_x$ has a lower melting temperature than silicon and it is more amenable to low-temperature processes such as annealing, dopant activation, and diffusion than silicon. Poly-Si$_{1-x}$Ge$_x$ offers the possibility of a MEMS mechanical material with properties similar to polysilicon, but the fabrication processing can be accomplished as low as 650°C. This will make poly-Si$_{1-x}$

Ge_x an attractive micromachining material for monolithic integration with micro-electronics, which requires a low thermal budget.[44]

Also, a surface micromachining process can be implemented utilizing poly-$Si_{1-x}Ge_x$ as the structural film, poly-Ge as the sacrificial film with a release etch of hydrogen peroxide when $x < 0.4$. Poly-Ge can be deposited as a highly conformable material that enables many MEMS structures.

3.7.3 DIAMOND

Diamond and hard amorphous carbon form a promising class of materials that have extraordinary properties, which promote new applications for MEMS devices. The various amorphous forms of carbon such as amorphous diamond (aD), tetrahedral amorphous carbon (ta-C), and diamond-like carbon (DLC) have hardness and elastic modulus properties that approach crystalline diamond, which has the highest hardness (~100 GPa) and elastic modulus (~1100 GPa) of all materials.[45] The appeal of this class of materials for MEMS designers is the extreme wear resistance, hydrophobic surfaces (i.e., stiction resistance), and chemical inertness. Recent progress has been achieved in the area of surface micromachining and mold-based processes[46,47] and a number of diamond MEMS devices have been demonstrated.[48,49] The use of diamond films in MEMS is still in the research stages. Recent progress in stress relaxation of the diamond films[50,51] at 600°C has been essential to the development of diamond as a MEMS material.

3.7.4 SU-8

EPON SU-8 (Shell Chemical) is a negative, thick, epoxy-photoplastic high aspect ratio resist for lithography.[52] This UV-sensitive resist can be spin coated in a conventional spinner to thicknesses ranging from 1 to 300 μm. Up to 2-mm thicknesses can be obtained with multilayer coatings. SU-8 has very suitable mechanical and optical properties and chemical stability; however, it has the disadvantages of adhesion selectivity, stress, and resists stripping. SU-8 adhesion is good on silicon and gold, but on materials such as glass, nitrides, oxides, and other metals the adhesion is poor. In addition, the thermal expansion coefficient mismatch between SU-8 and silicon or glass is large.

SU-8 has been applied to MEMS fabrication[52,53] for plastic molds or electro-plated metal micromolds. Also SU-8 MEMS structures have been used for micro-fluidic channels, and biological applications.[54]

3.8   CONCLUSIONS

Three categories of micromachining fabrication technologies have been presented; bulk micromachining, LIGA, and sacrificial surface micromachining.

Bulk micromachining is primarily a silicon-based technology that employs wet chemical etches and reactive ion etches to fabricate devices with high aspect ratio. Control of the bulk micromachining etches with techniques such as *etch stops* and material selectivity is necessary to make useful devices. Commercial applications utilizing bulk micromachining are available such as accelerometers and ink-jet nozzles.

LIGA is a fabrication technology that utilizes x-ray synchrotron radiation, a thick resist material and electroplating technology to produce high aspect ratio metallic devices.

Surface micromachining is a technology that uses thick films and processes from the microelectronic industry to produce devices. Surface micromachining employs two types of materials, a sacrificial material and a structural material, in alternating layers. A release process removes the sacrificial material in the last step in the process, which produces free function structural devices. Surface micromachining enables large arrays of devices since no assembly is required. Surface micromachining is also integratable with microelectronic for sensing and control. Two notable commercial applications of surface micromachining are the TI DMD and the Analog Devices ADXL accelerometers.

New materials are being developed to enhance MEMS applications. For example, silicon carbide is a hard, high-temperature material, which can withstand harsh environments. Silicon–germanium and diamond are materials that can be deposited at low temperatures, which enable increased MEMS process flexibility. SU-8 is an epoxy photo resin that can be used to produce high aspect ratio channels and molds.

REFERENCES

1. Sobel, D., *Longitude, The True Story of a Lone Genius Who Solved the Greatest Scientific Problem of His Time*, Penguin Books, New York, 1995.
2. Shockley, W., A unipolar field-effect transistor, *Proceedings of IRE*, 40, 1365, 1952.
3. Hoerni, J.A., Planar silicon transistors and diodes, *Proceedings of the IRE Electron Devices Meeting*, Washington, D.C., October 1960.
4. Hoerni, J.A., Method of Manufacturing Semiconductor Devices, U.S. Patent 3,025,589, issued March 20, 1962.
5. Feynman, R.P., There's plenty of room at the bottom, *Engineering and Science (California Institute of Technology)* February 1960, 22–36.
6. Feynman, R.P., There's plenty of room at the bottom, *JMEMS*, 1(1), March 1992.
7. Feynman, R.P., There's plenty of room at the bottom, http://nano.xerox.com/nanotech/feynman.html
8. Regis, E., *Nano: The Emerging Science of Nanotechnology*, Little Brown and Company, New York, 1995.
9. Maluf, N., *An Introduction to Microelectromechanical Systems Engineering*, Artech House, Inc., Norwood, MA, 2000.
10. Smith, C.S., Piezoresistive effect in germanium and silicon, *Physical Reviews*, April 4, 1954.
11. Nathanson, H.C., et al., The resonant gate transistor, *IEEE Transactions of Electronic Devices ED-14*, 117–133, 1967.
12. Petersen, K.E., Silicon as a mechanical material, *Proceedings of the IEEE*, 70(5), May 1982, 420–457.

13. Becker, E.W., et al., Fabrication of microstructures with high aspect ratios and great structural heights by synchrotron radiation lithography, galvanoforming, and plastic molding (LIGA process), *Microelectronic Engineering*, 4, 1986, 35.

14. Kovacs, G.T.A., Maluf, N.I., and Petersen, K.E., Bulk micromachining of silicon, *Proceedings of IEEE*, 86(8), 1536–1551, August 1998.

15. Shul, R.J. and Pearton, S.J., *Handbook of Advanced Plasma Processing Techniques*, Springer, New York, 2000.

16. US Patent 5,501,893, Method of Anisotropically Etching Silicon, Laermer, F., Schlp, A., Robert Bosch GmbH, issued March 26, 1996.

17. US Patent 2,749,598, Method of Preparing Electrostatic Shutter Mosaics, filed February 1, 1952, issued June 12, 1956.

18. Howe, R.T. and Muller, R.S., Polycrystalline silicon micromechanical beams, *Journal of Electrochemical Society: Solid-State Science and Technology*, 103(6), 1420–1423.

19. Fan, L.-S., Tai, Y.-C., and Muller, R.S., Integrated movable micromechanical structures for sensors and actuators, *IEEE Transactions of Electronic Devices*, 35(6), 724–730, 1988.

20. Pister, K.S.J., et al., Microfabricated hinges, *Sensors and Actuators A*, 33, 249–256, 1992.

21. Patrick, W., et al., Application of chemical mechanical polishing to the fabrication of VLSI circuit interconnections, *Journal of Electrochemical Society*, 138, 1778–1784.

22. Nasby, R.D., et al., Application of chemical mechanical polishing to planarization of surface micromachined devices, Solid State Sensor and Actuator Workshop, Hilton Head Is., SC, 48–53.

23. U.S. Patent 5,804,084, issued September 8, 1998, Use of Chemical Mechanical Polishing in Micromachining, Nasby, R.D., et al.

24. Yasseen, A., et al., Diffraction grating scanners using polysilicon micromotors, *Proceedings of the 8th International Conference on Solid — State Sensors and Actuators, and Eurosensors IX*, 1, Stockholm, Sweden, 206–209, 1995.

25. Legtenberg, R., Elders, J., and Elwenspock, M., Stiction of surface microstructures after rinsing and drying: model and investigation of adhesion mechanisms, *Proceedings of the 7th International Conference of Solid State Sensors and Actuators*, 198–201, 1993.

26. Fedder, G.K. and Howe, R.T., Thermal assembly of polysilicon microstructures, *Proceedings Microelectro Mechanical Systems '89*, 63–68.

27. Mulhern, G.T., Soane, D.S. and Howe, R.T., Supercritical carbon dioxide drying of microstructures, *Proceedings International Conference on Solid-State Sensors and Actuators (Transducers '93)*, Yokohama, Japan, 296–299, 1993.

28. Guckel, H., et al., Fabrication of micromechanical devices from polysilicon films with smooth surfaces, *Sensors and Actuators*, 20, 117–122, 1989.

29. Houston, M.R., Maboudian, R., and Howe, R.T., Self assembled monolayer films as durable anti-stiction coatings for polysilicon microstructures, *Proceedings Solid-State Sensor and Actuator Workshop, Hilton Head Is., SC, USA*, 42–47, 1996.

30. U.S. Patent 6,082,208, Method for Fabricating Five Level Microelectro Mechanical Structures and Five Level Micro Electro Mechanical Transmission Formed, Rodgers, M. S., et al., issued July 4, 2000.

31. SUMMiT™ (Sandia Ultra-planar, Multi-level MEMS Technology), Sandia National Laboratories, http://mems.sandia.gov.

32. Chau, K.H. and Sulouff, R.E., Technology for the high-volume manufacturing of integrated surface-micromachined accelerometer products, *Microelectronics Journal*, 29, 579–586, 1998.
33. Van Kessel, P.F., et al., A MEMS-based projection display, *Proceedings of the IEEE*, 86(8), August 1998.
34. Howe, R., Polysilicon integrated microsystems: technologies and applications, *Proceedings of Transducers 95*, 43–46, 1995.
35. Sullivan, J.P., et al., Developing a new material for MEMS: amorphous diamond, 2000 Fall MRS Meeting, November 27–December 1, 2000, Boston.
36. Yun, W., Howe, R., and Gray, P., Surface micromachined, digitally force balanced accelerometer with integrated CMOS detection circuitry, *Proceedings of IEEE Solid-State Sensor and Actuator Workshop '92*, p. 126, 1992.
37. Mehregany, M. and Zorman, C.A., SiC MEMS: opportunities and challenges for applications in harsh environments, *Thin Solid Films*, 355–356, 518–524, 1999.
38. Harris, G.L., *Properties of Silicon Carbide*, 1995.
39. Mehregany, M., et al., Silicon Carbide MEMS for Harsh Environments, *Proceedings of the IEEE,* 86(8), August 1998.
40. Fisher, G.R. and Barnes, P., *Philosophical Magazine*, B.61, 111, 1990.
41. Yasseen, A.A., Zorman, C.A., and Mehregany, M., Surface micromachining of polycrystalline sic films using microfabricated molds of SiO_2 and polysilicon. *Journal of Microelectromechanical Systems*, 8(3), 237–242, September 1999.
42. Rajan, N., et al., Fabrication and testing of micromachined silicon carbide and nickel fuel atomizers for gas turbine engines, *Journal of Microelectromechanical Systems*, 8(3), 251–257, September 1999.
43. Sedky, S., et al., Structural and mechanical properties of polycrystalline silicon germanium for micromachining applications, *Journal of Microelectromechanical Systems*, 7(4), pp. 365–372, December 1998.
44. Franke, A.E., Heck, J.M., and King, T.J., Polycrystalline silicon germanium films for integrated microsystems, *Journal of Microelectromechanical Systems*, 12(2), 160–171, April 2003.
45. Sullivan, J.P., et al., Developing a new material for MEMS: amorphous diamond, *Materials Research Society Symposium Proceedings*, 657, 2001.
46. Bjorkman, H., et al., Diamond replicas from microstructured silicon masters, *Sensors and Actuators*, 73, 24–29, 1999.
47. Ramesham, R., Fabrication of diamond microstructures from micro electro mechanical systems (MEMS) by a surface micromachining process, *Thin Solid Films*, 340, 1–6, 1999.
48. Bjorkman, H., et al., Diamond microstructures for optical microelectromechanical systems, *Sensors and Actuators*, 78, 41–47, 1999.
49. Shibata, T., et al., Micromachining of diamond film for MEMS applications, *Journal of Microelectromechanical Systems*, 9(1), 47–51, March 2000.
50. Friedmann, T.A., et al., Thick stress-free amorphous-tetrahedral carbon films with hardness near that of diamond, *Applied Physical Letters*, 71, p. 3820, 1997.
51. Friedmann, T.A. and Sullivan, J.P., Method of Forming a Stress Relieved Amorphous Tetrahedrally-Coordinated Carbon Film, U.S. Patent no. 6,103,305, issued Aug. 15, 2000.
52. Conradie, E.H. and Moore, D.F., SU-8 thick photoresist processing as a functional material for MEMS applications, *Journal of Micromechanical Microengineering*, 12, pp. 368–374, 2002.

53. Lorenz, H., et al., High-aspect-ratio, ultrathick, negative-tone near-UV photoresist and its applications for MEMS, *Sensors and Actuators A*, 64, 33–39, 1998.
54. Choi, Y., et al., High aspect ratio SU-8 structures for 3-D culturing of neurons, *2003 ASME International Mechanical Engineering Congress*, IMECE2003–42794.

4 Impact of Space Environmental Factors on Microtechnologies

M. Ann Garrison Darrin

CONTENTS

4.1 INTRODUCTION

Microelectromechanical systems (MEMS) devices used in space missions are exposed to many different types of environments. These environments include manufacturing, assembly, and test and qualification at the part, board, and assembly levels. Subsystem and system level environments include prelaunch, launch, and mission. Each of these environments contributes unique stress factors. An overview of these stress factors is given along with a discussion of the environments.

For space flight applications, microelectronic devices are often standard parts in accordance with NASA and Department of Defense (DoD)-generated specifications. Standard parts are required to be designed and tested for high reliability and long life through all phases of usage including storage, test, and operation. In contrast, there are no standard components in the MEMS arena for space flight application and no great body of knowledge or years of historical data and de-rating systems to depend on when addressing concerns for inserting devices in critical missions.

Civilian and military space missions impose strict design requirements for systems to stay within the allocations for size, weight, cost, and power. In addition, each system must meet the life expectancy requirements of the mission. Life expectancies of missions have continued to grow over the years from 6 months on early TIROS weather project to the current requirements of 30 years for the International Space Station (ISS). The Telstar 1 launched in 1962 had a lifetime of 7 months compared to Telstar 7 launched in 1999 with a 15+ year life expectancy. Albeit, the earlier Telstar weighed in at only 78 kg and cost US $6M compared to the 2770 kg Telstar 7 at a cost of US $200M. The geostationary operational environmental satellites (GOES) carry life expectancies greater than 5 years while current scientific satellites such as TERRA and AQUA have life expectancies greater than 6 years. Military-grade satellites such as Defense Satellite Communication System (DSCS) have design lives greater than 10 years.

To assure long-life performance, numerous factors must be considered relative to the mission environment when determining requirements to be imposed at the piece part (MEMS device) level. The high reliability required of all space equipment is achieved through good design practices, design margins (e.g., derating), and manufacturing process controls, which are imposed at each level of fabrication and assembly. Design margins ensure that space equipment is capable of performing its mission in the space environment. Manufacturing process controls are intended to ensure that a product of known quality is manufactured to meet the design requirements and that any required changes are made based on a documented baseline.

MEMS fall under the widely accepted definition of "part" as used by NASA projects; however, due to their often multifunctional nature, such as electrical and mechanical functions, they may well be better understood and treated as a component. The standard NASA definitions are:

- Part — One piece, or two or more pieces joined, which are not normally subjected to disassembly without destruction or impairment of designed use.
- Component — A combination of parts, devices, and structures, usually self-contained, which performs a distinctive function in the operation of the overall equipment.
- Assembly — A functional group of components and parts such as an antenna feed or a deployment boom.
- Subsystem — The combination of all components and assemblies that comprise a specific spacecraft capability.
- System — The complete vehicle or spacecraft made up of the individual subsystems.

4.2 MECHANICAL, CHEMICAL, AND ELECTRICAL STRESSES

4.2.1 THERMAL MECHANICAL EFFECTS

Spacecraft may receive radiant thermal energy from two sources: incoming solar radiation (solar constant, reflected solar energy, albedo) and outgoing long-wave radiation (OLR) emitted by the Earth and the atmosphere.[1]

High temperature causes adverse effects such as cracking, separation, wear-out, corrosion, and performance degradation on spacecraft system parts and components. These temperature-related defects may affect the electronic parts, the mechanical parts, and the materials in a spacecraft.

Although spacecraft environments rarely expose devices to temperatures below −55°C, a few spacecraft applications can involve extremely low temperatures. These cryogenic applications may be subjected to temperatures as low as −190°C. Cryogenic environments may be experienced by the electronics associated with solar panels or with liquid nitrogen baths used with ultrasensitive infrared detectors. The reliability of many MEMS improves at low temperatures but their parametric characteristics could be adversely affected. At such low temperatures many materials strengthen but may also become brittle. MEMS at cryogenic temperatures must be carefully selected. Evaluation testing is required for parts where cryogenic test data are not available.

It is important to evaluate the predicted payload environments to protect the system from degradation caused by thermal effects during ground transportation, hoisting operations, launch ascent, mission, and landing. The thermal effects on the spacecraft must be considered for each payload environment.

Spacecraft must employ certain thermal control hardware to maintain systems within allowable temperature limits. Spacecraft thermal control hardware including MEMS devices are usually designed to the thermal environment encountered on orbit which may be dramatically different from the environments of other phases of the mission. Therefore, temperatures during transportation, prelaunch, launch, and ascent must be predicted to ensure temperature limits will not be exceeded during these initial phases of the mission.[2]

The temperature of the spacecraft prelaunch environment is controlled by the supply of conditioned air furnished to the spacecraft through its fairing. Fairing air is generally specified as filtered air of Class 10,000 in a temperature range of 9 to 37°C and 30 to 50% relative humidity (RH).[3] The launch vehicle also controls the prelaunch thermal environment.

The design temperature range will have an acceptable margin that spacecraft typically require to function properly on orbit. In addition to the temperature range requirement, temperature stability and uniformity requirements can play an important role for conventional spacecraft hardware. The thermal design of MEMS devices will be subject to similar temperature constraints.

For the first few minutes, the environment surrounding the spacecraft is driven by the payload-fairing temperature. Prior to the fairing jettison, the payload-fairing temperature rises rapidly to 90 to 200°C as a result of aerodynamic heating. The effect of payload-fairing temperature rise may be significant on relatively low-mass MEMS devices if they are exposed. Fairing equipped with interior acoustic blankets can provide an additional thermal insulating protection.[2]

The highest ascent temperatures measured on the inside of the payload fairing have ranges from 27°C for Orbiter to 204°C for Delta and Atlas vehicles. For space flight missions, the thermal design for electronics is very critical since mission reliability can be greatly impacted. Systems are expected to operate continuously in orbit or in deep space for several years without performance degradation. For most low-power applications, properly designed heat conducting paths are sufficient to remove heat from the system. The placement of MEMS devices is therefore of great importance. The basic rule is that high power parts should not be placed too close to one another. This prevents heat from becoming concentrated in a localized area and precludes the formation of damaging "hot spots." However, some special high power boards require more intensive thermal management mechanisms such as ducting liquid cooling fluids through printed wiring assemblies and enclosures.

Aging effects of temperature are modeled after the Arrhenius or Eyring equations, which estimate the longevity of the subsystem. Similarly, the effects of voltage or power stress can be estimated using an inverse power model. From the microelectronic world comes a very mature understanding of the factors, such as the Arrhenius activation energy or the power law exponent, dependent on the part type being evaluated, and the expected dominant failure mechanism at the modeled stress level. However, the activation energy is based on electrochemical effects which may not be the predominant failure mode especially in the mechanical aspects of the MEMS device. Lack of an established reliability base remains a precautionary note when evaluating MEMS for space applications.

Accelerated stress testing can be used to activate latent failure mechanisms. The temperatures used for accelerated testing at the parts level are more extreme than the temperatures used to test components and systems. The latter temperatures exceed the worst-case predictions for the mission operating conditions to provide additional safety margins. High-temperature testing can force failures caused by material defects, workmanship errors, and design defects. Low-temperature testing can stimulate failures from the combination of material embrittlement, thermal contraction, and parametric drifts outside design limits.

Typical test levels derived from EEE parts include the following:

- High-temperature life test is a dynamic or static bias test usually performed between 125 and 150°C.
- Temperature–humidity testing is performed at 85°C and 85% RH (packaged).
- Temperature–pressure testing, also known as autoclave, is performed at 121°C at 15 to 20 psi (packaged).

Often, the space environment presents extreme thermal stress on the spacecraft. High-temperature extremes result from the exposure to direct sunlight and low temperature extremes arise because there is no atmosphere to contain the heat when not exposed to the sun. This cycling between temperature extremes can aggravate thermal expansion mismatches between materials and assemblies. Large cyclic temperature changes in temperature can cause cracking, separation, and other reliability problems for temperature sensitive parts. Temperature cycling is also a major cause of fatigue-related soldered joint failures.

For low-Earth orbit (LEO) and geosynchronous Earth orbit (GEO) satellites, the number and the temperature of thermal cycles experienced are dependent on the orbit altitude. For example, in a typical 550 km LEO, there would be approximately 15 eclipse cycles over a 24-h period. In a GEO, there would be only 90 cycles in a year with a maximum shadow time of 1.2 h per day. Trans-atmospheric temperature cycling depends on the orbit altitude and can have the same frequency as LEO; however, the amount of time in orbit is generally very short. Thermal cycling on planetary surfaces depends on the orbit mechanics in ascent acceleration relationship to the sun. For example, a system on the surface of Mars would endure a day or night cycle every 24.6 h. As Mars is 1.5 times farther away from the sun than the Earth, the sun's intensity is decreased by 43%. The lower intensity and attenuation due to the atmosphere on Mars limits the maximum temperature to 27°C. Temperature electronic assembly cycling is performed between high and low extremes (-65 to 125 or 150°C, typically).

4.2.2 MECHANICAL EFFECTS OF SHOCK, ACCELERATION, AND VIBRATION

Mechanical factors that must be considered are acceleration, random vibration, acoustic vibration, and shock. The effects of these factors must be considered during the launch phase, during the time of deployment of the system, and to a lesser degree, when in orbit or planetary trajectory. A folded or collapsed system or assembly is particularly sensitive to the effects of acoustic excitation generated during the launch phase. If the system contains large flat panels (e.g., solar panels), the effects of vibration and shock must be reviewed carefully since large flat surfaces of this type represent the worst-case condition.

Qualification at the component level includes vibration, shock, and thermal vacuum tests. Temperature effects precipitate most mechanically related failures; however, vibration does find some defects, which cannot be found, by temperature and vice versa. Data show that temperature cycling and vibration are necessary constituents of an effective screening program.

Acceleration loads experienced by the payload consists of static or steady state and dynamic or vibration loads. The acceleration and vibration loads (usually called load factors) are measured in "g" levels, "g" being the gravitational acceleration constant at sea level equal to 9.806 m/sec^2. Both axial and lateral values must be considered. For the Shuttle program, payloads are subjected to acceleration and vibration during reentry and during emergency or nominal landings (as well as the normal ascent acceleration and vibration-load events).

The vibration environment during launches can reach accelerations of 10 g at frequencies up to 1000 Hz. Vibration effects must also be considered in the design of electronic assemblies. When the natural frequency of the system and the forcing frequency coincide, the amplitude of the vibration could become large and destructive. Electronic assemblies must be designed so that the natural frequencies are much greater than the forcing frequencies of the system. In general, due to the low mass of MEMS devices, the effect of vibration will be minimal but assuredly must be considered with the packaging. For example, long wire bond leads have reached harmonic frequencies, causing failures during qualification tests.

Vibration forces can be stimulated by acoustic emissions. The acoustic environment of a spacecraft is a function of the physical configuration of the launch vehicle, the configuration of the propulsion system and the launch acceleration profile. The magnitude of the acoustic waves near the launch pad is increased by reflected energy from the launch pad structures and facilities. The first stages of a spacecraft (e.g., solid-rocket boosters) will usually provide a more demanding environment. The smaller the total vehicle size, the more stressed the payload is likely to be. The closer the payload is located to the launch pad, the more severe the acoustic environment will be.

Random vibration and multivibration tests (i.e., swept sine or frequency sine combined with random vibration) are typically performed. The use of vibration as a screen for electronic systems continues to increase throughout the industry (including airborne avionic, ground, military shipboard, and commercial applications).

Electronic assemblies in space applications must not degrade or fail as a result of mechanical shocks which are approximately 50 to $30,000g$ for 1.0 and 0.12 sec, respectively. To reduce effectively the negative effects of shock energy, electronic assemblies must be designed to transmit rather than absorb the shock. The assembly must therefore be stiff enough to achieve a rigid body response. Making individual electronic devices as low in mass as possible ensures that there is an overall increase in shock resistance of the entire assembly.

Commercial manufacturers of mass produced MEMS devices such as accelerometers for air bag deployment have incorporated shock and drop tests to their routing quality screens.

4.2.3 CHEMICAL EFFECTS

Chemical effects on MEMS devices are covered under three categories. These divisions are high-humidity environments, outgassing, and flammability. Moisture from high-humidity environments can have serious deleterious effects on the electronic assemblies particularly MEMS devices. Moisture causes corrosion, swelling, loss of strength, and affects other mechanical properties. To protect against moisture effects, electronic packages are typically hermetically sealed. However, many MEMS devices, especially those used for environmental sensors, cannot be hermetically sealed and require additional precautions. Systems are normally specified to operate in an environment of less than or equal to 50% RH. (A maximum of 50% RH is specified for the Space Shuttle.) Outgassing of moisture from sources such as wire insulation or encapsulants must be factored into the amount of humidity expected in an enclosed environment. Exposure during mission and launch is limited by the control of the environment. Prior to launch, the humidity of storage and processing must be controlled. Hermetic packaging schemes are preferred for space applications. The integrity of the package seal and the internal environment of the parts correlate directly with their long-term reliability. Moisture-related failure mechanisms might occur externally or internally to the packaged part. External moisture-related failure mechanisms include lead corrosion, galvanic effects, and dendrite growth. Internal moisture-related failure mechanisms can include metal corrosion or the generation of subtle electrical leakage currents, which disrupt the function of the device. The following factors are responsible for internal moisture-related failures: moisture, a path for the moisture to reach the active area, a contaminant, and for dendritic growth voltage. Space grade microcircuits, in contrast to MEMS devices, are protected by glassivating the die and controlling the sealing environment to preclude moisture and other contaminants. To be space qualified, a hermetic package requires a moisture content of no greater than 5000 ppm (by volume). This must be verified by performing an internal water vapor content check using residual gas analysis (RGA) in accordance with 1018.2 of MIL-STD-883. All space-qualified hermetic packages containing cavities receive a seal test to assure the integrity of the seal. Some space flight components, such as the computer of the Delta launch vehicle, are hermetically sealed assemblies. External to the parts, all assembled boards are conformally coated to reduce the chance for moisture or impurities to gain access to the leads, case, etc. Polymerics used in the conformal coating of assembled boards for NASA projects must comply with NASA-STD-8739.1 (formerly NHB 5300.4 (3J)). NASA has found the need to restrict certain materials in parts used for space flight. For instance, MIL-STD-975 prohibits the use of cadmium, zinc, and bright tin plating.

For outgassing requirements, an informal, but accepted, test specification used by all NASA centers is ASTM-E-595.[4] This specification considers the effects of a thermal vacuum environment on the materials. ASTM-E-595 does not set pass or fail criteria but simply lists the test results in terms of total mass loss (TML) and collected volatile condensable material (CVCM). The results are accumulated in the materials listings: NASA Reference Publication 1124 and MSFC-HDBK-527. The maximum acceptable TML and CVCM for general use are 1.0 and 0.10%, respectively. Materials used in near proximity or enclosed hermetically with optical components or surface sensors may require more stringent TML and CVCM percentages (such as TML < 0.50% and CVCM < 0.05%). Outgassing is of particular concern to EEE parts such as wire, cable, and connectors. Materials for space electronics must be able to meet a unique set of requirements. These are:

- Stability under high vacuum and thermal vacuum conditions
- Stability to the radiation of space (stability in high AO and UV conditions may also be required)
- Stability to sterilization conditions such as thermal radiation of outer space and ethylene oxide exposure
- Low outgassing under thermal vacuum conditions, nontoxicity of out gassed materials

4.2.4 ELECTRICAL STRESSES

Electrical stresses run the gamut from on-Earth damage as a result of electrostatic discharges through on-orbit damage due to degradation through radiation effects. Concerns for the prelaunch environment, launch, and postlaunch are addressed later in this chapter. The impact of radiation effects is addressed more fully in a dedicated chapter. The radiation issues are well worth an in-depth chapter as MEMS is a relatively new and emerging technology compared to microcircuits. For microelectronics there is a well-established knowledge base for space-grade parts. Unfortunately, there are no similar foundations for MEMS. Microelectronics for space are typically qualified to four standard total dose radiation levels, namely 3, 10, and 100 krads, and 1 megarad. Parts qualified to these levels are identified in MIL-M-38510 and MIL-PFR-19500 by the symbols M, D, R, and H, respectively. For the purposes of standardization, programs are encouraged to procure parts through the mentioned specifications using the designation, which most closely corresponds to their individual program requirements. The level of radiation hardness of a part must correspond to the expected program requirements. In addition, a safety margin (i.e., a de-rating factor) of 2 is frequently used. For example, if a system will be seeing a total dose level of 2 krads per year and the system is specified to operate for 5 years, then the individual part must either be capable of tolerating a total of 20 krads (10 krads × 2) or must be shielded so that it will not receive the total dose level of 2 krads per year. Any testing performed on actual MEMS devices is relatively recent. Commercial MEMS accelerometers such as the AD XL50 have been tested, and the IC component of the devices was found to be sensitive.[5,6] The author in one of these studies iterates the requirement that CMOS circuits in particular are known to degrade when exposed to low doses of ionizing radiation. Therefore, before MEMS can be used in the radiation environment of space, it is important to test them for their sensitivity to radiation ion-induced radiation damage.[6] In addition MEMS optical mirrors,[7] electrostatic, electrothermal, and bimorph actuators,[8] and RF relays[9] add to the rapidly growing database of components tested. In all fairness, these tests are performed on commercial grade MEMS as the concept of radiation-hardened space-qualified MEMS has yet to mature.

4.3 DESIGN THROUGH MISSION OPERATION ENVIRONMENTS

MEMS devices for space flight use are exposed to two application areas: design-through-prelaunch and launch-through-mission. The first phase includes the manufacture, qualification, integration, and test of the parts to the component level. The launch or mission environment includes the launch, lift-off, acceleration, vibration, and mission until the end-of-life (EOL).

The prelaunch period includes planning, procurement, manufacture, test, component assembly, and component acceptance testing. The procurement process for MEMS devices includes the fabrication run time and may well exceed the lengthy requirements of space grade microcircuits (48 to 70 weeks). Iterative runs must be considered when scheduling and planning for the incorporation of MEMS devices in space programs. Although vendors are claiming lead times for manufacturing consistent with the microcircuit world, the lack of high-volume manufacturing and the absence of low-cost packaging continue to keep most MEMS in a custom situation. Due to long lead times, devices spend a minimum of 10% of the prelaunch time span in the manufacture and test cycle; therefore, concerns about both handling and storage are of particular interest to space programs (based on the experiences in microelectronics). Board assembly and qualification take more than 20% of the prelaunch period. Integration and test at the board level takes approximately 6 to 18 months. This includes mechanical assembly, functional testing, and environmental exposure. Much time is spent in queuing for a mission. Factors such as budget negotiation and availability of the launch facilities and vehicle also contribute to the long time between program initiation and launch. It is not unusual for these time frames between initial plan and design to launch to stretch from 7 to 12 years as noted in Table 4.1. Proper handling control of MEMS devices during the prelaunch period is essential to avoid the introduction of latent defects that may manifest themselves in a postlaunch environment. Proper handling and storage require precaution to preclude damage from electrostatic discharge (ESD) and contamination. Temperature through test and storage should be maintained at $25 \pm 5°C$ and humidity should be held at $50 \pm 10\%$ RH. However, this requirement for ESD for the electronics runs counter to handling and storage precautions for MEMS devices. A chapter of this book is dedicated to handling and contamination control, and special storage requirements, which may well be required for MEMS devices in nonhermetic packaging.

Parts may degrade during the time between the manufacturing stage and the launch of the vehicle. This degradation generally proceeds at a much slower rate for nonoperating parts than for operating parts due to the lower stresses involved. Special precautions must be taken regarding humidity. Parts stored in a humid environment may degrade faster than operating parts that are kept dry by self-heating during operation. Keeping the parts in a temperature controlled, inert atmosphere can reduce the degradation that occurs during storage. Controls to prevent contamination are integral to good handling and storage procedures.

Most civilian contractors, and military space centers handle all EEE parts as if they were sensitive to ESD and have precautionary programs in place. These same precautions must be extended to MEMS devices once the devices have been singulated and released. NASA requirements for ESD control may be found in

TABLE 4.1
Time Span from Design Phase to Launch

Project	Initial Plan and Design	Launch	Duration (years)
TRMM	1985	1997	12
GRO or EGRET	1980	1991	11
COBE	1978	1989	11
ISTP	1985	1992–1993	8
TDRSS	1976	1983	7

TABLE 4.2
Cleanliness Requirements

Facility Type	Cleanliness Requirements in Parts per Million
Mechanical manufacturing	Not controlled
Electronic assembly	10,000
Electromechanical assembly	100
Inertial instrument	100
Optical assembly	100
Spacecraft assembly and test	100

NASA-STD-8739.7 ESD-control requirements are based on the requirements found in MIL-STD-1686, Electrostatic Discharge Control Program for Protection of Electrical and Electronic Parts, Assemblies and Equipment.

Manufacturing facilities consist of mechanical manufacturing, electronic manufacturing, spacecraft assembly and test, and special functions. Standard machine shops and mechanical assembly are part of the mechanical manufacturing facilities. In addition, plating and chemical treatment houses, adhesive bonding, and elevated treatment vendors are included. Aerospace facilities normally have operations performed under clean area conditions. In general, mechanical manufacturing steps are not performed in clean controlled areas. Certain assemblies such as electromechanical and optical components do need controlled clean rooms. Table 4.2 shows the different cleanliness requirements imposed in terms of particles per unit volume as defined in FED-STD-209. Cleanliness requirements are measured in particles (0.5 μm or larger) per cubic foot. Electronic part manufacturing facilities also require clean room environments for parts prior to sealing. Assembly of parts into the components and higher levels are normally performed under clean room (or area) influence of space environmental factors and NASA EEE parts selection and application conditions also. Assembly of spacecraft and test operations are often performed in large hangar bays. Depending on the particular instrument, special contamination controls may be required with optical equipment. Payload instruments that require cryogenic temperatures, RF isolation, or the absence of magnetic fields also require special handling.

4.4   SPACE MISSION-SPECIFIC ENVIRONMENTAL CONCERNS

The environmental concerns of the actual system mission are unique compared with those related to the test, prelaunch, and the launch environments. For instance, extreme vibrations and shock are not as prevalent during the mission as during test and take-off. On the other hand, radiation is definitely a major concern for systems operating in the mission environment, but there is little concern with radiation until the system leaves the Earth's atmosphere. The five mission-environmental factors that follow are: radiation, zero gravity, zero pressure, plasma, and atomic oxygen (AO), along with long-life requirements. These influences are reviewed in relation to their effects at the system and individual part levels.

A more in-depth discussion of the radiation environment is found in the chapter on space environment; however, some discussion of device level concerns is contained herein and would be applicable to device designer's incorporation of MOS components in their MEMS designs.

Commercial MEMS are designed to operate in our low radiation biosphere and the CMOS portions of the electronics can tolerate total radiation doses of up to 1 to 10 kRads. Terrestrial radiation levels are only about 0.3 rad/year so radiation damage is not normally an issue if you stay within the biosphere.[10]

There are primarily two types of radiation environments in which a system may be operated: a natural environment and a threat environment. Earth-orbiting satellites and missions to other planets operate in a natural environment. The threat environment is associated with nuclear explosions; this neutron radiation normally is a concern of non-NASA military missions. Irradiating particles in the natural environment consist primarily of high-energy electrons, protons, alpha particles, and heavy ions (cosmic rays). Each particle contributes to the total radiation fluence impinging on a spacecraft. The radiation effects of charged particles in the space environment cause ionization. Energy deposited in a material by ionizing radiation is expressed in "rads" (radiation absorbed dose), with 1 rad equal to 100 ergs/g of the material specified. The energy loss per unit mass differs from one material to another. Two types of radiation damage can be induced by charged particle ionization in the natural space environment: total dose effects and single event phenomena. In semiconductor devices, total dose effects can be time-dependent threshold voltage shifts, adversely affecting current gain, increasing leakage current, and even causing a loss of part functionality. A single-event phenomenon (SEP), which is caused by a single high-energy ion passing through the part, can result in either soft or hard errors. Soft errors (also referred to as single event upsets [SEUs]) occur when a single high-energy ion or high-energy proton causes a change in logic state in a flip-flop, register or memory cell of a microcircuit. Also, in low-power high-density parts with small feature sizes, a single heavy ion may cause multiple soft errors in adjacent nodes. Soft errors may not cause permanent damage. A hard error is more permanent. An example of hard error is when a single high-energy ion causes the four-layer parasitic silicon controlled rectifier (SCR) within a CMOS part to latch-up, drawing excessive current and causing loss of control and functionality. The part may burnout if the current is not limited. Single event latch-up (SEL) in CMOS microcircuits, single-event snapback (SES) in NMOS parts and single-event burnout (SEB) in power transistors are examples of hard errors that can lead to catastrophic art failures. Major causes of SED and latch-up are heavy ions. To valuate SED and latch-up susceptibility, the heavy-ion fluence is translated into linear energy transfer (LET) spectra. While the total dose radiation on a part may vary considerably with the amount of shielding between the part and the outside environment, the LET spectra (and hence the SED susceptibility) do not change significantly with shielding. SEU and latch-up problems are most critical for digital parts, such as memories and microprocessors, which have a large number of memory cells and registers. However, recent heavy-ion testing has shown that N-channel power MOSFETs are also susceptible to burnout caused by a single, high-energy heavy ion. A heavy ion passing through an insulator can sometimes result in a catastrophic error due to rupturing of the gate dielectric. This is known as single event gate rupture (SEGR) and it has been observed in power MOSFETs, SRAMs and EEPROMs. SEGR is a phenomenon that is presently being closely investigated by the space community. Microcircuits can be hard with respect to SED while being soft to the total dose effects, or vice versa.

In zero gravity, a significant reliability concern is posed by loose or floating particles during the process of manufacturing integrated circuits or discrete semi-conductor devices, loose conductive particles (e.g., solder balls, weld slag, flakes of metal plating, semiconductor chips, die attach materials, etc.) prior to sealing the package. In a zero-gravity environment, these particles may float about within the package and bridge metallization runs, short bond wires, and otherwise damage electronic circuitry. A thorough program of particle detection is necessary although the typical microcircuit programs may not be applicable to MEMS devices. Micro-circuits use a particle impact noise detection (PIND) Particle detection scheme (e.g., PIND screening). MIL-STD-883 and MIL-STD-750 both contain PIND test methods for testing microcircuits and discrete semiconductors, respectively. Both methods are required screens for space-level, standard devices in accordance with MIL-M-38510, MIL-PFR-19500, and MIL-STD-975. For MEMS devices having released structures such as cantilevers the use of a PIND test would fail good product, as the released structures would produce "chatter," negating the validity of the test. The use of particle capture test through stick tapes and other getter-type materials is encouraged. The inability to "blow off" particulate with an inert gas where release structures are present reinforces the need for an effective contaminant control program.

In space microgravity environments, atmospheres of hot, stagnant masses of gas can collect around sources of heat. Heat loss by unforced convection cannot occur without gravity. Heated masses of gas simply expand within the surrounding cooler and denser gaseous media. Heat sinks and fans can be used to prevent overheating in areas of anticipated heat generation. Unexpected heat producing events, such as an arc tracking failure of insulation or increasing power dissipation in a deteriorating capacitor, can rapidly lead to catastrophic failure by thermal runaway. Uncontrolled heating conditions can also lead to failure in low-pressure environments as heat loss by convection is effectively eliminated.

The postlaunch environment is one of near-zero atmospheric pressure. Atmos-pheric pressure changes as a function of altitude. The external pressure at high altitudes is minimal, thus the volume of existing and outgassed materials is forced to increase in accordance with Boyle's Law. The deep-space vacuum is less than 10^{-12} torr. Under these conditions, corrosive solids may sublimate and expand to cover exposed surfaces within the system. The corrosive power of these solids is enhanced by the fact that oxygen and other free ions are abundant in many orbits. The existence of free ions and active elements in the Earth's upper atmosphere makes it a much harsher environment than a laboratory on the Earth's surface. Two actions essential for enhancing the reliability of a satellite under such adverse conditions are: where possible use hermetically sealed parts and avoid the use of materials which outgas excessively or react to create corrosive material.

Outgassing of volatiles and toxic gases must be extremely low in the crew compartment areas. The maximum allowable levels for nonmetallics are defined in NASA specification MSFC-PA-D-67-l3. For manned space-flight (such as Apollo), conditions of 5 psi oxygen and 72 h of exposure, the total organics evolved must be less than 100 ppm.

To assure part performance in a zero-pressure environment, thermal vacuum testing is usually required at the component level. Zero-pressure environments cause more severe thermal stresses on parts. It was reported by Gibbel[11] that thermal or vacuum testing may yield a greater than 20°C temperature rise (at the high extreme) over a regular thermal or atmosphere test. The variance between how the piece part is tested and the environment in which the part will be used demonstrates the importance of temperature de-rating. Many times, extreme test temperatures are used to accelerate failure mechanisms. The near-perfect vacuum of the space environment provides little or no convective air cooling. All heat must be dissipated from the vehicle through radiation. Space-borne electronic equipment is cooled by conductive heat transfer mechanisms which transport dissipated heat to external radiating surfaces of the spacecraft. These conducting paths typically consist of thermally conductive pads, edge-connecting mechanisms, circuit-card fixtures, metal racks, and the system chassis.

The reduced pressure encountered in high-altitude operations can result in a reduced dielectric strength of the air in nonhermetically sealed devices. This permits an arc to be struck at a lower voltage and to maintain itself for longer, and may lead to contact erosion. Use of vented or nonhermetically sealed devices in high altitude or vacuum applications requires special precautions, such as additional de-rating.

In a low-pressure environment the likelihood of voltage flashover between conductors is increased. The voltage at which flashover occurs is related to gas pressure, conductor spacing, conductor material, and conductor shape. These relationships are plotted as Paschen's curves. Flashover resulting from corona discharge does not occur at voltages less than 200 V. Above that level, conductor separation, insulation, and conductor shapes must be carefully selected.

Within several hundred kilometers of the Earth, molecules in the upper atmosphere are ionized by solar ultraviolet and x-ray to form dense (up to 10^6 particles/cm^3) low-energy plasma. In this region, known as the ionosphere, plasma particles behave collectively because of the small range of individual particle influence (1 mm at shuttle orbit). Charged particles accumulate on spacecraft surfaces, creating differential charging and strong local electric fields. If a surface builds up sufficient electric potential, a high-energy discharge (arc) can blow away material and deposit it on optical or other sensitive equipment. The hot, thin plasma of the magnetosphere creates more devastating problems at the geosynchronous altitude. In the region above 1000 km, the electromagnetic influence of plasma particles extends over a kilometer or more. High-energy (greater than 100 keV) electron from plasma penetrates external spacecraft surfaces, accumulating inside on well-grounded conductors, insulators, and cables, causing strong electric fields and ultimately breakdown. Due to their high resistivities, dielectric surfaces can be charged to different potentials than metallic surfaces (which should be at spacecraft ground potential). Considering the effects of internal discharges is important when a system is expected to operate in an environment where penetrating radiation causes charging inside the system.

Internal discharges occur when ungrounded metal or dielectric surfaces collect enough charge from the plasma field so that the electric field generated exceeds the breakdown strength from the point of the deposited charge to a nearby point. Internal discharges have been suspected as the cause of a number of spacecraft performance anomalies. The conditions for discharging are dependent on the environment, the shielding provided by the spacecraft, the material, which is charging, and the geometry of the charged materials.

System response to internal charging depends on the location of the discharge and the sensitivity of the circuits. Charges that would go unnoticed on the exterior of a space system can be significant when they occur internally. Experiments on Long Duration Exposure Facility (LDEF)[12] have documented the phenomenon of spacecraft charging by plasma at low altitudes. The LDEF has been a wealth of information on the effects of the space environment.[13–17] LDEF was launched in 1984 and contained a package of 57 experiments placed in Earth orbit by the Space Shuttle for studying the effects of exposure to the environment of space. The LDEF was supposed to have been recovered after about 1 year. However, delays in the shuttle program meant that the package was not brought back until January 1990, just a few weeks before it would have reentered the atmosphere and been destroyed. One of the experiments measured long-term current drainage of dielectric materials under electric stress in space. Current leakage appeared to be much lower than predicted from ground simulations. The researchers believed that instead of gradual current drainage, instantaneous discharge to the space plasma reduced any excess charge. Carbon residue on the samples suggested breakdown of organic materials under the intense heat of an arcing discharge. The LDEF results suggested that comparing results from long-term space experiments and ground simulations was not fruitful. Simulating all space-environmental parameters during ground simulations is virtually impossible. In space, other environmental variables may alter or exacerbate plasma effects. This is an area of current research. Nonetheless, various options are available for testing and circumventing the effects of internal charging. For special missions, criteria can be generated that will eliminate or reduce internal discharge concerns.

The space station, orbiting at altitudes of 400 to 500 km, could lose considerable current to ambient plasma. Its solar arrays, 160 V cells connected end-to-end for high voltage and power efficiency, collect electrons from plasma, accumulating a substantial negative charge. To prevent the highly polarized station from losing large amounts of current, a plasma contactor generates a local high-density plasma to contact the ambient plasma, maintaining the system electric potential at zero.[18]

AO exists in significant amounts around low-Earth orbits and around Mars. AO is highly reactive and will react differently depending on the nature of the materials involved. AO effects were first detected during shuttle missions. Exposure to AO tends to cause metals to develop an oxide on their surface and polymers to lose mass and undergo a change in surface morphology. Due to their high reactivities with AO, polymers and other composites need to be protected. On an order of magnitude of scale, surfaces such as the solar arrays will be exposed to a stronger AO flux field than inboard components. The LEO range for AO exposure is 10^7 to 10^8 atoms/cm^3. Exposure to AO is a known detriment to Kapton$^{®}$ (DuPont High Performance Materials, Circleville, OH) wire as AO reduces the thickness of insulation materials and degrades their insulating properties. A thin, protective coating of silicon oxide is often used on Kapton solar array substrates for protection against AO threats.

4.5 CONCLUSION

This chapter is cursory and of an introductory nature giving merely an overview rather that handling any topic in depth. The consideration of inserting MEMS and microstructures in critical space flight programs must include the potential stresses that the piece, part, or component will be exposed to and each of their respective impact on the long-term survivability of the subsystem. In the reliability portion of this book there is a greater discussion on the combinations of stress factors from the various potential environments.

4.6 MILITARY SPECIFICATIONS AND STANDARDS REFERENCED

MIL-PFR-19500 General Specification for Semiconductors
MIL-M-38510 General Specification for Microelectronic Devices
MIL-STD-202 Test Methods for Electronic and Electrical Component Parts
MIL-STD-338 Electronic Design Reliability Handbook
MIL-STD-750 Test Methods for Semiconductor Devices
MIL-STD-883 Test Methods for Microelectronic Devices
MIL-STD-975 NASA Standard Electrical, Electronic, and Electromechanical (EEE) Parts List
MIL-STD-1540 (USAF) Test Requirements for Space Vehicles
MIL-STD-1541 (USAF) Electromagnetic Compatibility Requirements for Space Systems
FED-STD-209 Clean Room and Work Station Requirements, Controlled Environment

REFERENCES

1. James, B.F., The Natural Space Environment Effects on Spacecraft, in *NASA Reference Publication*, 1994.
2. Gilmore, D., *Spacecraft Thermal Control Handbook*. The Aerospace Press, Los Angeles, CA, 2002.
3. Loftus, J.P. and C. Teixeira, Launch systems, in Wertz, J.R. and W.J. Larson (eds), *Space Mission Analysis and Design*, 1999.
4. Standard test method for total mass loss and collected volatile condensable materials from outgassing in a vacuum environment, in *ASTM E595–93*, ANSI ASTM, 1999.
5. Lee, C.I., et al., Total dose effects on microelectromechanical systems (MEMS): accelerometers. *IEEE Transactions on Nuclear Science, Proceedings of the 1996 IEEE Nuclear and Space Radiation Effects Conference, NSPEC, July 15–19 1996*, 1996. 43(6 pt 1): p. 3127–3132.
6. Knudson, A.R., et al., Effects of radiation on MEMS accelerometers. *IEEE Transactions on Nuclear Science, Proceedings of the 1996 IEEE Nuclear and Space Radiation Effects Conference, NSPEC, July 15–19 1996*, 1996. 43(6): p. 3122–3126.
7. Miyahira, T.F., et al., Total dose degradation of MEMS optical mirrors. *IEEE Transactions on Nuclear Science*, 2003. 50(6): 1860–1866.
8. Caffey, J.R. and P.E. Kladitis, The effects of ionizing radiation on microelectromechanical systems (MEMS) actuators: electrostatic, electrothermal and bimorph. *17th IEEE International Conference on Micro Electro Mechanical Systems (MEMS)*. Maastricht MEMS, 2004 Technical Digest.
9. McClure, S.S., et al., Radiation effects in micro-electromechanical systems (MEMS): RF relays. *IEEE Transactions on Nuclear Science*, 2002. 49 (6): 3197–3202.
10. Janson, S., et al., Microtechnology for space systems, in *Proceedings of the 1998 IEEE Aerospace Conference. Part 1 (of 5), March 21–28 1998*, 1998. Snowmass at Aspen, CO, USA: IEEE Computer Society, Los Alamitos, CA.
11. Gibbel, M., *Thermal Vacuum vs Thermal Atmospheric Testing of Space Flight Electronic Assemblies*. The Gibbel Corporation, Montrose, CA.
12. Va, L.H., *LDEF Mission Document*. 1992, 1993, NASA.
13. Banks, B., M. Meshishnek, and R. Bourassa, LDEF materials, environmental parameters, and data bases. in *Proceedings of the LDEF Materials Workshop '91, November 19–22 1991*, 1992. Hampton, VA, USA: NASA, Washington, DC.
14. Kleis, T., et al., Low energy ions in the heavy ions in space (HIIS) experiment on LDEF. *Advances in Space Research*, 1996. 17(2): 163–166.
15. See, T.H., et al. LEO particulate environment as determined by LDEF, in *Proceedings of the 4th International Conference on Engineering, Construction, and Operations in Space, February 267 1990*, 1991. 19(1–4): 685–688.
17. Stevenson, T.J., LDEF comes home. *Materials Performance*, 1990. 29(10): p. 63–68.
18. Nama, H.K., Environmental interactions of the space station freedom electric power system. *Proceedings of the European Space Conference*, August 1991. ESA SP-320.

5 Space Radiation Effects and Microelectromechanical Systems

Stephen P. Buchner

CONTENTS

5.1 INTRODUCTION

The space environment presents a variety of hazards for spacecraft. Not only are there extremes of temperature and pressure to contend with, but the spacecraft must also withstand a constant onslaught of energetic ionized particles and photons that can damage both the spacecraft and its payload. Atomic oxygen (AO) poses a serious hazard because it corrodes materials with which it comes into contact, causing surface erosion and contamination of the spacecraft. High-energy photons

(UV, x-ray and gamma rays) that degrade the electrical, optical, thermal, and mechanical properties of materials and coatings are present. Charged particles also pose a danger to the spacecraft. For instance, plasmas are a significant hazard because they alter the spacecraft's electrical "ground" potential through the buildup of charge. After sufficient exposure, dielectrics may suddenly discharge, damaging sensitive electronic components in the process. Individual energetic ionized particles such as electrons, protons, alpha particles, and heavier ions are another hazard. They are able to penetrate the spacecraft's superstructure as well as the electronic, opto-electronic, and microelectromechanical systems (MEMS) devices contained on board. As they travel through matter, the particles collide with the atoms of the MEMS devices materials, which in turn, liberate charge and disrupt the lattice structure. Both of these effects contribute to the performance degradation of devices.

Radiation damage is sometimes gradual and at other times sudden. Gradual degradation is the result of cumulative radiation exposure, and is caused by total ionizing dose (TID) or displacement damage dose (DDD) or both. At some dose level the device may no longer function. For radiation sensitive parts, that dose may be quite small, whereas for radiation-hardened parts the level may be orders of higher magnitude. Sudden degradation sometimes occurs following the passage of a single particle through the device, and usually takes the form of a loss of data, disruption of normal operation, or even destructive failure. These effects are collectively known as single-event effects (SEEs) and have been the object of many investigations over the past two decades.

Over the years, these hazards have been responsible for numerous space mission failures. From reduced capability to total loss of the spacecraft, the associated financial losses have been significant.[1] Now that MEMS are being considered for space applications, particularly in microsatellites, which provide relatively little shielding against the harsh space environment, it is necessary to study how MEMS respond to all of the above-mentioned hazards.

5.1.1 THE SPACE RADIATION ENVIRONMENT

The radiation environment in space varies with both time and location, and models of particle flux include effects of the Sun, local magnetic fields, and galactic cosmic rays.

Radiation emitted by the Sun dominates the environment throughout the entire solar system. Solar emissions include both negatively charged electrons and positively charged ions that span the periodic table from hydrogen to uranium. These particles travel with velocities up to 800 km/sec.[2] Complicated processes that will not be discussed here are believed to be responsible for the electron emission from the Sun. To maintain electrical neutrality, the electrons "drag" positively charged particles with them as they speed away from the Sun. Positively charged solar emissions consist primarily of protons (85%) and alpha particles (14%), with the remaining 1% consisting of ions with atomic numbers greater than two. Both the relative and total numbers of solar particle emissions vary with time, exhibiting large increases during solar-particle events.

Characterizing space radiation environment requires knowledge of charge states and energies of the particles emitted by the Sun. In addition, the degree to which interactions between particles alter their charge states and energies as they travel through space must be determined. Electrons and ions spiral in opposite directions around the Sun's magnetic field lines in their journey away from the Sun. The resulting helical orbits are a function of the ions masses, charges, and velocities as well as the Sun's magnetic field strength. The particles emitted by the Sun form "solar wind." Solar wind is not constant, varying with both time and location. Temporal variations are due to changes in solar activity, whereas spatial variations are due to a number of factors, such as distance from the Sun, the effects of local magnetic fields, and to a lesser extent, interparticle scattering. Although the solar magnetic field decreases in strength with increasing distance from the Sun, the total magnetic field in the vicinity of certain planets, such as Earth and Jupiter, may be significantly greater because they contribute their own magnetic fields. Most of the particles streaming towards the Earth are deflected by the Earth's magnetic field. However, some become trapped in belts around the Earth where their densities are many times greater than in interplanetary space.

As already pointed out, the solar wind is not constant, fluctuating in intensity as a result of variable solar activity. Figure 5.1 shows that solar activity, as measured by the number of solar flare proton events, exhibits both long-term and short-term variations. Long-term variations are fairly predictable, consisting of periods of approximately 11.5 years. For 7 years the Sun is in its active phase characterized by an enhanced solar wind and an increase in the number of storms on the Sun's surface. Solar storms are either "coronal mass ejections" or "solar flares," both of

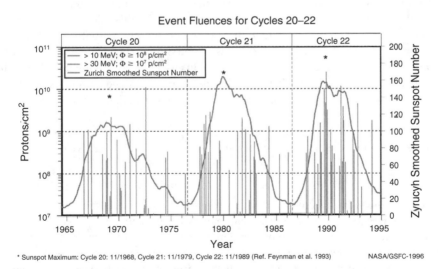

* Sunspot Maximum: Cycle 20: 11/1968, Cycle 21: 11/1979, Cycle 22: 11/1989 (Ref. Feynman et al. 1993)   NASA/GSFC-1996

FIGURE 5.1 Large solar proton events for cycles 20 to 22. The number of sunspots is superimposed on the graph.[2] (From J. Barth, Modelling Space Radiation Environments, *IEEE*, 1997.)

which produce large, short-duration increases in the solar wind. Solar storms can have a tremendous impact on the Earth's magnetic field. When enhanced solar wind associated with a solar storm reaches the Earth, it interacts strongly with the geomagnetic field, producing an intense electromagnetic pulse. The electromagnetic pulse can cause considerable damage not only to space hardware, but also to Earth-based infrastructure, as evidenced by the failure in 1989 of the electrical power grid serving Canada and the northeastern states of the U.S. Enhanced solar wind can also "pump up" the radiation belts by injecting large numbers of particles.

The distribution of particles in the solar wind generally decreases with increasing energy. At the same time, however, great variations in the particle energy spectra have been observed from one solar-flare event to another. Measurements indicate that 10 GeV is the upper energy limit of particles in the solar wind. Short-duration flux increases of up to five orders of magnitude have been observed near Earth following a solar event identified as a coronal mass ejection.[3]

The Sun's quiet phase typically lasts 4 years, and is characterized by a diminished solar wind and a reduction in the frequency of storms. During both the active and quiet phases of the Sun the occurrence of solar storms is random and, therefore, impossible to predict with certitude. Figure 5.1 illustrates that major solar storms occur randomly during each solar cycle. Predictions of how many solar storms to expect during a mission require the use of probabilistic techniques and can only be stated within certain confidence levels.[4] In general, long-term average predictions are more reliable than short-term predictions.

Galactic cosmic rays (GCRs), whose origins are believed to be outside the solar system, most likely in supernova millions of light years away, also contribute to the radiation environment. Although the GCR flux is relatively low, GCRs consist of fully ionized atoms, some of which have energies in the TeV range, making them capable of penetrating most spacecraft as well as the Earth's magnetosphere. Solar wind, whose direction is opposite to that of the cosmic rays, partially attenuates the cosmic ray flux. Therefore, during times of maximum solar activity, when solar wind is at its most intense, the cosmic ray flux is reduced. Figure 5.2 shows the relative fluxes of the nuclei that make up the cosmic rays and illustrates that there are very few cosmic rays with nuclear charge greater than that of iron ($Z = 28$). A detailed description of the radiation environment in space is beyond the scope of this book. Only a brief summary of the major aspects will be included here, and the interested reader is referred to the literature for a more comprehensive exposition.[2,5]

5.1.2 EARTH ORBITS

Predicting the radiation environment experienced by a spacecraft in orbit around the Earth requires knowledge of orbital parameters, such as apogee, perigee, and angle of inclination as well as launch date and mission duration. Some orbits are relatively benign from a radiation exposure point-of-view, whereas others are quite severe. For example, a spacecraft in a low-Earth equatorial orbit (LEO), where the radiation environment is relatively benign, would be expected to survive for many years, whereas in medium-Earth orbit (MEO), where the radiation belts are at their most

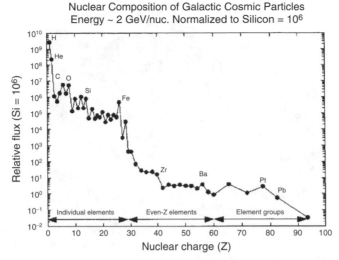

FIGURE 5.2 Relative abundances of galactic cosmic ray ions in interplanetary space.[2] (From J. Barth, Modelling Space Radiation Environments, *IEEE*, 1997.)

intense, the spacecraft might survive for only a few days. This dependence on orbit is a result of the complex structure of the Earth's magnetic field, which determines the shape of the radiation belts and attenuates the flux of solar particles and cosmic rays.

The magnetic field experienced by LEO spacecraft is dominated by the Earth's geomagnetic field, which may be assumed to be a bar magnet (dipole) located within the Earth. The axis of the bar magnet is tilted by 11° with respect to the Earth's axis of rotation and is also displaced from the Earth's center. The geomagnetic field, which, to first order, is independent of azimuthal angle (latitude), does vary significantly with both altitude and longitude. At a distance of about 5 Earth radii is the "shock" region where the solar wind and the geomagnetic fields interact strongly. Because magnetic field lines cannot cross, those from the Sun and the Earth "repel" each other and the solar wind is redirected around the Earth. This effectively shields the Earth from direct exposure to most solar particle radiation. On the Earth's "dark" side, solar wind has the shape of a cylinder with its axis directed along a line extending from the Sun through the Earth. The distortion on the "dark" side of the Earth extends to more than 100 Earth radii and is the region where particles are injected into the radiation belts.[2]

An important consequence of the interaction between the solar wind and the Earth's magnetic field is the presence of radiation belts, known as van Allen belts. These radiation belts are regions containing high fluxes of charged particles surrounding the Earth (and other planets with magnetic fields, such as Jupiter). For the Earth, there is an inner belt of mostly protons and electrons located at approximately 1.5 Earth radii in the equatorial plane, and an outer belt dominated by electrons at approximately 5 Earth radii. Figure 5.3 shows the two belts around

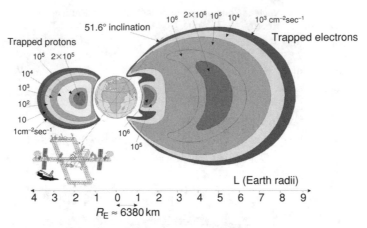

FIGURE 5.3 Artist's impression of the radiation belts. The protons on the left are separated in the figure from the electrons on the right. (From C. Dyer, Space Radiation Environment Dosimetry, NSREC Short Course, *IEEE*, 1998.)

the Earth. They take the form of "jug handles," approaching closer to the Earth's surface near the North and South Poles. Heavy ions are also present in the belts, but at much lower fluxes. Also, the outer belt, though dominated by electrons, is not devoid of protons. Some protons in the belts have energies of hundreds of MeV, making them very penetrating and, therefore, difficult to shield against. Most electrons in the belts have energies below 10 MeV, so shielding on the spacecraft is much more effective. Figure 5.4a shows the energy distributions as a function of altitude for protons and Figure 5.4b that for electrons. The highest energy protons and electrons have their maximum concentrations at about 1.5 Earth radii. To avoid the high radiation exposure levels, most spacecraft orbits avoid this region.

As a result of the displacement of the Earth's magnetic axis with respect to the center of the Earth, the magnetic field in the South Atlantic is much weaker, allowing protons and electrons to reach lower altitudes than at other locations on Earth. This produces the well-known South Atlantic Anomaly (SAA) where the radiation belts extend down to very low altitudes. Most spacecraft in LEO with large inclination will pass through the SAA where they will accumulate most of their radiation dose. Another characteristic of the Earth's magnetic field is that the magnetic field lines at the North and South Poles are perpendicular to the Earth's surface and connected to those emanating from the Sun. Therefore, the geomagnetic field does not deflect cosmic rays and solar particles from the North and South Poles. As a result, there are large fluxes of protons and heavy ions over both poles. Enhanced particle fluxes associated with solar storms are first apparent on Earth at the poles and signal future enhanced particle fluxes in the belts. Spacecraft in orbits that pass over the poles will be directly exposed to high particle fluxes during solar storms.

Given that the ions in the radiation belts originate primarily in the Sun, it is not surprising that the Sun's activity also affects the structure of those radiation belts. In

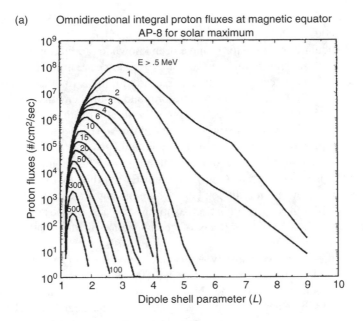

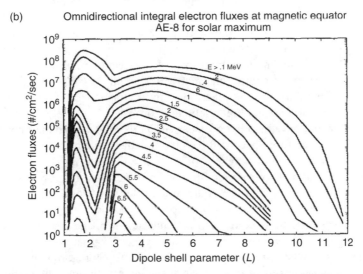

FIGURE 5.4 (a) Variation of omnidirectional integral proton flux with distance from the surface of the earth at the magnetic equator.[2] (From J. Barth, Modelling Space Radiation Environments, *IEEE*, 1997.) (b) Variation of omnidirectional integral electron flux with distance from the surface of the earth at the magnetic equator.[2] (From J. Barth, Modelling Space Radiation Environments, *IEEE*, 1997.)

particular, solar storms compress the belts on the side facing the Sun, forcing them to lower altitudes while at the same time populating them with additional charged particles. Particularly intense storms have been known to produce an extra radiation belt that lasts for several months in the "slot" region between the inner and outer electron belts. At distances greater than 5 Earth radii, the azimuthal component of the Earth's magnetic field is highly nonuniform and its mathematical description is very complex. The shifting of the geomagnetic fields as a result of solar storms will modify the radiation environment experienced by a spacecraft in Earth orbit; particularly those close to the edges of the radiation belts.

Most Earth orbits fall into one of three categories — LEO, highly elliptical orbit (HEO), and geostationary orbit (GEO). Medium-Earth orbits (MEOs) are generally avoided because they are in radiation belts where the high-radiation fluxes severely limit mission lifetimes. The radiation exposure in each of these orbits is very different due to the combined effects of geomagnetic shielding and the presence of the radiation belts.

LEOs in the equatorial plane typically have an altitude of only a few hundred kilometers (300 km for the Space Shuttle) and, therefore, spend most of their time below the radiation belts. At that height they are also shielded against solar particles and cosmic rays by the Earth's magnetosphere. As the angle of inclination increases, the orbits pass through the "horn" regions of the belts located at high latitudes. There the belts dip down closer to the Earth's surface and the particle flux is enhanced. The SAA is part of the southern "horn" region, and spacecraft in LEO regularly pass through it, obtaining a significant boost to their total radiation exposure. For orbit inclinations close to 90°, spacecraft pass near the magnetic poles where the magnetosphere is ineffective at shielding against solar particles and cosmic rays. Therefore, low altitude and low inclination orbits are much more benign than high altitude and high inclination orbits.

HEOs typically have their apogee near GEO (36,000 km) and their perigee near LEO (300 km). Therefore, spacecraft pass through the radiation belts twice per orbit where they experience high fluences of protons and electrons. Beyond the belts, spacecrafts are exposed for extended periods of time to cosmic rays and particles expelled during solar storms. HEOs are among the most severe from a radiation standpoint.

Spacecraft in GEO are exposed to the outer edges of the electron belts and to particles originating in cosmic rays and solar events. The magnetosphere provides some shielding against cosmic rays and solar particles, but storms on the Sun can compress the magnetosphere and reduce the effective particle attenuation. Components that are not well shielded will acquire a significant dose from the relatively low-energy electrons in the belts and from solar storms. In addition, the energy spectrum at GEO is considerably "harder" than in LEO because of the presence of high-energy galactic cosmic rays.

Launch date and mission duration must be factored into any calculation of radiation exposure in Earth orbit, particularly for orbits with high angles of inclination that approach the polar regions. For example, if launch date and mission duration occur entirely during a period of low solar activity where the Earth's geomagnetic fields extend to higher latitudes, the exposure to solar particles in the polar regions will be reduced. This will be somewhat compensated by the enhanced cosmic ray flux. Nevertheless, the total radiation exposure will not be as severe as during a period of high solar activity.

5.1.3 INTERPLANETARY SPACE

Missions in interplanetary space consist of a number of phases, each with a different radiation environment. The first phase typically involves placing the spacecraft into geosynchronous orbit, which requires it to pass through the heart of the Earth's radiation belts, sometimes more than once. During this first phase, the spacecraft will accumulate a significant radiation exposure. The spacecraft then spends some time in GEO before being injected into its interplanetary orbit. Depending on how long it spends in GEO, the spacecraft and the components on board could experience a substantial total dose due to the electrons in the outer region of the second electron radiation belt. Once the spacecraft has left GEO and is traveling in interplanetary space, it is exposed to radiation from the Sun and from cosmic rays. The total radiation flux is then due to solar activity, the distance between the Sun and the spacecraft, and cosmic ray contributions. The major danger to spacecraft lies in solar particle events that, as noted previously, can lead to an increase in the radiation flux to which a spacecraft is normally exposed by many orders of magnitude.

5.1.4 PLANETARY MISSIONS

The radiation environment around other planets varies greatly, depending on the strength of their magnetic fields and their distance from the Sun. For instance, the Earth's moon has no magnetic field and the radiation exposure there does not differ significantly from that in interplanetary space. In contrast, Jupiter's magnetic field is much stronger than that of the Earth. In addition, Jupiter's moons orbit within the intense radiation belts, requiring any mission to one of Jupiter's moons to use parts with a high degree of immunity to radiation and, if necessary, to shield the parts as well. Mars has a very weak magnetic field, which offers little "shielding" against cosmic rays and solar particles. The absence of a magnetic field around Mars also means that spacecraft in orbit around Mars will not encounter radiation belts, such as those on Earth.

5.2   RADIATION EFFECTS

Before a device can be used in space it must be qualified to ensure that it will survive the rigors of the space environment. Radiation qualification is one of many different qualification procedures that must be performed.[6] Others include temperature, pressure, and vibration. In the absence of specific guidelines for qualifying MEMS devices for a radiation environment, radiation test engineers make use of standard radiation qualification procedures that have been developed for microelectronics.

A radiation qualification procedure consists of a series of steps to ascertain whether a part will operate properly in a radiation environment. The first step is to define the environment by calculating its temporal and spatial compositions, that is, fluxes, energies, and masses of the ions. Computer models, such as Space Radiation®, CREME96, and SPENVIS are available for predicting the flux of each radiation component as a function of both location and time. The programs require information such as launch date, mission duration, and orbital parameters, such as perigee, apogee, and inclination.

The second step involves determining the level of shielding provided by the spacecraft superstructure, by any boxes housing the parts, and by packaging. The above programs are able to calculate how isotropic shielding modifies the radiation environment at the device level. Figure 5.5 is an example of such a calculation. It shows how the deposited radiation dose decreases with aluminum shielding thickness for a 5-year mission in GEO. However, in those cases where the shielding is not isotropic, more versatile programs, such as GEANT4 that employ ray tracing, must be used. Not only does shielding reduce the particle flux at the device location, it also modifies the energy spectrum, attenuating low-energy particles preferentially over high-energy particles. This is important because the degree of device degradation depends not only on the particle type and flux but also the energies of the particles actually striking the device.

Next, the failure modes of the device must be identified and the dependence on radiation characteristics determined. For those cases where radiation test data already exists for the failure modes identified, calculations are performed to determine whether the devices will survive the mission given the parameters of the radiation environment determined in step two.

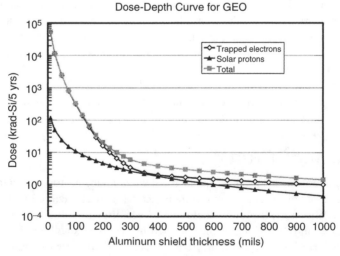

FIGURE 5.5 Dose–depth curve for geosynchronous orbit. (From J. Barth, Modelling Space Radiation Environments, *IEEE*, 1997.)

Where no test data exist, radiation testing of parts identical to those intended for space is the next step. The parts chosen for testing should have the same date and lot codes as those selected for the mission because it is well known that performance degradation during and following exposure to radiation is very device- and process-dependent. Ground testing involves the use of particle (proton or heavy ion) accelerators for SEE and displacement damage testing and radioactive sources (Co^{60}) or x-rays for total ionizing dose testing.[7,8] The kinds of degradation are identified and their dependence on particle fluence and deposited energy measured to quantify the degradation. That information is then used to predict the operation of the device in the charged particle environment of interest.

Finally, subsystem and system-level analyses must be undertaken to determine how the specific device degradation affects the overall spacecraft performance. Some radiation-induced effects may have no adverse effects on the system, whereas others may cause system failures. In those cases where the effects are pernicious, one can adopt any one of a host of measures that have been used successfully to mitigate them. Such measures might include the use of "cold spares" or extra shielding for devices that are sensitive to TID, or protecting data with "error-detecting-and-correcting" codes in devices found to be SEE sensitive. When such measures are not possible, the device should be discarded and an alternate one used in its place.

5.2.1   SPACE RADIATION INTERACTION WITH MATERIALS AND DEVICES (IONIZATION)

This section deals primarily with radiation damage by charged particles, including electrons, protons, and heavy ions ($Z > 2$). Most of the investigations of radiation damage have been in electronic, opto-electronic, and optical devices. Those results will be applied to the case of radiation damage in MEMS.

The first step is to investigate the interactions between incoming charged particles and the materials (metals, dielectrics, and semiconductors) used in the manufacture of MEMS. This requires knowledge of the particles' masses and energy distributions as well as of the properties, species, and density of the materials through which they pass.

When radiation interacts with materials it liberates bound charge, breaks chemical bonds, and displaces atoms from their equilibrium positions. These effects have been investigated for a long time and are quite well understood. Mechanical properties, such as density, brittleness or stress, are largely unaffected by the typical particle fluences encountered in space, and are ignored. In contrast, electrical properties of materials are greatly affected by radiation. Charge generation and displacement of atoms are known to alter the electrical properties of materials to such an extent that the performance of devices, such as transistors, may become severely degraded.[9] Studies of charged particle interaction with various materials will be used to draw general conclusions concerning radiation effects in MEMS.

Charged particles traveling through matter scatter off atoms, losing energy and slowing down in the process. The primary interaction involves Coulomb scattering off electrons bound to constituent atoms. Those electrons acquire sufficient energy to break free from the atoms. As the liberated electrons (known as delta rays) travel away from the generation site, they collide with other bound electrons, liberating them as well. The result is an initially high density of electrons and holes that together form a charge track coincident with the ion's path. The initial diameter of the track is less than a micron, but in a very short time — on the order of picoseconds — the electrons diffuse away from the track and the initial high charge density decreases rapidly.

The energy lost by an ion and absorbed in the material is measured in radiation absorbed dose or rad(material). One rad(material) is defined as 100 ergs of energy absorbed by 1 g of the material. Thus, for the case of silicon, the rad is given in terms of how much energy is absorbed per gram of silicon, or rad(Si). Absorbed dose may be calculated from Bethe's formula, which gives the energy lost per unit length via ionization by a particle passing through material,[10] as shown in the following equation:

$$-\frac{dE}{dx} = \frac{4\pi e^4 z^2}{m_o v^2} NZB(m_o, v, I) \qquad (5.1)$$

In the equation, v and z are the velocity and charge of the incoming particle, N and Z are the number density and atomic number of the absorber atoms, m_o is the electron mass and e is the electron charge. I is the average ionization potential, which is determined experimentally and depends on the type of material. For silicon $I = 3.6$ eV, whereas for GaAs $I = 4.8$ eV. $B(m_o, v, I)$ is a slowly varying function of v so that the energy lost by an ion traveling through material is greatest for highly charged (large Z) incoming particles with low energy (small v).

A normalized form of this equation, independent of material density, is obtained by dividing the differential energy loss by the material density (ρ) and is termed linear energy transfer (LET), and is the metric used by most radiation test engineers in the following equation:

$$\text{LET} = \frac{1}{\rho} \frac{dE}{dx} \qquad (5.2)$$

Figure 5.6 shows a plot of dE/dx as a function of energy for a number of different ions passing through silicon. At low energies the LET increases with increasing energy until a maximum is reached after which the LET decreases with increasing energy. Therefore, a high-energy particle traveling through matter loses energy, and as its energy decreases its LET increases, with the result that energy is lost at an ever-increasing rate. The density of charge in the track mirrors that of the LET. Near the end of the track is the Bragg peak where the amount of energy lost increases significantly just before the charged particle comes to rest. Figure 5.7 shows how the LET changes with depth for a 2.5 MeV helium ion in silicon. The charge density along the track is proportional to the LET at each point.

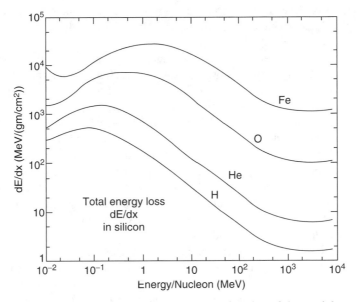

FIGURE 5.6 Energy lost per unit length in silicon as a function of the particle energy.

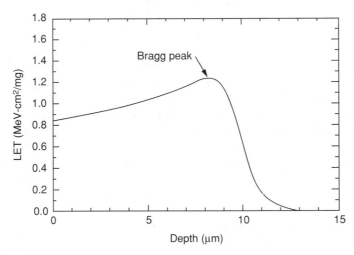

FIGURE 5.7 LET as a function of depth for a 2.5 MeV He ion passing through silicon. (From E. Petersen, Single Event Upsets in Space: Basic Concepts, NSREC Short Course, *IEEE*, 1983.)

How MEMS respond to radiation is determined, in part, by where the charge deposition occurs. For instance, ions passing through metal layers in MEMS generate additional electrons, but because they constitute a tiny fraction of the electrons already present in the metal, they have no effect on MEMS performance.

In contrast, ions passing through insulators and semiconductors are capable of generating sufficient charge to cause noticeable radiation effects in devices such as transistors and diodes.

Charge generated in insulators may become trapped at sites where they can reside for a long time. Their presence distorts the local electric fields and can affect the density of carriers in the semiconductor near the interface. For instance, positive charge trapped in the oxides used in the construction of a transistor will attract electrons in the semiconductor to the interface. The increased concentration of electrons at the field-oxide or semiconductor interface may lead to increased leakage currents in the transistor, whereas positive charge trapped in the transistor's gate oxide may prevent the transistor from switching on and off, thereby causing functional failure.

The amount of trapped charge is a function of the TID, which increases with exposure. Therefore, in space where devices are continuously exposed to radiation, there is a steady increase in the amount of trapped charge that is first observed as an increase in the leakage current and eventually a failure to operate.

TID effects in MEMS can originate in either the electronic or mechanical parts of the device, or both. Whatever the origin, the essential requirement is that charge be trapped in an insulator and that the trapped charge distort the existing electric field to such an extent that the operation of the device is affected.

Electrons and holes generated by energetic ions passing near or through a semiconductor metallurgical (n/p) junction will be separated by the associated electric field. Charge separation disturbs the electrical potential across the junction, and that voltage disturbance may propagate through the circuit to other nodes. When the voltage disturbance occurs in a latch or a memory, the information stored there may be nondestructively altered. The change in the state of the latch is known as a single-event upset (SEU). It is called a SEU because a single particle interacting with the material liberates sufficient charge to cause the effect. Of the many different kinds of single event effects, those that occur when charge is deposited in the semiconductor part of a device include single-event upset, single-event latchup, single-event snapback, single-event transient, and single-event burnout. In some cases, charge deposited in the gate oxide of a power MOSFET will lead to single-event burnout. These types of effects are expected to occur in the electronic circuits of MEMS but are unlikely to occur in the mechanical parts.

5.2.2   SPACE RADIATION INTERACTION WITH MATERIALS AND DEVICES (DISPLACEMENT DAMAGE)

Particle radiation may also interact with the atomic nuclei of the materials through which they pass. Those interactions consist of either elastic or inelastic nuclear scattering events. In either case, the atomic nuclei of the constituent atoms recoil and move away from their normal lattice sites, thereby disrupting the regular crystal lattice, and producing vacancies and interstitials.[11] Vacancies in semiconductors are usually electrically active whereas interstitials are not. Electrically active sites act as either short-lived traps or recombination centers for free carriers. Such traps reduce minority carrier lifetimes and doping levels, causing certain devices, such as bipolar transistors and LEDs, to suffer from degraded performance. Although nuclear interactions also occur in metals and insulators, their effects are typically not detectable. Thus, MEMS that contain bipolar devices or LEDs may be expected to degrade via displacement damage.

At extremely high levels of displacement damage, bulk material properties, such as stiffness, could be affected. This will be evident in MEMS devices that rely on the values of these bulk properties for proper operation. For example, changes in a bulk material property such as stiffness would modify the degree of flexibility of silicon layers used in comb drives that form part of MEMS engine.[12] Levels of radiation exposure for most space missions, except perhaps those to Jupiter, are several orders of magnitude lower than what would be necessary to have a noticeable effect on the bulk material properties and may largely be ignored.

5.2.3　RADIATION TESTING OF MEMS

Radiation testing of MEMS can be accomplished by following well-established procedures developed for radiation testing electronic and photonic devices. SEE testing is usually accomplished with heavy ions and protons at accelerators. TID susceptibility is most conveniently measured with gamma rays in a Co^{60} cell or with x-rays. DD is typically produced with protons at accelerators, as well as with neutrons in reactors or at accelerators. Parts are exercised either during (for SEE) or following (for TID and DD) irradiation to ascertain how they respond to the radiation.

One issue relevant for MEMS is that of ion range. Heavy ions available at most accelerators have relatively short ranges in material — at the most a few hundred microns. In some MEMS the radiation sensitive parts are covered by material, such as in the case of digital mirror devices, where a transparent glass covers the mechanical part. Removal of the glass destroys the mirror so that testing must be performed at those accelerators with sufficient energy for the ions to penetrate the overlying material. Particle range is not a problem for protons or gamma ray exposures.

5.3　EXAMPLES OF RADIATION EFFECTS IN MEMS

MEMS are unique from a radiation-effects point of view because they contain electronic control circuits coupled with mechanical structures, both of which are potentially sensitive to radiation damage. The electronic circuits in MEMS are either CMOS or bipolar technologies that are known potentially to exhibit great sensitivity to radiation damage. It is not at all obvious that radiation doses that produce measurable changes in performance in electronic circuits will have any effect on mechanical structures; however, they can.

The first commercial MEMS tested for radiation sensitivity was an accelerometer exposed to an ion beam.[13] By using a small aperture it was possible to confine the beam to the area of the chip containing only the mechanical structure. Significant changes in performance were noted following moderate particle fluences. The radiation damage was attributed to charge generated in an insulating layer that was part of the mechanical structure. The charge altered the magnitude of the applied electric field, which, in turn, changed the acceleration reading. Subsequent tests of other MEMS devices, such as a RF switch, a micromotor and a digital mirror device, also revealed radiation damage originating in insulating layers incorporated in the mechanical structure. These results suggest a common theme for radiation effects in MEMS that depend on sensing electric fields across insulators in the mechanical portions, that is, charge deposited in insulating layers of MEMS modifies existing electric fields in those layers, and the system responds by producing an erroneous output.

The responses to radiation exposure of four different MEMS will be discussed in detail. They include an accelerometer, a comb drive, a RF relay, and a digital mirror device. In all cases the radiation damage is attributable to charge generated in insulators that cause unwanted mechanical displacements. Inspection of these four different MEMS confirms that there are no conceivable ways for SEE to occur in the mechanical parts. Thus, no SEE testing was done.

5.3.1 Accelerometer

The first MEMS device subjected to radiation testing was a commercial accelerometer (ADXL50) used primarily in the automotive industry for deploying air bags during a collision.[13] Because of their small size, light weight, and low power consumption, MEMS accelerometers also have applications in space, such as in small autonomous spacecraft that are part of NASA's New Millennium Program (NMP).

Figure 5.8 shows the construction of the ADXL50. It consists of two sets of interdigitated fingers. One set is stationary (y and z) and the other (x) is connected

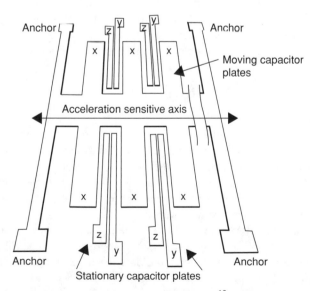

FIGURE 5.8 Construction of the ADXL50 accelerometer.[13] (From F. Sexton, Measurement of Single Event Phenomena in Devices and ICs, NSREC Short Course, *IEEE*, 1992.)

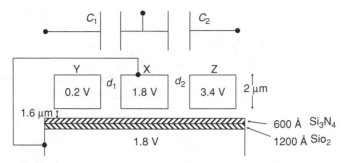

FIGURE 5.9 Cross-sectional view of the ADXL50.[13] (From A. Knudson, The Effects of Radiation on MEMS Accelerometers, *IEEE*, 1996.)

to a spring-mounted beam that moves when the device experiences a force due to acceleration along the length of the beam. Figure 5.9 is a cross-sectional view of the ADXL50 showing the beams suspended above the silicon substrate covered with thin layers of Si_3N_4 and SiO_2. The operation of the device has been described in a previous publication.[13] A distance d_1 separates beams X and Y that form the two "plates" of capacitor C_1, whereas d_2 separates X and Z that form the "plates" for capacitor C_2. Movement of beam X changes both d_1 and d_2. That causes both C_1 and C_2 to change. Figure 5.10 shows the circuit used to measure the changes in capacitance. An internal oscillator applies two separate square wave signals to beams Y and Z. Since the two signals are 180° out of phase, the output voltage from the sensor is zero because $C_1 = C_2$. However, when the part is accelerated,

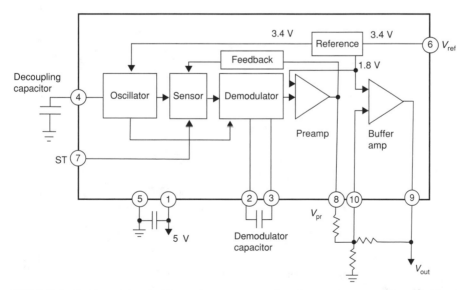

FIGURE 5.10 Electronic circuit used to measure the changes in capacitance.[13] (From A. Knudson, The Effects of Radiation on MEMS Accelerometers, *IEEE*, 1996.)

beam X moves relative to beams Y and Z so that $C_1 \neq C_2$. The result is an AC voltage on X, which is demodulated and compared with a reference voltage in the buffer amplifier. The difference between the two voltages is a measure of the acceleration and appears at the device's output. Beam X is electrically tied to the substrate to prevent the arms from bending down towards the substrate in the presence of a voltage difference between the beam X and the substrate. This effect would lead to an erroneous voltage reading on the output.

The first experiment involved irradiating the entire device with 65 MeV protons and monitoring the outputs of the preamplifier (V_{pr}) and of the buffer amplifier (V_{out}). Proton irradiation caused both V_{pr} and V_{out} to change, but in opposite directions. Furthermore, the dose rate had a significant effect on both the magnitude and direction of change. These results were not too surprising given that the ADXL50 contained CMOS control circuits that are known to be radiation-sensitive.

With an aperture placed over the accelerometer to cover the electronic circuit and expose only the mechanical part to ion beam irradiation, it was possible to determine whether the mechanical part also responded to radiation. Figure 5.11 shows that V_{out} decreases exponentially with cumulative fluence. The decrease does not depend on dose rate. Additional experiments with protons indicate that the magnitude of the decay depends only slightly on whether the device was on or off.

These results suggest that charge trapping in either the SiO_2 or Si_3N_4 layers is responsible for changes in V_{out}. Ionizing particles passing through the insulators generate charge that may become trapped in the insulators and modify the existing

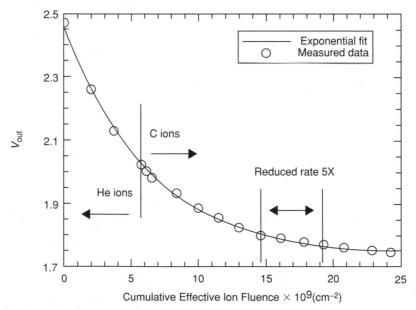

FIGURE 5.11 Change in the output voltage V_{out} as a function of particle fluence.[13] (From A. Knudson, The Effects of Radiation on MEMS Accelerometers, *IEEE*, 1996.)

electric fields between the fingers. That could cause one set of fingers to move relative to the other. The result is a change in the capacitance between the two sets of interdigitated fingers that results in a change in the output voltage.

The proposed mechanism of charge generation and trapping in the insulators causing a shift in V_{out} was confirmed by testing another accelerometer (ADXL04) that contained a conducting polycrystalline silicon layer on top of the insulators. That layer was electrically connected to the moveable set of fingers. The conducting layer effectively screens out any charge generated in the insulators, so that the mechanical part of the device should exhibit no radiation-induced changes. Irradiation of the device with protons confirmed that there was no change in V_{out}. Mathematical modeling also confirmed that charge trapping in the insulators could cause an offset in V_{out}.[14] Another investigation showed that very high doses of radiation actually caused the device to lock up and stop operating, presumably by bending the beams to such an extent that they made contact with the substrate.[15]

5.3.2 MICROENGINE WITH COMB DRIVE AND GEARS

MEMS microengines have been designed and built by Sandia National Laboratories that could be used for a variety of space applications.[12] A microengine consists of two comb drives moving perpendicular to each other and linkage arms connecting them to a small drive gear rotating about a shaft. The mechanical and electrical performances of the microengine components following exposure to various forms of radiation, including x-rays, electrons, and protons, were evaluated. Performance degradation, in the form of limited motion and "lockup" were observed, but only at very high exposure levels. This relative immunity to radiation was designed into the devices by incorporating a polysilicon layer that, when grounded, screened out any radiation-generated charge trapped in the Si_3N_4 or SiO_2 insulating layers covering the silicon substrate. This is completely analogous to the ADXL04 accelerometer discussed in the previous section.

Figure 5.12 shows the structure of the comb drive that is responsible for driving the machine. It is, in effect, a reciprocating linear electrostatic drive. Application

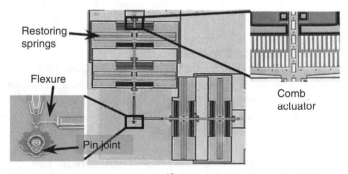

FIGURE 5.12 MEMS comb drive and gear.[12] (From A. Knudson, The Effects of Radiation on MEMS Accelerometers, *IEEE*, 1996.)

and removal of bias between the two sets of interdigitated teeth cause them to move back and forth in a direction parallel to the long dimension of the teeth. Two sets of comb drives located such that their linear movements are perpendicular to each other are used to drive a cog connected to an axle. During movement, the comb is subjected to both adhesive and abrasive wear, as well as to microwelding and electrostatic clamping. These failure modes are the result of the very small spacing between the two sets of interdigitated fingers and between the comb fingers and the substrate. Trapped charge could cause the two sets of fingers to make contact with one another or to make contact with the substrate. The much larger tooth-to-substrate capacitance suggests that the buildup of charge will be much more effective in bending the teeth towards the substrate. Because it is important to prevent this from happening, a grounded polysilicon layer was deposited on the substrate below the comb teeth, and any radiation-induced charge trapped in the Si_3N_4 or SiO_2 layer below the polysilicon layer could be screened from the comb teeth. Permitting the comb fingers to bend down and make contact with the substrate would lead to the enhanced likelihood of abrasion, microwelding, and electrostatic clamping.

The magnitude of the charge trapped in the oxide was obtained by measuring the capacitance between the comb and the substrate following each radiation exposure. Radiation-induced wear in the comb was obtained by measuring the resonant operating frequency spectrum of the micromotor: the maximum of the resonant frequency spectrum decreases with wear. Radiation effects in the gear drive were quantified by measuring the reduction in the rotation rate of the gear with radiation dose. During irradiation, three different bias configurations were used — all pins floating, all pins grounded, or all pins biased in a particular configuration.

Experimental results indicated that the configuration in which all the pins were grounded is the one in which the microengine was the least sensitive to radiation-induced changes. For instance, the gear rotation rate decreased only slightly, while the resonant frequency response for the grounded comb drive did not change for x-ray doses between 3 and 100 Mrad (SiO_2). Figure 5.13 shows a large shift in the capacitance versus voltage curves for the comb drive, indicating a large buildup of radiation-induced charge in the insulating layers. Despite the large buildup of charge in the Si_3N_4 or SiO_2 layers, the grounded polysilicon layer was effective in shielding the associated electric field and preventing the comb fingers from bending down and making contact with the substrate.

Electron-beam irradiation of grounded comb drives caused lockup at a fluence of 10^{14}/cm^2 (14.4 Mrad [SiO_2]) an order of magnitude larger than for a floating comb drive. Similarly, the resonant frequency of the floating comb drive decreased between electron fluences of 1 and 3×10^{13}/cm^2 whereas no change in resonant frequency was measured for the grounded device. Microengines, containing a ground polysilicon layer, exhibited no degradation in motion when exposed to electrons up to a fluence of 4×10^{16}/cm^2 (5.76 Grad [SiO_2]).

Proton beam irradiation of an operating comb drive had no effect on the motion until a dose of 10^{13} protons/cm^2 at which the comb drive locked up. At this high

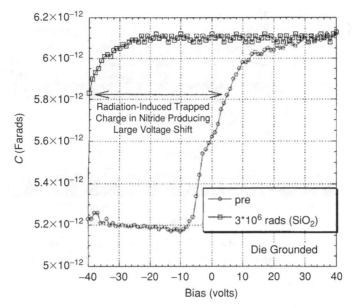

FIGURE 5.13 Capacitance as a function of voltage for a capacitor indicating the presence of trapped charge.[12] (From L.P. Schanwald, Radiation Effects on Surface Micromachines Combdrives and Microengines, *IEEE*, 1998.)

level it is possible that displacement damage effects cause fatigue in the polysilicon spring attached to the one end of the comb drive.

In summary, microengines can operate with little radiation effects in a typical space environment provided the devices are designed with a polysilicon layer deposited on top of the Si_3N_4 or SiO_2 layers that can be connected to ground to shield the mechanical parts from the effects of the trapped charge, thereby greatly extending the useful life of the MEMS engine.

5.3.3 RF RELAY

Compact, low-loss RF switches manufactured using MEMS technology are commercially available and are potentially useful for a variety of applications in space, such as for electronically scanned antennas for small satellites. Because RF switches must be able to operate in a radiation environment, NASA's Jet Propulsion Laboratory radiation tested two similar RF switches that differed only in the location of an insulating layer.[16] The switch with the insulator between capacitor metal plates proved to be significantly more sensitive to radiation damage than the switch with the insulator outside the capacitor plates.

Figure 5.14 shows the design of the two switches. Application of a voltage greater than the activation voltage (V_{act}) to the upper capacitor plates at each end of the switch, forces the two metal plates together, thereby "closing" the switch. Upon removal of the bias, the two contacts separate and the switch is in its "open"

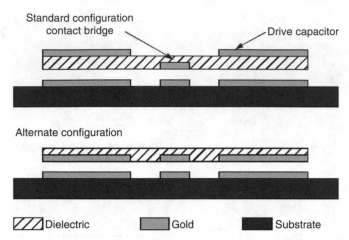

FIGURE 5.14 Construction of two standard RF switches: Contact Bridge and Drive Capacitor.[16] (From L.P. Schanwald, Radiation Effects on Surface Micromachines Combdrives and Microengines, *IEEE*, 1998.)

position. The switches have slightly different structures: switch A contains an insulating layer between the two metal capacitor plates, whereas switch B does not. The switches were made on GaAs substrates with a dielectric thickness of 2 μm. V_{act} was 60 V and the gap between the metal plates was 3.5 μm when open and 0.8 μm when closed.

The parts were exposed to gamma rays in a Co^{60} source. During irradiation a constant electrical bias was applied; in some cases the top metal plates were biased positive relative to the bottom plates, whereas in others the bias was the reverse. The activation voltage (V_{act}) was measured following incremental doses of radiation.

Figure 5.15 shows V_{act} as a function of dose for switch A. Under positive bias, V_{act} increased approximately linearly with dose. Under negative bias, V_{act} shifted in the negative direction and appeared to degrade more rapidly with dose. Annealing for 3 days under no bias caused a slight recovery (3 V) in V_{act}. Unbiased devices showed no measurable degradation with dose. No significant degradation up to a dose of 150 krad (GaAs) was found for switch B.

Previous studies of radiation damage in accelerometers suggest that the buildup of charge in an insulator alters the magnitude of an electric field applied across that insulator. In the case of the RF switch, the trapped charge in the insulator either reduces or increases V_{act}, depending on the charge distribution in the dielectric. V_{act} becomes more positive for both bias configurations if the charge produces a positive V_{act}. On the other hand, V_{act} becomes more negative for both bias configurations when V_{act} is negative. In fact, V_{act} in the two bias configurations are always opposite, one increasing and the other decreasing in magnitude. No radiation-induced changes in V_{act} were observed for switch B.

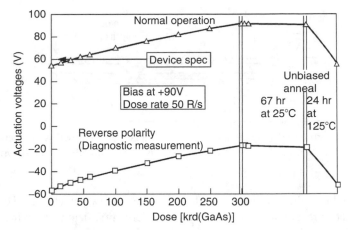

FIGURE 5.15 V_{act} as a function of dose for switch A.[16] (From L.P. Schanwald, Radiation Effects on Surface Micromachines Combdrives and Microengines, *IEEE*, 1998.)

A calculation of the dependence of V_{app} on dose shows that a much smaller dose is required to produce a given offset voltage than observed here. To account for the smaller V_{app}, when the top electrode is negatively biased, the authors suggested thermal emission of electrons from the insulator that are collected at the bottom electrode. Alternatively, when the top electrode is positively biased, electrons are thermally emitted from the bottom electrode and captured by the insulator.

In summary, it is possible to design MEMS switches that show little radiation sensitivity. The source of the radiation degradation is an insulating layer in which radiation-induced charge can be trapped. By removing the insulating layer to a region where there is little electric field, the radiation sensitivity of a MEMS RF switch can be minimized.

5.3.4 DIGITAL MIRROR DEVICE

Two structurally different types of digital mirror devices (DMDs) have been tested for their TID responses using gamma rays in a Co^{60} source at the Jet Propulsion Laboratory.[17] DMDs consist of arrays of tiny mirrors that assume one of two positions, depending on the magnitude of an applied electric field. In one configuration, a mirror directs an incident beam of light in a particular direction designated as "on." In the other configuration obtained by changing the applied electric field, a mirror directs the beam in a different direction designated as "off." The metallic mirror elements in DMDs are one of two electrodes. The second electrode is typically on the surface of the silicon substrate separated from the mirror elements either by an insulator, air, or vacuum.

Figure 5.16 shows the structure of a membrane-based device manufactured by Boston Micromachines Corporation. Deflection of the membrane is achieved by applying a voltage of at least 30 V between the surface electrode and the polycrystalline silicon electrode covering the insulator on top of the silicon substrate. The

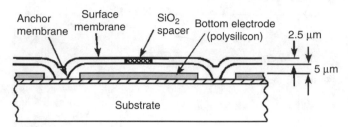

FIGURE 5.16 Structure of membrane-based device manufactured by Boston Micromachines Corporation.[17] (From S. McClure, Radiation Effects in MEMS: RF Relays, *IEEE*, 2003.)

figure also shows that there is no insulator between the two electrodes. The absence of an insulator results in very little change in deflection depth following radiation testing. Figure 5.17 shows that there is no change in the deflection depth as a function of total dose up to 3 Mrad(Si). The results are independent of whether or not the devices were irradiated under bias and confirm that the DMD device with no insulating layer between the two electrodes is relatively immune to radiation degradation.

The second device tested also consists of deformable mirrors and was manufactured by the Jet Propulsion Laboratory in conjunction with Pennsylvania State University. It is not commercially available. A piezoelectric membrane, comprised of a layer of lead zirconium titanate (PZT) deposited on silicon nitrite, is anchored at two opposite edges to silicon posts. At the center and on top of the membrane is an indium post that supports a thin silicon layer, which is the mirror membrane.

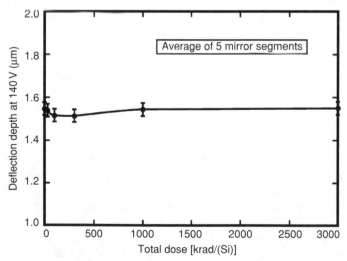

FIGURE 5.17 Plot of deflection depth as a function of dose.[17] (From T.F. Miyahara, Total Dose Degradation of MEMS Optical Mirrors, *IEEE*, 2003.)

Application of a voltage across the PZT causes it to flex and, in so doing, it deforms the mirror membrane, which turns the DMD off. Exposure to an ionizing radiation dose of 1 Mrad(Si) causes unbiased mirrors to deflect by about 5% and biased mirrors to deflect by about 10%. The degradation is due to charge trapped in the silicon nitrite on which the PZT is deposited and not to degradation of the PZT itself.

The results of TID testing of DMDs show once again that the radiation sensitivity of MEMS may be reduced by eliminating dielectric layers between electrodes, because any charge generated in the dielectrics will modify the electric field applied between the two electrodes, thereby causing an erroneous reading.

5.4 MITIGATION OF RADIATION EFFECTS IN MEMS

Reducing the sensitivity of MEMS to radiation effects is possible but may be quite challenging given that MEMS are manufactured using normal silicon processing steps, some of which are not necessarily compatible with radiation immunity. Most commercial-off-the-shelf devices are well known for being radiation intolerant because they do not use the special processing techniques required for obtaining radiation hardened devices. For instance, hardening a device to TID requires the avoidance of both high-temperature anneals and exposure to gases such as hydrogen. However, using an approach termed "hardening by design" makes it possible to harden commercial processes to both TID and SEE.[18]

One way to mitigate the effects of the trapped charge produced by the radiation in MEMS is to modify the design to include a grounded conducting plane over the insulator to shield the mechanical part from the effects of the trapped charge.[13] This approach proved effective for the accelerometer. Another approach is to place the insulator in a position where charge trapped in it will have no effect on the device, as in RF switches.

In summary, it should be possible to use special processing and design approaches to harden MEMS against radiation damage, but so far the only approach that has been tried and proved to be effective is the avoidance of dielectric layers between metal electrodes used for applying electric fields to the MEMS.

5.5 CONCLUSION

The performance of MEMS devices in space depends, critically, on the characteristics of the radiation environment. The environment, in turn, depends on spacecraft location and time. Characterizing the environment requires knowledge of such factors as launch date, mission duration and orbit, as well as the amount of shielding provided by the spacecraft. Fortunately, there are models for solar activity, cosmic ray intensity, and fluxes in the radiation belts surrounding planets that simplify calculations of total radiation exposure. By combining the calculated levels of radiation exposure with the results of ground testing, it will be possible to assess whether MEMS will meet mission requirements.

REFERENCES

1. M. Lauriente and A.L. Vampola, Spacecraft Anomalies due to Radiation Environment in Space, presented at NASDA/JAERI 2nd International Workshop on Radiation Effects of Semiconductor Devices for Space Applications, Tokyo, Japan, 21st March 1996.
2. J. Barth, Modeling space radiation environments, IEEE Nuclear and Space Radiation Effects Short Course, Snowmass, CO, 21st July 1997.
3. J. Feynman and S.B. Gabriel, On space weather consequences and predictions, *Journal of Geophysical Research* Vol. 105, No. A5, pp. 10,543–10,564 (May 2000).
4. M.A. Xapsos, G.P. Summers, J.L. Barth, E.G. Stassinopolous, and E.A. Burke, Probability model for cumulative solar proton event fluences, *IEEE Transactions on Nuclear Science* Vol. 47, pp. 486–490 (June 2000).
5. E. Daly, Space weather and radiation effects, Radiation and its effects on components and systems short course, Grenoble, France, 10th September 2001.
6. C. Poivey, Radiation hardness assurance for space systems, 2002 IEEE Nuclear and Space Radiation Effects Conference Short Course Notebook, Phoenix, AZ, 15th July 2002.
7. D.M. Fleetwood and H.A. Eisen, Total-dose radiation hardness assurance, *IEEE Transactions on Nuclear Science* Vol. 50, pp. 552–564 (June 2003).
8. R.A. Reed, J. Kinnison, J. Pickel, S. Buchner, P.W. Marshall, S. Kniffin, and K.A. LaBel, Single event effects ground testing and on-orbit rate prediction methods: the past, present, and future, *IEEE Transactions on Nuclear Science* Vol. 50, pp. 622–634 (June 2003).
9. T.P. Ma and P.V. Dressendorfer, *Ionizing Radiation Effects in MOS Devices and Circuits*, John Wiley and Sons, Hoboken, NJ, 1989.
10. G.F. Knoll, *Radiation Detection and Measurement*, John Wiley and Sons, Hoboken, NJ, p. 32, 1989.
11. P.W. Marshall and C.J. Marshall, Proton effects and test issues for satellite applications, IEEE Nuclear and Space Radiation Effects Conference Short Course, Norfolk, VA, 12th July 1999.
12. L.P. Schanwald, J.R. Schwank, J.J. Sniegowski, D.S. Walsh, N.F. Smith, K.A. Petersen, M.R. Shaneyfelt, P.S. Winokur, J.H. Smith, and B.L. Doyle, Radiation effects in surface micromachined comb drives and microengines, *IEEE Transactions on Nuclear Science* Vol. 45, pp. 2789–2798 (December 1998).
13. A.R. Knudson, S. Buchner, P. McDonald, W.J. Stapor, A.B. Campbell, K.S. Grabowski, D.L. Knies, S. Lewis, and Y Zhao, The effects of radiation on MEMS accelerometers, *IEEE Transactions on Nuclear Science* Vol. 43, pp. 3122–3126 (December 1996).
14. L.D. Edmonds, G.M. Swift, and C.I. Lee, Radiation response of a MEMS accelerometer: an electrostatic force, *IEEE Transactions on Nuclear Science* Vol. 45, pp. 2779–2788 (December 1998).
15. C.I. Lee, A.H. Johnson, W.C. Tang, C.E. Barnes, and J. Lyke, Total dose effects on micromechanical systems (MEMS): accelerometers, *IEEE Transactions on Nuclear Science* Vol. 43, pp. 3127–3132 (December 1996).
16. S. McClure, L. Edmonds, R. Mihailovich, A. Johnson, P. Alonzo, J. DeNatale, J. Lehman, and C. Yui, Radiation effects in micro electro mechanical systems (MEMS): RF relays, *IEEE Transactions on Nuclear Science* Vol. 49, pp. 3197–3202 (December 2002).
17. T.F. Miyahira, H.N. Becker, S.S. McClure, L.D. Edmonds, A.H. Johnson and Y. Hishinuma, Total dose degradation of MEMS optical mirrors, *IEEE Transactions on Nuclear Science* Vol. 50, pp. 1860–1866 (December 2003).

18. R.C. Lacoe, CMOS scaling, design principles and hardening-by-design methodologies, Nuclear and Space Radiation Effects Conference Short Course Notebook, Monterey, CA, 21st July 2003.

6 Microtechnologies for Space Systems

Thomas George and Robert Powers

CONTENTS

6.1 INTRODUCTION TO SPACE TECHNOLOGY DEVELOPMENT

The "maturing" of advanced micronanotechnology (MNT) concepts for space applications faces a very similar dilemma similar to that faced in the commercial world.[1,2,3] NASA has pioneered a means of evaluating the maturity of new technologies, known as the technology readiness level (TRL) scale that has now found widespread use in government and industry. As shown in Table 6.1, the TRL scale ranges from levels 1 through 9, with levels 1 to 3 being at the so-called "low-TRL," that is basic research into demonstrating the proof-of-concept. Levels 4 to 6 correspond to "mid-TRL" development, which is the reliable demonstration of subsystems based on the new technologies, and finally, levels 7 to 9 (high-TRL) correspond to successful utilization of these technologies at the system or subsystem level in NASA's space missions. A large majority of the exciting MNT

TABLE 6.1
Technology Readiness Level (TRL) Scale

TRL1	Basic principles observed and reported
TRL2	Technology concept and application formulated
TRL3	Analytical and experimental critical function and/or characteristic proof-of-concept
TRL4	Component and/or breadboard validation in laboratory environment
TRL5	Component and/or breadboard validation in relevant environment
TRL6	System or subsystem model or prototype demonstration in a relevant environment (ground or space)
TRL7	System prototype demonstration in a space environment
TRL8	Actual system completed and "flight qualified" through test and demonstration (ground or flight)
TRL9	Actual system "flight proven" through successful mission operations

Source: NASA/JPL.

concepts are at the low TRL stage, sometimes referred to as the "technology push" stage, with the daunting challenge of having to bridge the "TRL gap"[4] to successfully transition to the high-TRL space applications or "technology pull" stage. The TRL gap, sometimes referred to as the "valley of death" in the commercial sector, therefore represents an order-of-magnitude increase in effort (and consequently funding) that is required to make the transition to high TRL. The primary reason that most new technologies fail to bridge the TRL gap is that because of their relatively low maturity, they do not have a compelling mission "pull" to drive further system-level development.

Another important consideration is that space applications only need components and systems in relatively minuscule volumes compared to the consumer market. However, the performance requirements for these technologies are no less stringent, and in most cases, much more so than for consumer products. Thus, the system development costs are considerable since a sufficiently large body of laboratory test data has to be generated in order to conclusively demonstrate the reliability of the new technology. Additionally, there is also a more subtle perception barrier to be overcome. This involves the generation of sufficient "space heritage" via actual space flights of the new system. Carried to the extreme, this perception barrier leads to the conundrum that a new technology cannot fly unless it has flown before! NASA has recognized the impact of the space heritage barrier as a major obstacle impeding the infusion of new technologies into its missions. This recognition has led to the setting up of programs such as the New Millennium Program (NMP) that are aimed specifically to provide flight demonstration opportunities for new technologies. These flights, however, are few and far between, and are also generally restricted to technologies that are already at a high level of maturity (TRL 4 and above).

6.2 HIGH TRL SUCCESS STORIES

The following is a description of a few MNT-based devices and instruments that have successfully transitioned either to space mission development or are currently at a very high level of technology maturity. This list is by no means comprehensive but serves to show that the applications for MNT in space are numerous and varied. In each case, the key factors that were responsible for the successful technology infusion have also been identified.

6.2.1 "SPIDER WEB" BOLOMETERS FOR HERSCHEL SPACE OBSERVATORY AND PLANCK SURVEYOR MISSIONS

NASA and the European Space Agency (ESA) are jointly developing the PLANCK Surveyor Mission and the Herschel Space Observatory, both scheduled for launch in 2007. The PLANCK Surveyor will carry on board a high-frequency instrument (HFI),[5] which will map the entire sky in six frequency bands ranging from 100 to 857 GHz. HFI will be used to probe the cosmic microwave background (CMB) anisotropy and polarization. The spectral and photometric imaging receiver (SPIRE)[6] will be an imaging photometer and spectrometer for ESA's Herschel Space Observatory. SPIRE will be used to conduct deep extragalactic and galactic imaging surveys as well as spectroscopy of star-forming regions. It contains a three-band imaging photometer with bands in the range of 570–1200 GHz, and an imaging Fourier transform spectrometer (FTS) covering the 450–1500 GHz range. Both HFI and SPIRE depend on "spider web" bolometer detectors operating at temperatures between 0.1 and 0.3 K.

The spider web bolometer detector[7] was developed at the Jet Propulsion Laboratory as shown in Figure 6.1, and rapidly made the transition from a low

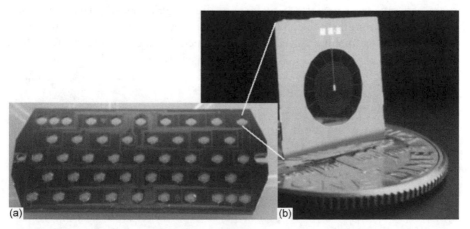

FIGURE 6.1 (a) An array of microfabricated "Spider Web" bolometers. (b) Magnified view of a single detector showing the spider web suspension for the rectangular thermistor chip mounted in the center of the device. (*Source*: NASA/JPL.)

TRL "push" technology to a mission-enabling "pull" technology. Thus, this highly sensitive detector (noise equivalent power ~ 10^{-18} W/rt-Hz at 100 mK) is the first "success story" for JPL-developed microelectromechanical system (MEMS) technologies. The device consists of a high-purity, neutron transmutation doped (NTD), single crystal Ge thermistor chip mounted on a "spider web" suspension comprising metallized, suspended SiN filaments. The spider web structure has several advantages: (a) it provides a large area for microwave absorption; (b) it has low heat capacity; (c) it provides excellent thermal isolation for the NTD chip from the surrounding environment; and (d) it also has a low cross section for cosmic rays. The detection mechanism consists of the NTD chips measuring the local temperature rise due to the absorbed microwave radiation.

6.2.2   MEMS-BASED SUN SENSOR

Sun sensors are used commonly as part of the attitude control systems of spacecraft. JPL has developed a miniaturized sun sensor with a mass of less than 30 g and with power consumption less than 20 mW.[8] The device in Figure 6.2 consists of a focal-plane array photodetector above which a microfabricated, silicon chip with several hundred small apertures is mounted. The focal plane captures the image of the aperture array upon illumination by the sun. The orientation of the spacecraft with respect to the sun is then computed to accuracies of better than a few arcminutes by analysis of the resultant image on the focal plane detector. The simplicity and robustness of the device have made it a candidate technology for the Mars Surface Laboratory mission to be launched in 2009.

6.2.3   MEMS VIBRATORY GYROSCOPE

MEMS-based miniature gyroscope development has become a very active area of research and development for a number of research groups around the world. The main performance parameter used for classifying gyroscopes is the angular bias stability or the minimum uncertainty in rotation rate as a function of the time over which the measurements are averaged or integrated. For inertial grade performance,

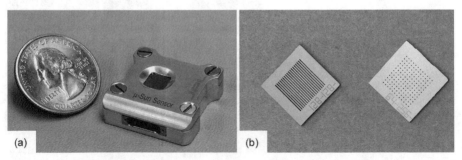

FIGURE 6.2 MEMS-based sun sensor device. (a) Fully assembled device consisting of the (b) microfabricated silicon mask mounted over a focal plane array detector. (*Source*: NASA/JPL.)

that is, for spacecraft navigation applications, the requirements are for angular bias stabilities to be in the range of 0.001–0.01°/h. The JPL-developed postresonator gyroscope (PRG) holds the world record for the performance of MEMS gyroscopes, at 0.1°/h.[9] Although, not yet meeting the stringent requirements for spacecraft navigation, by virtue of its low mass and power consumption, the PRG is being considered for incorporation into inertial measurement units that are augmented by other attitude measurement devices such as miniature star trackers or GPS receivers (for low-Earth orbit [LEO] applications).

Figure 6.3 shows the PRG, consists essentially of a two degree-of-freedom (DOF), planar resonator arrangement, which is "rocked" about an in-plane axis using capacitive actuation electrodes. The gyroscope senses rotation, also capacitively, by measuring the Coriolis-coupled vibration about the orthogonal in-plane axis. Thus, for optimum performance it is very important for the Coriolis-coupled, in-plane resonance modes to have very high-quality factors (low mechanical energy loss) and be "degenerate," that is, be closely matched in frequency (for maintaining linearity with feedback control). Further development in device design, materials choices, and fabrication processes is underway to enhance the performance of these gyroscopes.

The PRG can be classified as a mid TRL (~ TRL 4) technology. Therefore, the development strategy being pursued is to capture "niche applications" on the path to full-scale implementation in space missions. This gyroscope is being considered

FIGURE 6.3 Exploded view of the PRG. Rotation about the central post is sensed electrostatically via capacitive electrodes. The post is mounted on a layer containing in-plane orthogonal resonators. The post or resonator assembly is suspended over a substrate containing an arrangement of multiple electrodes for actuation, sensing and tuning the frequencies of the resonance modes. The gyroscope operates by "rocking" the post about an in-plane axis and consequently sensing the Coriolis force-generated oscillation about the orthogonal in-plane axis. (*Source*: NASA/JPL.)

initially for microspacecraft applications in which the severe constraints on the size, mass, and power consumption preclude non-MEMS solutions.

6.2.4 MEMS Microshutter Arrays for the James Webb Space Telescope

This space application represents an excellent example in which the only viable solution is a MEMS device. Thus, without the 175 × 384 array of densely packed microfabricated shutters[10,11] allowing the simultaneous selection of over 200 imaged celestial objects, the near-infrared multiobject spectrometer (NIRMOS) instrument would not be possible. The NIRMOS is an important part of the instrument suite for the James Webb Space Telescope. It operates in the 0.6 to 5.0 μm wavelength range with a 3.6 × 3.6 in. field of view (FOV) as shown in Figure 6.4.[11] Each individual shutter is approximately 100 × 200 μm in size and subtends 0.2 × 0.4 in. within this FOV. The microshutter approach has several advantages over micromirror arrays namely, possibility of high contrast between open and closed states, interchangeability of transmissive geometry with a fixed mechanical slit (backup solution) and elimination of the need for flatness of the mirror surface. The MEMS microshutter arrays are being developed for NASA/ESA by NASA's Goddard Space Flight Center (GSFC).

6.2.5 Carbon Nanotube-Based Thermal Interface

The Hubble Space Telescope (HST) is soon expected to have its fourth servicing mission. Installation of new and high-power instruments in the HST's aft shroud section is expected to generate excessive waste heat. A capillary-pump loop (CPL)

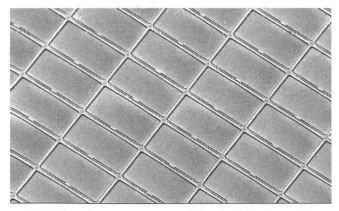

FIGURE 6.4 Scanning electron micrograph of 200 × 100 μm sized, hinged microshutters forming part of a 175 × 384 array. The microshutter array is the enabling component for the NIRMOS for the James Webb Space Telescope. Each SiN shutter is hinged about a torsion bar and is rotated downwards using magnetic actuation. Once lowered, they can be electrostatically clamped as required, to allow light from a selected celestial object to pass through into the spectrometer. (*Source*: M. J. Li et al., Microshutter arrays for near IR applications, *SPIE Proceedings*, SPIE Vol. 4981, 2003, pp. 113–121. With permission.)

technology will be implemented to transport the waste heat towards the aft shroud's exterior, where it is radiated into outer space. The primary challenge with this type of cooling scheme is the development of an efficient thermal interface between the CPL and the external radiator.

This heat transfer challenge provides a unique opportunity for the first use of carbon nanotubes (CNTs) in space.[12] The CNTs will be incorporated within a thermal interface kit for the HST instrument system as the only materials choice that satisfies the following requirements. The interface should withstand the harsh vacuum and radiation environment of space, have good electrical isolation, be mechanically compliant, be abrasion tolerant for manual assembly by astronauts, and not contaminate the spacecraft or its instruments. Competing solutions based on polymer materials have the needed flexibility, however, these are either not good thermal conductors or could contaminate the HST by outgassing high-vapor pressure compounds. CNTs on the other hand are flexible (10% linear elasticity), strong (Young's modulus greater than 1 TPa) and with high thermal conductivity (theoretical axial conductivity of 6000 W/m-K). The CNT-based thermal interface is a joint development of NASA's GSFC and the Ames Research Center (ARC) in collaboration with the Applied Physics Laboratory (APL) at the Johns Hopkins University. ARC and APL are aligning arrays of CNTs, and optimizing their characteristics for the needs of the HST project.[13] GSFC will be responsible for the testing and integration of the novel thermal interface. The CNT-based thermal interface will be a 6-square-inch, copper-backed thermal conductor, with approximately 4 billion, 40-μm-tall nanotube "bristles."

6.2.6   RF MEMS Switch

Rockwell Science Center (RSC) developed a MEMS-based RF switch that was flown successfully on two PICOSAT missions.[14] The RSC devices were metal-contacting switches that were fabricated using low-temperature surface-micromachining techniques. RF MEMS switches are an exciting alternative to the conventional semiconductor field effect transistor-based switches, since they overcome several of the shortcomings of semiconductor switches. Among other advantages, RF switches have low mass, low power and small size, low RF insertion loss, high isolation, and high intermodulation product. Additionally, micromechanical switches are inherently radiation tolerant and robust for space applications. As shown in Figure 6.5,[14] the RSC switch is a microrelay consisting of a metal shunt bar that is suspended over a gap in the RF conductor line. Contact is made by electrostatically attracting the shunt bar down to the RF conductor by means of voltage applied to two drive capacitors attached on either side of the shunt bar. The mechanical restoring force is provided by cantilevered silicon dioxide springs. The RSC switch has a low insertion loss of ~ 0.2 dB for the range between dc and 40 GHz. It also has very good isolation of greater than 60 dB at DC and ~ 25 dB at 40 GHz. Actuation voltages are generally around 80 V with settling times following on or off and off or on transitions being about 10 μs. The relay materials

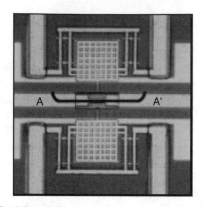

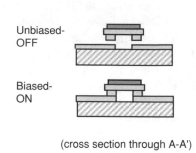

(cross section through A-A')

FIGURE 6.5 Scanning electron micrograph of the Rockwell Science Center RF MEMS switch. On the right are cross-sectional schematic views of the switch in the "off" and "on" states. (*Source*: J. Jason Yao et al., MEM system radio frequency switches, *Smart Material and Structures*, 10, Institute of Physics Publishing (2001) pp. 1196–1203. With permission.)

have shown no fatigue after 60 billion cycles and "hot" switching at ~ 1 mA has been demonstrated for tens of millions of cycles.

Given the significant advantages of MEMS switches for space RF transceiver systems, this device was an excellent candidate for a LEO technology demonstration flights via the PICOSAT missions (see below). Two PICOSAT-based flight demonstrations of the RSC RF switches were conducted: The first in February 2000 and the second in September 2001. In both of these missions the RF switches were not part of the functional RF communication system but comprised the test payload. The mission objective to actuate the switches while in orbit was successfully accomplished on both missions. No detectable degradation in performance from the baseline performance prior to launch was found for the RSC MEMS RF switches.

6.2.7 MICROCHEMICAL SENSORS

NASA's Glenn Research Center (GRC) has spearheaded the development of miniaturized chemical sensors based on MEMS and nanomaterials technologies.[15] GRC's most successful technology is a microfabricated hydrogen sensor that won the 1995 R&D 100 award. It has been successfully demonstrated on the STS-95 and STS-96 missions as a point contact sensor for the detection of hydrogen fuel leaks. Fuel leaks have led to the grounding of the Space Shuttle while on the launch pad. No commercial sensors were available that operated satisfactorily for the detection of hydrogen over a wide range of partial pressures, and that could detect the presence of hydrogen in inert environments (He purged environments) or in air. Commercially available sensors often needed oxygen to operate or needed the presence of moisture. The GRC hydrogen sensor remains highly sensitive in both inert and oxygen-bearing environments and can operate over a wide concentration range of hydrogen as shown in Figure 6.6. Since it is a microfabricated device, it has

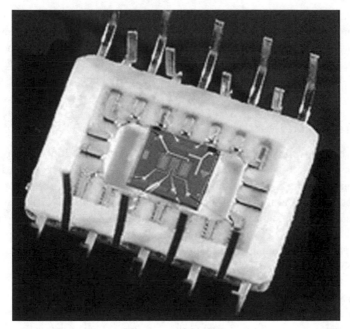

FIGURE 6.6 Optical micrograph of the NASA Glenn Research Center Hydrogen Sensor. The device consists of Pd alloy-based resistor and metal-oxide-semiconductor hydrogen sensors. Also incorporated on the chip are a microfabricated heater and temperature sensor for thermal control of the sensor. (*Source*: NASA Glenn, www.grc.nasa.gov/WWW/RT1999/5000/5510hunter.html.)

low mass, size, and power, and can be integrated with miniaturized electronics for signal processing and temperature control. The GRC sensor has also been delivered to the X-33 and X-43 projects and has been baselined for use in the water processing and oxygen generator on the International Space Station. The GRC chip contains two Pd-alloy-based hydrogen sensors. These are a resistor and a metal-oxide-semiconductor device. Also integrated within the chip are a resistive heater and a temperature sensor for controlling the thermal environment of the sensor.

6.2.8   MEMS VARIABLE EMITTANCE CONTROL INSTRUMENT

This MEMS-enabled instrument described in detail elsewhere in this book contains a MEMS shutter array radiator[16] that allows tunable control of the radiative properties of spacecraft skins. The project is led by NASA GSFC in partnership with The Johns Hopkins University Applied Physics Laboratory and Sandia National Laboratories. The technology is based on an array of micromachined, hinged shutters that can be opened or closed using MEMS comb drives (maximum operating voltage: 60 V), thus presenting a variable emittance surface to the outside environment for the spacecraft. Each shutter is a 1.77 × 0.88 mm rectangular surface. The entire shutter array contains a total of 2592 such shutters (36 chips, each containing 72 shutters) and is assembled within a 9 × 10 × 3 cm enclosure. The instrument is scheduled to fly on a NASA NMP ST5 technology demonstration flight in May 2005. The key factors that led to the selection of this technology for the demonstration flight are the simplicity and robustness of the core technology, the mission-enabling nature of the technology (can enable 20-kg class satellites), and the strong technical team.

6.2.9 TUNNELING INFRARED SENSOR ON THE SAPPHIRE SATELLITE

The University CubeSat Project[17] represents a rapid and low-cost approach to testing new technologies in a LEO space environment. Pioneered by Prof. Bob Twiggs, head of the Space Systems Development Laboratory at Stanford University, this is an exciting movement that has spread to several universities worldwide. The CubeSat development is closely related to the PICOSAT satellite development described below. Prof. Twiggs's group has had a long history of experience in launching and operating nano- and picosatellites. One of the SSDL satellites named SAPPHIRE[18] carried a MEMS-based tunneling infrared sensor (TIS) payload. The TIS was used as a horizon detector on SAPPHIRE.

Kenny et al.[19] developed the TIS initially at JPL, following up with further development of it at Stanford University. They modified the pneumatic infrared detector invented in 1947 by Marcel Golay[20] using MEMS-based silicon micromachining techniques and utilized quantum mechanical electron tunneling as the displacement transducer. As shown in Figure 6.7, the TIS consists of a stack of three silicon chips. The top two chips enclose a volume of air that expands upon absorption of infrared energy. The enclosed cavity is bounded by a flexible, metallized, silicon nitride membrane, which forms one of the electrodes for the tunneling transducer. The second tunneling electrode is a metallized silicon micromachined tip on the bottom chip. This conductive tip is surrounded at its base by a larger, planar electrode. The purpose of this planar electrode is to electrostatically attract the membrane to within electron-tunneling distance (~ 1 nm) of the tip. Once tunneling is initiated, the distance is maintained constant using feedback electronics. The infrared signal is subsequently measured by the change in the bias voltage on the planar electrode when the membrane moves outwards towards the tip. The TIS is highly sensitive and falls within the general class of uncooled infrared detectors.

6.2.10 FREE MOLECULE MICRO-RESISTOJET

The Free Molecule Micro-Resistojet (FMMR)[21] was developed by the Air Force Research Laboratory (AFRL) in collaboration with the University of Southern California and JPL. This novel MEMS-based micropropulsion device is based on resistively heating molecules within a Knudsen flow regime (Knudsen number ~ 1) in order to increase their kinetic energy as they exit the thruster, and thereby impart momentum to the spacecraft. The design is extremely simple: The propellant is solid ice at an ambient temperature of 245 K with a vapor pressure of 50 Pa. The water molecules pass through 100-μm-wide silicon-micromachined slots,

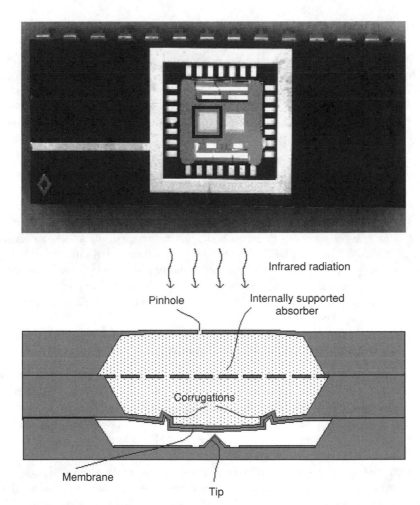

FIGURE 6.7 Tunneling infrared sensor flown on a Stanford University satellite called SAPPHIRE. The sensor was used as a horizon detector. The MEMS-based sensor combines the principles of the Golay cell infrared sensor and scanning tunneling microscopy. The three-chip device (bottom cross-sectional view) consists of two chips enclosing a small volume of air. Infrared absorption causes the enclosed air to expand, pushing out the lower membrane. The membrane movement is sensed by a quantum-mechanical tunneling electrode tip. The top view shows a 2-pixel device (red square marks a single 1.5 × 1.5 mm pixel). (*Source*: NASA/JPL.)

resistively heated to 600 K. The thrust levels per slot are in the range of tens of μN, and can be adjusted by either changing the ambient temperature of the propellant or by changing the dimensions of the slot. These extremely low thrusts could be utilized for spacecraft attitude control or for precise station-keeping applications. Such low-level, precision thrust is needed for ensembles of spacecraft involved in long-baseline, space-based interferometry missions such as in the Terrestrial Planet Finder mission aimed at detecting planets orbiting distant stars. Components of the FMMR device have been successfully flown on low-altitude rockets. The Propulsion Directorate at the Edwards Air Force Base in California packaged the FMMR as part of the Traveler I suborbital experimental payload for launch on a suborbital vehicle during the fall of 2003. Traveler I is a joint mission between the directorates' Aerophysics Branch, Microcosm Inc. of El Segundo, California, and the University of Southern California's microsatellite program.

6.3 TECHNOLOGY DEVELOPMENT PIPELINE

The above examples represent a very small subset of the broad spectrum of MNT-based systems that have potential space applications. In order to advance the maturity of a larger number of new technologies for space applications, a coherent strategy has to be put in place for creating a smoothly functioning "technology development pipeline." For each of these technologies, the two most important issues to be addressed are: bridging the mid-TRL gap and the acquisition of space heritage cheaply and rapidly. In the discussion below, novel solutions are proposed for each of these issues.

6.3.1 TECHNOLOGY MATURATION TEAM APPROACH

The primary challenge facing MNT developers and sponsors seeking to increase the efficiency of the technology "harvesting" process is how to bring together the various communities involved in space technology development in order to create a continuous technology development pipeline. A possible solution suggested by George and Powers[4] lies in the creation of a "TRL maturation team" (TMT), composed of representatives from the high and low TRL communities. They proposed that such a team should be formed at the early stages of low TRL development, essentially immediately after a new concept has been selected for funding. The importance of creating the TMTs after funding decisions for low TRL concepts have been made was to avoid coloring the initial technology selection process in any way with high TRL pragmatism.

During the low TRL development phase, the high TRL team members essentially have an advisory role, guiding the inventor away from technological dead ends that could stop the technology from transitioning to the ultimate system level aerospace application. An important consideration is that design changes are far cheaper and more cost-effective at low TRL than after the technology has matured in a direction that is not well aligned with the end application. Also, during this phase, the high TRL members become intimately acquainted with the emerging technology and its various nuances, so that they can anticipate many challenges they have to face during the ultimate system development. The TMT's role becomes increasingly important once the proof-of-concept for the technology has been successfully demonstrated. A crucial juncture in the development cycle is of course, the mid TRL development — the point in time and funding at which the

TRL gap manifests itself. The reason mid-TRL development is such a dreaded phase is because a successful transition to high TRL depends on several technological and programmatic factors that have to come together in the correct sequence. The ultimate objective for the TMT at the mid-TRL stage is to change the character of the new technology from "push" to "pull" — and thereby create a customer demand. The TMT approach can potentially increase significantly the number of new technologies transitioned and the overall efficiency of the transition process. Recently, there has been recognition at the international level of this need to bring together the disparate communities involved in aerospace technologies under one roof.[22]

6.3.2 LOW-COST, RAPID SPACE FLIGHT

A novel solution developed to overcome the "TRL Gap" problem has been to fly MNT-based devices at the low TRL stage of development. It is hoped that such flight demonstrations will generate the necessary space heritage required for future NASA, military, and commercial spacecrafts. By having space flights at the low-TRL stage, one can either "screen" the technology for space-worthiness or alternatively, build in the requisite robustness, far more cheaply and cost-effectively, than at higher TRLs. Screening space-suitable devices at an early stage in the development cycle avoids wastage of effort and investment over several years into technological "dead-ends." On the other hand, design changes are often necessary to make MNT devices and systems comply with the form, fit, and functional requirements of space missions. These design changes could be identified and implemented based on lessons learned from the space flight experiment.

The primary limitations to obtaining space heritage for new technologies are the limited flight opportunities that are available and the conservatism of mission managers to the infusion of technologies not tested in space. This risk-averse conservatism is understandable since an average space mission costs several hundreds of millions of dollars and therefore has to have a low probability of technology-related failure. Therefore, until recently, the only option for new technologies was to conduct extensive reliability testing in terrestrial laboratories, and if possible, by simulating the expected space environment.

An important innovation that makes the testing of new technologies in space competitive with terrestrial laboratory testing is the development of the low-cost, rapid-launch PICOSAT spacecraft. The PICOSAT has also spawned the worldwide, university-based CubeSat program mentioned above. The PICOSAT is an invention of the Aerospace Corporation[23] and is developed primarily as a rapid-launch, low-cost platform for testing new technologies and mission architectures in LEO. The MEMS technology group at JPL[24] partnered with the Aerospace Corporation, under sponsorship from the Defense Advanced Research Projects Agency (DARPA) and AFRL to develop a 1 kg class ($10 \times 10 \times 12.5$ cm) PICOSAT spacecraft. The 10×10 cm cross-section for the satellite and the type of spring-loaded launcher developed for ejecting the PICOSAT has been adopted as the standard by the CubeSat program. Once released in orbit, the PICOSAT is designed to be fully autonomous,

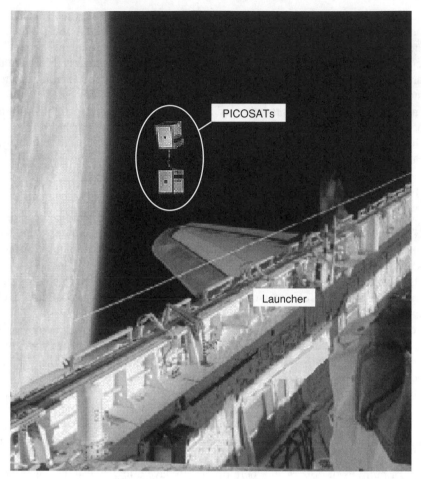

FIGURE 6.8 A pair of PICOSATS launched from the cargo bay of the space shuttle during the STS-113 mission in December 2002. Each PICOSAT carried a three-axis inertial measurement assembly consisting of MEMS gyroscopes and accelerometers. (*Source*: NASA/JPL.)

and can communicate directly with ground stations on Earth. Its low mass and size allow taking advantage of the numerous opportunities to fly secondary payloads on Earth-orbiting missions, in some cases by replacing the "ballast" that would otherwise be flown. The PICOSAT spacecraft is amenable to testing a wide range of MNT devices and systems including those developed for inertial guidance, micropropulsion, RF communication, and microinstrumentation. The most recent flight of the PICOSAT was on the Space Shuttle (STS-113) in December 2002. Figure 6.8 shows a pair of PICOSATs being released into LEO from the cargo bay of the Space Shuttle.

6.4 CONCLUSION

Given launch costs ranging from approximately $10,000/pound for LEO to as much as $1M/pound for deep space missions, there is no question that the high degree miniaturization afforded by MNT-based devices and systems is key to enabling the faster, better, and cheaper missions of tomorrow. Although there exists a large diversity of space-related applications, there is an equally large diversity of MNT, thus ensuring that nearly every MNT-based solution is guaranteed to find a home in future space missions. However, the infamous "mid-TRL gap" represents the single biggest obstacle to the infusion of a much broader range of micro- and nanotechnologies, than the lucky few that have been selected to date. For rapid and cost-effective infusion of MNT into space applications, a coordinated technology development approach, via a TMT-like mechanism proposed above, is essential. Furthermore, cheap and rapid access to space testing via novel spacecraft platforms such as the PICOSAT or CubeSats will ensure that the "infant mortality" rate of MNT remains low.

REFERENCES

1. Beatty, C.C., A chronology of thermal ink-jet structures, in: Technical Digest of the Solid-State Sensor and Actuator Workshop held June 2–6, 1996 on Hilton Head Island, South Carolina. Published by the Transducer Research Foundation, Cleveland, Ohio, pp. 200–204, 1996.
2. Payne, R.S., S. Sherman, S. Lewis, and R.T. Howe, Surface micromachining: from vision to reality to vision [accelerometer], Solid-State Circuits Conference, 1995. Digest of Technical Papers. 42nd ISSCC, 1995 IEEE International, pp. 164–165, 358, 1995.
3. Hornbeck, L.J., Digital light processing update: status and future applications, *Proceedings of SPIE*, v. 3634, pp. 158–170, 1999.
4. George, T. and R.A. Powers, Closing the TRL gap, *Aerospace America*, v. 41(8), pp. 24–26, 2003.
5. Yun, M., J. Beeman, R. Bhatia, J. Bock, W. Holmes, L. Husted, T. Koch, J. Mulder, A. Lange, A. Turner, and L. Wild, Bolometric detectors for the planck surveyor, *Proceedings of SPIE*, v. 4850, pp. 136–147, 2003.
6. Griffin, M., B. Swinyard, and L. Vigroux, SPIRE — Herschel's Submillimetre Camera and Spectrometer, *Proceedings of SPIE*, v. 4850, pp. 686–97, 2003.
7. Turner, A.D., J.J. Bock, J.W. Beeman, J. Glenn, P.C. Hargrave, V.V. Hristov, H.T. Nguyen, F. Rahman, S. Sethuraman, and A. L. Woodcraft, Silicon nitride micromesh bolometer array for submillimeter astrophysics, *Applied Optics*, v. 40, pp. 4921–4932, 2001.
8. Liebe, C.C. and S. Mobasser. MEMS based sun sensor, in: Aerospace Conference, 2001, *IEEE Proceedings*, v. 3, pp. 3/1565–3/1572, 2001.
9. Tang, T.K., R.C. Gutierrez, C.B. Stell, V. Vorperian, G.A. Arakaki, J.Z. Wilcox, W.J. Kaiser, J.T. Rice, W.J. Li, I. Chakraborthy, K. Shcheglov, Packaged silicon MEMS vibratory gyroscope for microspacecraft, in: Micro Electro Mechanical Systems, 1997. *MEMS'97, Proceedings, IEEE, Tenth Annual International Workshop*, pp. 500–505, 1997.
10. Moseley, S.H., R. Arendt, R.A. Boucarut, M. Jhavala, T. King, G. Kletetschka, A.S. Kutyrev, M. Li, S. Meyer, D. Rapchun, R.F. Silverberg, Microshutter arrays for the JWST near infrared spectrograph, *Proceedings of SPIE*, v. 5487, pp. 645–652, 2004.

11. Li, M.J., A. Bier, R.K. Fettig, D.E. Franz, R. Hu, T. King, A.S. Kutyrev, B.A. Lynch, S.H. Moseley, D.B. Mott, D.A. Rapchun, R.F. Silverberg, W. Smith, L. Wang, Y. Zheng, and C. Zinke, Microshutter arrays for near-infrared applications on the James Webb space telescope, *Proceedings of SPIE*, v. 4981, pp. 113–121, 2003.
12. Powell, D., Nanotechnology in space, *Nanotech Briefs* (www.nanotechbriefs.com), v. 1, pp. 6–9, 2003.
13. Sample, J.L., K.J. Rebello, H. Saffarian, and R. Osiander, Carbon nanotube coatings for thermal control, Ninth Intersociety Conference on Thermal and Thermomechanical Phenomena in Electronic Systems, v. 1, pp. 297–301, 2004.
14. Yao, J.J., C. Chien, R. Mihailovich, V. Panov, J. DeNatale, J. Studer, X. Li, A. Wang, and S. Park, Microelectromechanical system radio frequency switches in a picosatellite mission, *Smart Materials and Structures*, v. 10, pp. 1196–1203, 2001.
15. Hunter, G.W., C.C. Liu, and D. Makel, D. Microfabricated chemical sensors for aerospace applications, *MEMS Handbook*, CRC Press LLC, ed. M. Gad-el-Hak, Ch. 22, 2001.
16. Osiander, R., S.L. Firebaugh, J.L. Champion, D. Farrar, and M.A.G. Darrin Microelectromechanical devices for satellite thermal control, *IEEE Sensors Journal of Microsensors Microacuators: Technology and Applications*, v. 4 (4), pp. 525, 2004.
17. The official website of the International CubeSat Project is http://cubesat.calpoly.edu/
18. The official website for the Stanford SAPPHIRE Satellite is http://ssdldelta.stanford.edu/squirt1/sapphire_overview.html
19. Kenny, T.W., J.K. Reynolds, J.A. Podosek, E.C. Vote, L.M. Miller, H.K. Rockstad, and W.J. Kaiser, Micromachined infrared sensors using tunneling displacement transducers, *Reviews on Scientific Instrument*, v. 67 (1), pp. 112–128, 1996.
20. Golay, M.J.E. Theoretical considerations in heat and infra-red detection, with particular reference to the pneumatic detector, *Reviews on Scientific Instruments*, v. 18, p. 347, 1947.
21. Ketsdever, A.D., D.C. Wadsworth, and E.P. Muntz, Predicted performance and systems analysis of the free molecule micro-resistojet, *Progress in Astronautics Aeronautics*, v. 187, pp. 167–183, 2000.
22. URL 3: The CANEUS organization has been set up to bring together the full spectrum of technology developer communities involved in aerospace technology development. More details are provided at the CANEUS website: http://www.CANEUS.org
23. Janson, S.W., H. Helvajian, and E.Y. Robinson, The Concept of 'Nanosatellite' for Revolutionary Low-Cost Space System, Paper No. IAF-93-U.5.573, 44th IAF Congress, Graz, Austria, 1993.
24. George, T., Overview of MEMS/NEMS technology development for space applications at NASA/JPL, *SPIE*, v. 5116, pp. 136–148, 2003.

7 Microtechnologies for Science Instrumentation Applications

Brian Jamieson and Robert Osiander

CONTENTS

7.1 INTRODUCTION

Within the last decade, public support for large-scale space missions has slowly decreased and there has been a strong incentive to make them "faster, better, cheaper." Reducing the development time for space instruments can have the advantage of having the latest, most capable technology available, but it runs the risk of reducing the reliability through lack of testing. The major cost saver for space missions is a reduction of launch cost by reducing the weight of the spacecraft and the instrument. Microelectromechanical systems (MEMS) provide an opportunity to reduce the weight of the scientific instruments.

The science instrument is, apart from commercial and government communications satellites, the most important aspect of the spacecraft. There are a number of insertion points for MEMS into scientific instruments based on the advantages of microsystems. One example is thermal transport. The small size of a MEMS device, and the small features with high aspect ratios allow mechanically strong structures to be built with very small thermal conduction paths and small thermal capacities. Such devices can be used in microbolometers and allow the detection and imaging of particles and electromagnetic radiation from x-rays to mm-waves with very high resolution. The technology allows small shutters or mirrors to be built, which can block or deflect light from a single pixel in a telescope such as the James Webb Space Telescope (JWST), the designated replacement for the Hubble Space Telescope (HST). The small dimensions also allow for building ultrasmall plasma detectors and mass spectrometers with sufficient electric fields at very small supply voltages.

Science instruments can be divided into different groups based on the mission. For earth and solar sciences, and in some respects, planetary and deep space missions, the detection, investigation, and mapping of electromagnetic fields, particle distributions, and gravitational fields are important. Instruments to be employed are plasma and ion detectors, magnetometers, and accelerometers. There are a number of MEMS designs and prototype systems available for these instruments.[1,2] For the observation of stars and planetary emissions, telescopes and spectrometers are of importance. Here MEMS instruments can be used as the detector (e.g., a bolometer), or can improve the operation of the telescopes as in the case of the JWST.[3,4] For planetary exploration, MEMS instruments can help reduce the size and weight of planetary landers. For these applications, instruments such as hygrometers, seismometers, mass spectrometers, and micromagnetic resonance systems such as the magnetic resonance force microscope have been designed and fabricated and could be used for robotic and human exploration.[4,5] A further set of instruments can be applied for human space exploration, all those which monitor and measure the environmental conditions within the spacecraft or habitat. The applications cover the range of all medical diagnosis instrumentations, environmental monitors such as oxygen detectors, monitors for soil quality in space-based growth chambers, etc. This chapter will provide an overview of instruments with the capability or development goal to be used in spacecraft applications.

7.2   ELECTROMAGNETIC FIELD AND PARTICLE DETECTION FOR SPACE SCIENCE

MEMS-based detectors for electromagnetic fields and particles are expected to be important for future planetary and deep space missions, and their use in Earth-orbiting satellites is planned for the near future.[1] A mission concept which relies totally on the basic advantages of MEMS instruments — light weight, batch-processes, inexpensive instruments, and satellites in large numbers — is the mapping of ion distributions or magnetic fields. This goal can only be achieved with a large number of microsatellites, which can simultaneously map the fields at different positions in space. One example for such a mapping mission is the magneto-spheric constellation mission, MagCon.[6] It consists of a constellation of 50 small satellites distributed in the domain of the near-Earth plasma sheet. The mission will answer the fundamental question of how the dynamic magnetotail stores, transports, and releases matter and energy. An artist's concept of the mission is shown in Figure 7.1. Another science mission planned in the near future is the Geospace Missions Network, which is part of the "Living with a Star" (LWS) Space Weather Research Network, consisting of constellations of small satellites located in key regions around the Earth to measure downstream effects of the solar wind.[7] Figure 7.2 shows artist's concepts of different missions for satellites which carry magnetometers as well as ion and neutral particle detectors. The required size and mass restrictions provide a great opportunity for insertion of MEMS instruments.

7.2.1 PLASMA PARTICLE SPECTROMETERS

One of the keys to the solar–terrestrial interaction is the temporal and spatial distribution of ions, electrons, and neutral particles in the space surrounding Earth and between the Earth and Sun. An example is the Ion and Neutral Mass Spectrometer (INMS) on the Cassini Spacecraft, a direct sensing instrument that analyzes charged particles (like protons and heavier ions) and neutral particles near Titan and Saturn to learn more about their atmospheres. The Cassini INMS is intended also to measure the positive ion and neutral environments of Saturn's icy satellites and rings. Another example is the plasma experiment for planetary exploration (PEPE), which is a space plasma, energy, angle, and mass or charge spectrometer now taking data aboard the Deep Space 1 (DS1) spacecraft. These

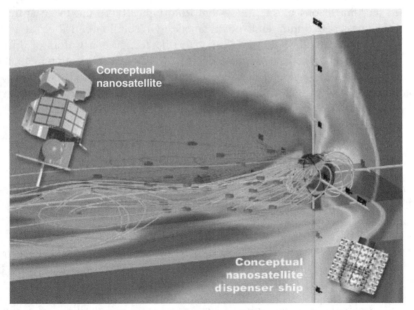

FIGURE 7.1 Artist's concept of the Magnetospheric Constellation Mission. (*Source*: NASA, http://stp.gsfc.nasa.gov/missions/mc/mc.htm.)

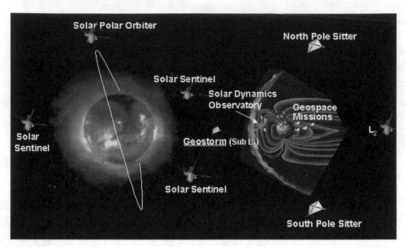

FIGURE 7.2 Schematic of the different missions for "Living with a Star." (*Source*: NASA, http://lws.gsfc.nasa.gov/overview2.htm.)

instruments are large (20 kg for PEPE), require a large amount of electrical power, are expensive, and could not be easily implemented into 20 to 50 small satellites. A good overview on plasma spectrometers, how they work, and the drive to make them smaller (thereby making the missions less expensive) is given by Young.[8] One of the first micromachined designs was used by Stalder et al., who used micromachining techniques to generate an array of Bessel boxes.[9] They have reduced the dimensions of such a system from typically 10 cm to an array of 4 with a thickness of 2.6 mm, and shown an energy resolution of 1.2 eV at 100 eV with an acceptance half-angle of 14°. For this device, silicon wafers of different thickness were wet-etched and bonded together. A different fabrication method was used by Enloe et al., who fabricated an electrostatic analyzer out of laminated, photolithographically etched stainless steel.[10] The analyzer worked without charge multiplication and was about 5 cm × 5 cm in size, with 1920 individual analyzer elements. The acceptance angle was 5°, with an energy resolution of 0.66 eV at 10 to 30 eV ion energies.

A similar analyzer design was used at the Johns Hopkins University Applied Physics Laboratory (JHU/APL) and NASA Goddard Space Flight Center (GSFC) for a flat plasma spectrometer to fly on the Air Force Academy's Falcon Sat 3 mission in 2005.[1] This instrument, including sensor-head-array, printed circuit board with amplifier array electronics, power supply, and chassis has been designed and built to occupy a volume of approximately 200 cm^3 in a 0.5 kg, 300 mW package. The sensor head as shown in Figure 7.3 consists of an array of five identical spectrometer modules, each with a different fixed field-of-view (FOV) consisting of a collimator, electrostatic analyzer, energy selector masks, microchannel plates, and anode plate for detection. Ions enter the instruments via the collimator, which serves to select the entrance angle of the incident particles. It is comprised

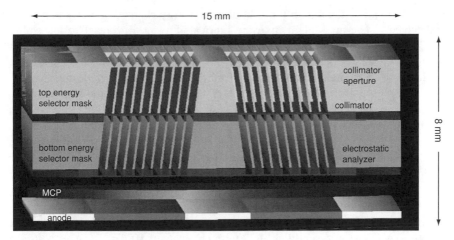

FIGURE 7.3 Cross-section of the FlaPS sensor-head-array (± 8 pixel elements shown). (*Source*: JHU/APL.)

of single-crystal silicon die with an array of 50-μm wide and 4.2-mm long channels, as shown in Figure 7.4a. These die are bonded so that each channel is in the center of an array of 200-μm wide channels that have been micromachined in CuBe using electrical discharge machining (EDM), as shown in Figure 7.4b. Each of the five pixels defining the sensor-head-array was micromachined at selected angles with respect to the normal plane of incidence to achieve a maximum FOV of $\pm 8°$. The total thickness of the collimator is 2.75 mm, which with the input channels results in a 1° acceptance angle and a transmission of 11% per detector. Before entering and exiting the electrostatic analyzer, the particles encounter entrance and exit apertures which act as energy selector masks. For a given electric field in the electrostatic analyzer, only particles of a given energy pass through both apertures. The design

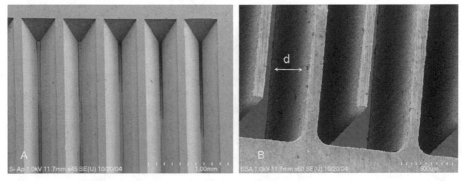

FIGURE 7.4 (a) FlaPS analyzer entrance aperture and (b) EDM machined analyzer electrodes. (*Source*: JHU/APL.)

does not provide a direct path for light or high energy charged particles to pass and be detected.[10] The mask elements were fabricated using deep reactive ion-etching (DRIE) and anisotropic wet-etch techniques. Processing was performed on silicon-on-insulator (SOI) wafers taking advantage of the 1-μm thick buried oxide layer in providing electrical isolation of the electrostatic analyzer region from the rest of the device. A highly anisotropic plasma etch of the handle-side of the SOI die generating 400-μm wide and 300-μm deep slits was followed by a device-side wet-etch to generate 20-μm wide and 50-μm deep slits. This, and oxide removal in only the aperture areas resulted in entrance and exit aperture mask elements for each pixel. With the geometry specified above, the energy resolution is about 5%. The electrostatic analyzer is almost identical to the collimator, made from copper beryllium using EDM machining at different angles in each of the detectors.

The final two elements of the FlaPS[1] instrument head (excluding the control electronics) are the microchannel plates for signal amplification and the anode for detection. A Chevron assembled MCP with channels of 10 μm diameter is used in this instrument. On the anode, one anode per pixel is patterned on a single ceramic substrate with Cr, Cu, Ni, or Au conductors and plated through vias for bonding to each of the five preamplifier discriminator circuits located on the amplifier array electronics board mounted below. An appealing extension of the basic FlaPS design is to distribute an array of analyzers around a satellite or onto a spinning satellite, with a common high-voltage supply and microchannel plate holder, allowing wide ranges of directions to be measured.

7.2.2 MAGNETOMETERS AND ELECTRIC FIELD DETECTORS

The determination of planetary electric and magnetic fields and their interactions with the solar wind and other charged particles have been an important focus for past space missions and future mission planning. Magnetic measurements, such as those carried out by MagSat and Oersted, are essential for the maps used in satellite orientation and navigation, as well as for geophysical mapping of Earth's field. In addition, magnetometers are used for navigation and attitude control, which will be discussed in Chapter 10. For the Oersted mission, the fluxgate magnetometer's noise level was in the order of about 100 pT at 1 Hz; for deep space mission, sensitivities below these levels are desirable. Most instruments are fluxgates, which are the most suitable vector magnetic field sensors besides SQUIDs in this range.[11,12] Initial attempts to micromachine fluxgates have resulted in sensitivities of the order of around 100 nT.[13–17] Disadvantages of fluxgate sensors are offset due to the magnetic cores, limited dynamic range, and relatively low frequency ranges. Other miniature magnetometers are based on magnetoresistance,[18] or giant magnetoresistance,[19] which can achieve about 10 nT sensitivities at very high frequencies and relatively low noise levels. Two types of magnetometers have been successfully micromachined, one based on the torque[20–22] or magnetostriction[23,24] created by a ferromagnetic material, and the other based on the Lorentz force.[2,24] The disadvantages of the first type are the integration of ferromagnetic materials into the fabrication process. This may interfere with the promise of batch production at traditional foundries. In addition, all magnetometers, which use ferromagnetic materials have a limited dynamic range and the variation in magnetization requires a calibration process. An interesting approach for space application, where large current-carrying supply lines can change the magnetic environment around the magnetic boom, would be the use of such a magnetometer with remote interrogations.[25] Lorentz force-based magnetometers promise a high dynamic range with a zero offset and wide linearity. They are based on the measurement of the deflection of a MEMS structure with an AC or DC current flowing in it. One example is the JPL device,[26] that uses DC current and measures the static deflection of a membrane with conductors using a tunneling current as the transduction method. The sensitivities of this device are in the order of mT. A more sensitive magnetometer has been designed at JHU/APL,[2,24] based on a resonating "xylophone" bar, a few hundred microns long and supported at the nodes where an AC current is supplied. At the resonance frequencies, Qs for these devices in vacuum are in the order of 50–100k, and small fields can generate a large magnitude of deflection. Devices etched photolithographically from CuBe with lengths of a few millimeters have been used to measure magnetic fields with sensitivities as low as 100 pT/Hz$^{1/2}$ using optical beam deflection as the transduction method. Figure 7.5 shows a device surface micromachined in polysilicon using the MUMPs process. The sensitivity of theses devices is limited by the current-carrying capability of the polysilicon supports as well as the integration of the transduction into the device. An improvement has been achieved by using a complementary metal oxide semiconductor (CMOS) process or a silicon on sapphire (SOS) CMOS process.[27] While the mechanical properties of the resonating device are somewhat degraded, the use of multiple metal layers and the integration of the control electronics as well as the capacitive readout onto the same die improve the performance. Major advantages of the Lorentz force magnetometers are the wide dynamic range, since the signal is the

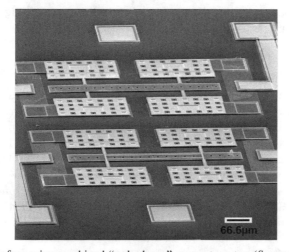

FIGURE 7.5 Surface micromachined "xylophone" magnetometer. (*Source*: JHU/APL.)

product of the current in the magnetometer, which can be chosen depending on the field to be detected, limited only by the current-carrying capability of the material, and the absence of any offset other than the detection limit of the Brownian motion of the resonator itself. It also can be used to detect AC magnetic fields with the same narrow bandwidth and sensitivity.[28]

The measurement of electric fields in space is important to investigate wave processes in space plasma. To our knowledge, the only micromachined device reported for measurements of electric fields for microsatellites is based on a split Langmuir probe, consisting of two conductive plates in a small distance.[12] Such a prototype was tested on board of the Prognoz-10 satellite.

7.3   TELESCOPES AND SPECTROMETERS

The development of optical MEMS components during the telecom boom of the late 1990s, has provided building blocks for a new generation of space-based optical devices. Micromachined silicon slits and apertures provide a high degree of precision for critical optical paths, and have been used in space flight dual slit spectrometers. A MEMS Fabry–Perot (FP) interferometer has been developed at NASA GSFC,[29] and additional spectrometers with surface micromachined grating structures controlled via small MEMS motors have been reported.[30] More dramatically, microoptoelectromechanical systems (MOEMS) can deflect certain image areas to a spectrometer, can block other areas, or can be used to correct for optical aberrations in the telescope or the instrument. An example is the Near Infrared Spectrograph (NIRSpec) for the JWST, planned for launch in 2009, which will have MOEMS devices as an integral part of the instrument.[31–34]

Another application, and one that is relatively well established, is in bolometers. Here, the small pixel size enabled by MEMS and the resulting small thermal capacities allow for integration of large arrays of very small bolometric devices which can be used to detect radiation from the millimeter wave range all the way up to x-rays and particles.[35]

7.3.1   THE JAMES WEBB SPACE TELESCOPE NEAR-IR SPECTROGRAPH

The study of galaxy formation, clustering, chemical abundances, star formation kinematics, active galactic nuclei, young stellar clusters, and measurements of the initial mass function of stars (IMF) requires a near-infrared spectrograph. The NIRSpec for the JWST (in earlier publications referred to as Next Generation Space Telescope or NGST) will be the spectrograph in the wavelength range of 0.6 to 5 μm, providing three observation modes with a FOV of ~3.4 × 3.4 arcmin in the current design. In the R~1000 modes, NIRSpec provides users of JWST with the ability to obtain simultaneous spectra of more than 100 objects in a >9 square arcminute FOV. Three gratings cover the wavelength range from 1 to 5 μm, and the spectrograph will take advantage of a MEMS shutter system to enable users to observe hundreds of different objects in a single FOV. The European

Space Agency (ESA) will be providing the NIRSpec instrument, and NASA will provide the detectors and the MEMS aperture mask as part of their contribution to JWST. Two approaches were initially proposed in 1996 to NASA for the NIRSpec, one using a MEMS micromirror array[36,37] and one using a MEMS shutter array.[31–34] The requirements for both mirror and shutter arrays are very strict: The size of each pixel was to be 100×100 μm, with a fill factor better than 80% and a contrast better than 2000:1, expandable to an array size of 1800×1800 square elements, operating at 40K. Two mirror microarray technologies were considered, one developed at NASA GSFC[38] and one at Sandia National Laboratories (SNL).[39] A scanning electron microscope (SEM) photograph of the SNL micromirrors is shown in Figure 7.6. The mirrors are made using SNL's SUMMiT V process as seen in Chapter 3. An advantage of the mirror design is that the drive and selection electronics can be hidden under the mirrors. However, in order to improve the image quality and the contrast, the mirror needs to be fabricated with different materials, gold on silicon in this case, which causes stress when cooled down to cryogenic temperatures. This results in required distortion of the image quality and may cause de-lamination of the gold coating itself.

The shutter approach was selected because of its better contrast and image quality, since no reflective surface is involved, and scattered light from the edges is predominantly reflected back away from the spectrograph. A major challenge for this approach is to integrate the actuation mechanism as well as the single shutter control — each shutter needs to be uniquely addressed — within less than 20% of the entire area. The JWST "microshutter array" is a programmable aperture which

FIGURE 7.6 A micromirror developed by SNL, pictured in the tilted position. Each mirror is about 100 μm in width.[1] (Courtesy of Sandia National Laboratories, http://www.sandia.gov/media/NewsRel/NR1999/space.htm.)

is used to select the light from multiple objects for transmission to the infrared spectrograph. The array consists of 250,000 individually addressable 100×200 μm shutters, which are magnetically actuated 90° out of plane and then electrostatically latched. A deep reactive ion etched (DRIE) silicon frame supports the shutters and provides interconnects and electrodes for latching individual shutters. Figure 7.7 shows SEM images of the 100×100 μm shutters of the proto type as shown in Figure 7.8 made from single crystal silicon. For the NIRSpec 100×200 μm shutters will be used. All shutters are slightly magnetized and are opened by scanning a permanent magnet over the array. Selected apertures are held open electrostatically via application of a potential between the shutter and an electrode on the wall. After the magnet has passed, the resilience of the hinges flips the remaining shutters close. Light shields fabricated onto the frame prevent light from passing around the edges. The shutters are fabricated on 4-in. SOI wafers and the completed dies are flip-chip bonded to a silicon substrate which contains the drive electronics. Four adjoining substrate assemblies produce a complete flight array, which must withstand launch conditions and be operated at cryogenic temperatures.

At present the MEMS shutter design is finalized and initial flight prototypes have been fabricated and tested under operating conditions.[40,41] The final flight

FIGURE 7.7 Scanning electron microscope image of the Si microshutter shutter blade, which is suspended on a torsion beam that allows for a rotation of 90°. The torsion beam is suspended on a support grid. While actuated with a probe tip in this image, the blades will be actuated magnetically in the JWST–NIRSpec. (*Source*: NASA GSFC.)

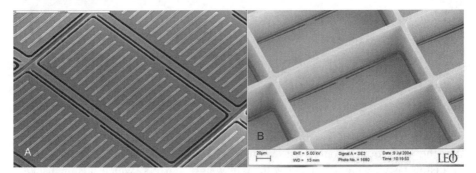

FIGURE 7.8 Scanning electron microscope image of the front (A) and back (B) side of the JWST microshutter array (*Source*: NASA GSFC.)

FIGURE 7.9 Optical transmission image of a pattern written with the microshutter array (*Source*: NASA GSFC.)

device is scheduled to be delivered for instrument integration in 2007, with a projected launch date in 2009. As a risk mitigation path, a macroscopic slit array is fabricated in parallel.[42]

7.3.2 ADAPTIVE OPTICS APPLICATIONS

A similar application is the use of dense arrays of MEMS mirrors in adaptive optics for space telescopes. In this case the requirements for the mirror motion are more stringent: they need to be positioned continuously and not just toggled between two positions. On terrestrial telescope applications, adaptive optics compensate for atmospheric turbulence during observations. In principle, very faint objects can be imaged during long exposures, provided there is a bright "reference beacon" nearby to allow the AO system to analyze the atmospheric effects. It is conceivable that the same optics could be used in space-based applications to replace high-precision heavy-weight mirrors with light-weight mirrors, which are themselves adaptive or are corrected via adaptive optics. One principle for such a mirror array has been developed at Boston University[43-48] and is commercially available from Boston Micromachines.[49] The device offers a displacement of 2 μm with no hysteresis, and surface finishes of highly reflective gold or aluminum coating of 30 nm RMS. A similar device has been designed and fabricated by Vdovin et al.,[50-52] which also uses an electrostatic membrane mirror. This device has been demonstrated at the Air Force Research Laboratory (AFRL).[53-56]

Two other concepts, flexure-beam micromirror devices (FBMD) and axial-rotation micromirror devices (ARMD), have been developed at the AFRL and SNL.[57,58] These devices are fabricated in SNL's four-level planarized polysilicon process (SUMMiT V, see Chapter 3). Although square FBMDs are sufficient for most applications, the same size array of ARMDs demonstrates significantly improved performance since this device combines tilting and piston deflection. The tilting of the ARMD mirror surface, in addition to its piston deflection, allows for a closer adherence to the curvature of typical wavefront aberrations.

FIGURE 7.10 Photograph of assembled Fabry–Perot tunable filter. (*Source*: NASA GSFC.)

A similar program proposed by JPL is the Advanced Segmented Silicon Space Telescope (ASSiST), which utilizes thin silicon wafers as the building blocks of highly segmented space telescope primary mirrors.[59-61] Using embedded MEMS actuators operating with high bandwidth control, this technology can achieve diffraction-limited image quality in the 3–300 μm wavelength range. The use of silicon wafers as cryogenic mirror segments is carried forward considering a point design of a future NASA ORIGINS mission. Individual segments of the ASSiST consist of 1-mm thick, 300-mm diameter silicon wafers with 10-mm deep frames, assembled into 3-m diameter rafts. This achieves considerable reductions in primary mirror mass through the elimination of a heavy back plane support structure. Rather, they exploit the micromachining capabilities of silicon processing technology to achieve sophisticated control of a highly segmented mirror using high-bandwidth, high-stroke MEMS actuators, which will ultimately be built directly into the mirror segment, resulting in an integrated optics package. Thus, a single segment can perform the traditional light-focusing function of a telescope as well as the control functions, and quite possibly the space deployment functions.

7.3.3 SPECTROMETER APPLICATIONS

The size of spectrometers, especially infrared spectrometers, has been rapidly reduced in recent years due to uncooled IR detectors with ultrasmall pixel size and modern micromachining techniques.[62] Infrared spectrometers are some of the most important instruments since most molecules show a characteristic "fingerprint" spectrum within this range. A reduction in size for these instruments will have a major impact on space-based observations, as well as for terrestrial sensors for chemical and biological agent detection. One example is a Fabry–Perot (FP)-based interferometer.[63] A FP interferometer or etalon consists of two flat, parallel, semitransparent plates coated with films of high reflectivity and low absorption. The pass band of the etalon is determined by the separation between the plates, which is generally varied using piezoelectric actuators. For any large aperture wide field telescope, low-resolution FPs are an ideal option for narrow-band imaging as opposed to linear or circular variable interference filters as they ease size require-ments on filter wheels and offer flexibility in choice of spectral resolution. Tunable filters on space telescopes will require operation at cryogenic temperatures, where piezo actuators alone do not provide sufficient translation to tune the etalon over the desired orders of interference without becoming large and cumbersome. In addition, low-resolution infrared etalons require cavity spacings on the order of a micron. Figure 7.10 and Figure 7.11 show a photograph and the schematic of a FP interfer-ometer design developed at NASA GSFC.[29]

In this design, the mechanism is fabricated in two sections that are assembled into the final FP filter. The stationary mirror structure consists of a micromachined 350-μm thick silicon wafer coated with a multilayer dielectric (MLD) in the aperture. The moving mirror structure is also machined from a 350-μm thick silicon wafer and is identically coated with MLD over its aperture. Its reflector is attached

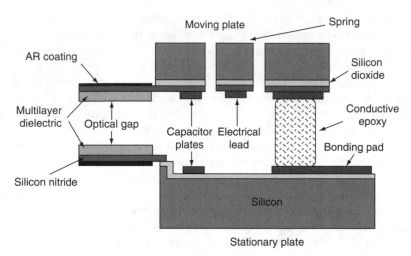

FIGURE 7.11 A cross-section of the outer edge of a Fabry–Perot filter. (*Source*: NASA GSFC.)

to a moving inner annulus suspended from an outer fixed annulus by silicon leaf springs. The moving plate is joined to the fixed plate with conductive epoxy for mechanical alignment and electrical connection for the moving plate's electrodes. The two mirrors consist of thin silicon nitride membranes with high-reflectance MLD coatings on their gap-facing surfaces and antireflection (AR) coatings on their outward-facing surfaces.

The inner annulus is suspended on three leaf springs designed to allow scanning of the FP gap. Three gold capacitance pads deposited onto each of the moving and fixed plates form three equally-spaced electrostatic actuation and measurement pairs. A DC (~35 V) bias across these pads generates an attractive force that works against the restoring force of the spring. Micromachined FP tunable filters are an enabling component for wide-field imaging spectroscopy and optics components for a wide range of hyperspectral imaging sensor systems.

Another approach for a MEMS infrared interferometer is the use of programmable diffraction gratings.[30,64] A commercial product of this kind is sold by Silicon Light Machines.[65] Small ribbons, which constitute an optical grating, are actuated electrostatically to change the grating constant and therefore the transmission or reflection spectrum of the device. An interesting application for such a device is in correlation spectroscopy,[30] where a spectrum of interest is programmed into the grating and correlated with the received thermal infrared radiation to detect and identify substances such as chemical agents or pollutants in the environment. MEMS fabrication has also been used in the design of a millimeter-wave Fourier transform spectrometer.[66] In this case, the quasi-optical arrangement of a Fourier transform infrared (FTIR) system was replaced with a MEMS-based, high-impedance coplanar waveguide (CPW) line loaded with RF switches that produced a linear variable time delay line. This technology is extensively described in Chapter 8, under MEMS devices for communications.

7.3.4   MICROMACHINED BOLOMETERS

Bolometers are an important application for MEMS devices in infrared spectrometry. Most IR detectors require cryogenic cooling, bolometers can be used at ambient temperatures and are almost wavelength independent. While bolometers are used as detectors from microwaves to the visible spectrum, but visible MEMS fabrication has given this technology a new dimension. We now can fabricate bolometers as mechanical structures, which are the size of a wavelength, with thermal masses so small that even the smallest amount of absorbed energy is detectable in arrays with standard video array sizes. In a commercially available imaging array from Sarnoff Corporation, bimetallic cantilevers deflect upon absorption and change the capacity of the respective pixel.[67] The bi-material cantilever deflects upon absorption and changes the capacity in this pixel. The small dimensions of MEMS technology allow the bimetallic cantilever to be thermally insolated from the substrate with a very thin element and to have such a low thermal mass that the absorbed energy creates a temperature change large enough to measurably deflect the cantilever. Other bolometer designs developed for satellite-based infrared imaging use active and reference detectors arranged in Wheatstone bridge configurations.[3,68,69] The energy absorbed in the optical stack formed by the materials changes the temperature and therefore the resistance of the active pixel.

The same approach can be used not only for infrared radiation, but also for other radiation such as x-rays. NASA GSFC has been working on a high-resolution x-ray spectrometer for the Constellation–X mission.[35] The spectrometer is micromachined and consists of a Bi or Cu multilayer absorber for stopping and thermalizing the incident x-rays, an e-beam evaporated Mo or Au proximity bilayer with sputtered Nb leads for sensing the resultant temperature rise, and a silicon nitride membrane to provide a weak thermal link to the thermal sink so that the calorimeter can return to its equilibrium temperature. The x-ray spectrometers have achieved resolutions of about 28 eV at 3.3 keV x-rays. MEMS are an enabling technology for these position sensitive spectrometers, which require small sizes for resolution as well as for small thermal capacities.

7.4   MEMS SENSORS FOR *IN SITU* ANALYSIS

All of the scientific spacecraft instruments discussed so far are essentially remote sensing devices, measuring photons, fields, or particles incident upon an orbiting spacecraft or space telescope. Equally important is the ability to measure the chemical composition or other properties of a sample encountered on a planet's surface or in its atmosphere. Robotic spacecraft carrying mass spectrometers, for example, have been used in the exploration of Venus, Mars, Jupiter, the Moon, the comet Halley, and most recently Saturn and its moon Titan. Other devices such as x-ray spectrometers, x-ray fluorescence and diffraction instruments, nuclear magnetic resonance force microscopes, and scanning electron microscopes have been either flown or proposed for use in a planetary exploration mission to identify the composition of planetary samples *in situ*. In all cases, existing spacecraft instruments are quite large and consume a lot of power. Miniaturization would allow these instruments to be incorporated onto small multiple entry probes, autonomous rovers, and sample handling systems such as robotic arms, booms, and drills. Accordingly, MEMS is an attractive technology for developing highly miniaturized versions of these instruments, if they can maintain the performance of existing space flight instruments. In addition, new instruments based on technologies such as lab-on-a-chip have been proposed to provide the ability to carry out analytical chemistry in a miniature, integrated package.

7.4.1 MICROMACHINED MASS SPECTROMETERS

A mass spectrometer consists of a sample handling system, an ion source, a mass filter, and a detector. After being introduced to the instrument by the sample handling system, atoms in gaseous, solid, or liquid states are ionized by electron bombardment, electrospray ionization, laser ablation, or other methods. The ions are then separated by their charge to mass ratio in a mass filter. Common mass filters include: magnetic sectors, in which ions of different masses are deflected differentially in a magnetic field; quadrupoles and ion traps, which are scanning devices in which ions of a particular mass exhibit stable trajectories at a given RF frequency; and time-of-flight, in which ions of constant initial kinetic energy but different mass are separated by their flight times due to their differences in velocity. Work on MEMS-based mass spectrometers has been reported, including magnetic, quadrupole, ion trap and time-of-flight mass filters.[70–79] In all cases, instrument performance has fallen far short of the requirements for a space flight mass spectrometer, and the need for additional research and development in this area is clear.

7.4.2 MAGNETIC RESONANCE FORCE MICROSCOPY

Nuclear magnetic resonance is a very sensitive way to detect the presence of water, and therefore is a desirable instrument on any explorer mission. There has been a recent push to develop imaging magnetic resonance microscopes to be able to measure spin distributions and identify molecules. These methods are based on magnetic resonance force microscopy, where the force applied by the spins rotating in an RF field on a micromachined resonant cantilever beam with a magnetic particle is measured via interferometric techniques. Such instruments could be potentially built entirely on a MEMS or microelectronics platform and used in space exploration as element detectors for landers.[5,80–82]

7.5 CONCLUSION

While it is difficult to imagine the instrumentation for future spacecraft that will be enabled or improved by the integration of MEMS, it is obvious from the examples that it is already being done, and that there are devices that can be inserted into space systems as well as devices that have already been designed and fabricated for specific missions. In many cases, the fast development technology and fabrication capability allow systems and instruments to be designed and fabricated that could not have been thought of a few years ago. This development requires strong interaction between the space scientist and the engineer, who can use a toolbox of new capabilities of microsystems to generate new instruments.

REFERENCES

1. Wesolek, D.M. et al., A micro-machined flat plasma spectrometer (FlaPS), *Proceedings of SPIE* **5344**, 89, 2004.
2. Wickenden, D.K. et al., Micromachined polysilicon resonating xylophone bar magnetometer, *Acta Astronautica* **52** (2–6), 421, 2003.
3. Jerominek, H. et al., 128 × 128 pixel uncooled bolometric FPA for IR detection and imaging, *Proceedings of SPIE* **3436**, 585, 1998.
4. Tang, T.K., MEMS for space applications, *Proceedings IEEE 25th International Silicon-on-Insulator Conference* 67, 1999.
5. George, T. et al., MEMS-based force-detected nuclear magnetic resonance spectrometer for *in situ* planetary exploration, *2001 IEEE Aerospace Conference Proceedings* **1**, 1273, 2001.
6. Magnetospheric constellation mission, MC, NASA, http://stp.gsfc.nasa.gov/missions/mc/mc.htm#overview
7. Geospace Missions Network, NASA, http://lws.gsfc.nasa.gov/overview2.htm
8. Young, D.T., Space plasma particle instrumentation and the new paradigm: Faster, cheaper, better, in *Measurement techniques in space plasmas*, Pfaff, R.F., Borovsky, J.E., and Young, D.T., (eds), American Geophysical Union, Washington, D.C., 1998, 1.
9. Grunthaner, F.J. et al., Micromachined silicon-based analytical microinstruments for space science and planetary exploration, 706, 1994.
10. Enloe, C.L. et al., Miniaturized electrostatic analyzer manufactured using photolithographic etching, *Review of Scientific Instruments* **74** (3), 1192, 2003.
11. Ripka, P., New directions in fluxgate sensors, *Journal of Magnetism and Magnetic Materials* **215**, 735, 2000.
12. Korepanov, V., Electromagnetic sensors for microsatellites, *Proceedings of IEEE Sensors 2002* **1**, 1718, 2002.
13. Kawahito, S. et al., Fluxgate magnetic sensor with micro-solenoids and electroplated permalloy cores, *Sensors and Actuators A: Physical* **43** (1–3), 128, 1994
14. Liakopoulos, T.M. and Ahn, C.H., Micro-fluxgate magnetic sensor using micromachined planar solenoid coils, *Sensors and Actuators A: Physical* **77** (1), 66, 1999.
15. Kawahito, S. et al., Micromachined solenoids for highly sensitive magnetic sensors, 1077, 1991.
16. Gottfried-Gottfried, R. et al., Miniaturized magnetic field sensor system consisting of a planar fluxgate sensor and a CMOS readout circuitry, *Proceedings of the International Conference on Solid-State Sensors and Actuators* **2**, 229, 1995.
17. Gottfried-Gottfried, R. et al., A miniaturized magnetic-field sensor system consisting of a planar fluxgate sensor and a CMOS readout circuitry, *Sensors and Actuators A: Physical* **54** (1–3), 443, 1996.
18. Lenz, J.E. et al., A high-sensitivity magnetoresistive sensor, 114, 1990.

19. Hill, E.W. et al., Giant magnetoresistive magnetometer, *Sensors and Actuators A: Physical* **59** (1–3), 30, 1997.
20. Yee, J.K., Yang, H.H., and Judy, J.W., Dynamic response and shock resistance of ferromagnetic micromechanical magnetometers, *Proceedings of the 15th IEEE International Conference on Micro Electro Mechanical Systems (MEMS)* 308, 2002.
21. Latorre, L. et al., Micromachined CMOS magnetic field sensor with ferromagnetic actuation, *Proceedings of SPIE* **4019**, 398, 2000.
22. Beroulle, V. et al., Micromachined CMOS magnetic field sensors with low-noise signal conditioning, *Proceedings of the IEEE Micro Electro Mechanical Systems (MEMS)* **2002**, 256, 2002.
23. Kistenmacher, T.J. et al., Design and properties of a thin-film, MEMS-based magnetostrictive magnetometer, *Materials Research Society Symposium — Proceedings* **444**, 75, 1997.
24. Givens, R.B. et al., High sensitivity, wide dynamic range magnetometer designed on a xylophone resonator, *Applied Physics Letters* **69** (18), 2755, 1996.
25. Vasquez, D.J. and Judy, J.W., Zero-power magnetometers with remote optical interrogation, *Proceedings of the 17th IEEE International Conference on Micro Electro Mechanical Systems (MEMS)* 109, 2004.
26. Miller, L.M. et al., M-magnetometer based on electron tunneling, *Proceedings of the 9th IEEE Micro Electro Mechanical Systems (MEMS) Workshop* 467, 1996.
27. Tejada, F. et al., Surface micromachining in silicon on sapphire CMOS technology, *Proceedings — IEEE International Symposium on Circuits and Systems* **4**, 2004.
28. Givens, R.B. et al., Heterodyne detection of alternating magnetic fields with a resonating xylophone bar magnetometer, *Applied Physics Letters* **74** (10), 1472, 1999.
29. Mott, D.B. et al., Micromachined tunable Fabry–Perot filters for infrared astronomy, *Proceedings of SPIE* **4841**, 578, 2002.
30. Sinclair, M.B. et al., A MEMS-based correlation radiometer, *Proceedings of SPIE* **5346**, 37, 2004.
31. Li, M.J. et al., Fabrication of microshutter arrays for space application, *Proceedings of SPIE* **4407**, 295, 2001.
32. Moseley, S.H. et al., Programmable 2-dimensional microshutter arrays, *Proceedings of SPIE* **3878**, 392, 1999.
33. Mott, D.B. et al., Magnetically actuated microshutter arrays, *Proceedings of SPIE* **4561**, 163, 2001.
34. Zheng, Y. et al., Microshutter arrays for near-infrared applications on the James Webb space telescope, *Proceedings of SPIE* **4981**, 113, 2003.
35. Tralshawala, N. et al., Design and fabrication of superconducting transition edge x-ray calorimeters, *Proceedings 8th International Workshop on Low Temperature Detectors* **444**, 188, 2000.
36. Connelly, J.A. et al., Alignment and performance of the infrared multi-object spectrometer, *Proceedings of SPIE* **5172**, 1, 2003.
37. MacKenty, J.W. et al., IRMOS: an infrared multi-object spectrometer using a MEMS micro-mirror array, *Proceedings of SPIE* **4841**, 953, 2002.
38. Winsor, R. et al., Optical design for an infrared multi-object spectrometer (irmos), *Proceedings of SPIE* **4092**, 102, 2000.

39. Walraven, J.A. et al., Failure analysis of polysilicon micromirror arrays, Conference *Proceedings from the International Symposium for Testing and Failure Analysis* 283, 2002.

40. Zamkotsian, F., Gautier, J., and Lanzoni, P., Characterization of MOEMS devices for the instrumentation of next generation space telescope, *Proceedings of SPIE* **4980**, 324, 2003.

41. Zamkotsian, F. et al., MEMS-based slit generator for NGST-NIRMOS: modeling and characterization, *Proceedings of SPIE* **4850**, 527, 2002.

42. Erickson, D.A., Design of a mechanically actuated reconfigurable slit mask (MARS) for the NGST near IR spectrograph, *Proceedings of SPIE* **4850**, 517, 2002.

43. Bifano, T. et al., Micromachined deformable mirrors for adaptive optics, *Proceedings of SPIE* **4825**, 10, 2002.

44. Bifano, T.G. et al., Microelectromechanical deformable mirrors, *IEEE Journal on Selected Topics in Quantum Electronics* **5** (1), 83, 1999.

45. Perreault, J.A. et al., Manufacturing of an optical quality mirror system for adaptive optics, *Proceedings of SPIE* **4493**, 13, 2002.

46. Perreault, J.A. et al., Adaptive optic correction using microelectromechanical deformable mirrors, *Optical Engineering* **41** (3), 561, 2002.

47. Perreault, J.A., Bifano, T.G., and Martin Levine, B., Adaptive optic correction using silicon based deformable mirrors, *Proceedings of SPIE* **3760**, 12, 1999.

48. Reimann, G. et al., Compact adaptive optical compensation systems using continuous silicon deformable mirrors, *Proceedings of SPIE* **4493**, 35, 2002.

49. Boston Micromachines, http://www.bostonmicromachines.com

50. Sakarya, S., Vdovin, G., and Sarro, P.M., Technological approaches for fabrication of elastomer based spatial light modulators, *Proceedings of SPIE* **4983**, 334, 2003.

51. Sakarya, S., Vdovin, G., and Sarro, P.M., Spatial light modulators based on micromachined reflective membranes on viscoelastic layers, *Sensors and Actuators A: Physical* **108** (1–3), 271, 2003.

52. Vdovin, G.V. et al., Technology, characterization, and applications of adaptive mirrors fabricated with IC-compatible micromachining, *Proceedings of SPIE* **2534**, 116, 1995.

53. Dayton, D. et al., MEMS adaptive optics: field demonstration, *Proceedings of SPIE — The International Society for Optical Engineering* **4884**, 186, 2002.

54. Dayton, D. et al., Air Force research laboratory MEMS and ICM adaptive optics testbed, *Proceedings of SPIE* **4825**, 24, 2002.

55. Dayton, D. et al., Demonstration of new technology MEMS and liquid crystal adaptive optics on light astronomical objects and satellites, *Optics Express* **10** (25), 1508, 2002.

56. Gonglewski, J. et al., MEMS adaptive optics: field demonstrations, *Proceedings of SPIE* **4839**, 783, 2002.

57. Comtois, J. et al., Surface-micromachined polysilicon moems for adaptive optics, *Sensors and Actuators A: Physical* **78** (1), 54, 1999.

58. Michalicek, M.A., Bright, V.M., and Comtois, J.H., Design, fabrication, modeling, and testing of a surface-micromachined micromirror device, American Society of Mechanical Engineers, Dynamic Systems and Control Division (Publication) DSC 57–2, 981, 1995.

59. Dekany, R. et al., Advanced segmented silicon space telescope (ASSiST), *Proceedings of SPIE* **4849**, 103, 2002.

60. Yang, E.-H., Dekany, R., and Padin, S., Design and fabrication of a large vertical travel silicon inchworm microactuator for the advanced segmented silicon space telescope, *Proceedings of SPIE* **4981**, 107, 2003.

61. Yang, E.-H., Wiberg, D.V., and Dekany, R.G., Design and fabrication of electrostatic actuators with corrugated membranes for MEMS deformable mirror in space, *Proceedings of SPIE* **4091**, 83, 2000.

62. Daly, J.T. et al., Recent advances in miniaturization of infrared spectrometers, *Proceedings of SPIE* **3953**, 70, 2000.

63. Barry, R.K. et al., Near IR Fabry–Perot interferometer for wide field, low resolution hyperspectral imaging on the next generation space telescope, *Proceedings of SPIE* **4013**, 861, 2000.

64. Butler, M.A. et al., A MEMS-based programmable diffraction grating for optical holography in the spectral domain, Technical Digest — International Electron Devices Meeting IEDM 2001, 909, 2001.

65. Silicon Light Machines, http://www.siliconlight.com

66. Barker, N.S., Shen, H., and Gernandt, T., Development of an integrated millimeter-wave Fourier transform spectrometer, *Proceedings of SPIE* **5268**, 61, 2004.

67. Sarnoff Corporation, http://www.sarnoffimaging.com/technologies/uncooled_ir.asp

68. Jerominek, H. et al., Micromachined uncooled VO_2-based IR bolometer arrays, *Proceedings of SPIE* **2746**, 60, 1996.

69. Saint-Pe, O. et al., Study of an uncooled focal plane array for thermal observation of the Earth, *Proceedings of SPIE* **3436**, 593, 1998.

70. Holland, P.M. et al., Miniaturized GC/MS instrumentation for *in situ* measurements: micro gas chromatography coupled with miniature quadrupole array and Paul ion trap mass spectrometers, *Proceedings of the SPIE* **4878**, 1, 2003.

71. Peddanenikalva, H. et al., A microfabrication strategy for cylindrical ion trap mass spectrometer arrays, *Proceedings of IEEE Sensors* **1**, 651, 2002.

72. Siebert, P. et al., Surface microstructure/miniature mass spectrometer: processing and applications, *Applied Physics A: Materials Science and Processing* **67** (2), 155, 1998.

73. Siebert, P., Petzold, G., and Muller, J., Processing of complex microsystems: a micro mass spectrometer, *Proceedings of the SPIE* **3680**, 562, 1999.

74. Sillon, N. and Baptist, R., Micromachined mass spectrometer, *Proceedings of 11th International Conference on Solid State Sensors and Actuators—Transducers '01* **1**, 788, 2001.

75. Taylor, S., Gibson, J.R., and Srigengan, B., Miniature mass spectrometry: implications for monitoring of gas discharges, *Sensor Review* **23** (2), 150, 2003.

76. Taylor, S., Tindall, R.F., and Syms, R.R.A., Silicon based quadrupole mass spectrometry using microelectromechanical systems, *Journal of Vacuum Science & Technology B (Microelectronics and Nanometer Structures)* **19** (2), 557, 2001.

77. Tullstall, J.J. et al., Silicon micromachined mass filter for a low power, low cost quadrupole mass spectrometer, *Proceedings IEEE Eleventh Annual International Workshop on Micro Electro Mechanical Systems* 438, 1998.

78. Wiberg, D. et al., LIGA fabricated two-dimensional quadrupole array and scroll pump for miniature gas chromatograph/mass spectrometer, *Proceedings of SPIE* **4878**, 8, 2002.

79. Yoon, H.J. et al., The test of hot electron emission for the micro mass spectrometer, *Proceedings of the SPIE* **4408**, 360, 2001.

80. Chabot, M.D. et al., Single-crystal silicon triple-torsional micro-oscillators for use in magnetic resonance force microscopy, *Proceedings of SPIE* **4559**, 24, 2001.

81. Choi, J.-H. et al., Oscillator microfabrication, micromagnets, and magnetic resonance force microscopy, *Proceedings of SPIE* **5389**, 399, 2004.
82. Goan, H.-S. and Brun, T.A., Single-spin measurement by magnetic resonance force microscopy: effects of measurement device, thermal noise and spin relaxation, *Proceedings of SPIE* **5276**, 250, 2004.

8 Microelectromechanical Systems for Spacecraft Communications

Bradley Gilbert Boone and Samara Firebaugh

CONTENTS

8.1 INTRODUCTION

The communications subsystem is responsible for reception and demodulation of signals sent up from the ground station (uplink) as well as transmission of signals back to the ground station (downlink). The system is also responsible for any communication with other satellites. The uplink signal consists of commands and range tones, which are signals first transmitted by the ground station, and then received and retransmitted by the satellite. The delay is used to determine the satellite's distance from the station. In addition to range tones, the downlink signal includes telemetry for spacecraft status and any payload data. The downlink signal is usually coherent in phase with the uplink signal, which allows for Doppler shift detection of spacecraft velocity.

The signal frequency range for ground to satellite communications is from 0.2 to 50 GHz, depending on the application. Intersatellite links sometimes use 60 GHz signals. Uplink and telemetry downlink data rates are typically less than 1 kbit/sec, and are transmitted using low-bandwidth, widebeam antennas.[1] When payload data requires a higher transmission rate, high-gain, directional antennas are used. These antennas need to be steered either mechanically or electrically. Mechanical steering places additional demands on the attitude determination and control subsystem, which must balance the reaction forces caused by antenna movement. For more detailed information on the communications subsystem, interested readers should consult Morgan and Gordon.[2] Applications for MEMS in spacecraft communications systems include routing switches, phase shifters, electrically steerable antenna, higher performance filters for transmitter or receiver circuits, and scanning mirrors for intersatellite optical communications.

Optical communication links offer many advantages over microwave links. In particular, free space laser systems can provide narrow beam widths and high gains with much smaller hardware. High gains allow for much higher data rates, on the order of Gbps for sufficiently close link ranges, for example, near terrestrial space. Because of the significant attenuation of optical frequencies by the atmosphere, optical links are most easily employed for intersatellite communications, which is particularly attractive for crosslinks within satellite constellations.[1]

Optical communication hardware is well suited to small satellites. The flight mass of an optical communications subsystem is typically 55 to 65% of that of a conventional microwave subsystem.[3] This derives from the use of low-mass detectors and semiconductor laser diodes, and fiber amplifier or lasers, many of which were developed for the terrestrial fiber optics communications market.[4] However, macroscale electromechanical beam steering subsystems make up a significant fraction of the mass of these systems. This is where MEMS offer a solution in optical communications for many aerospace applications.[5]

8.2 MEMS RF SWITCHES FOR SPACECRAFT COMMUNICATION SYSTEMS

Microwave and RF MEMS are especially applicable to commercial communication satellites, where communication systems make up the payload as well as are part of the satellite bus.[6] These systems require many switches for signal routing and redundancy. In the past, they have been implemented by large electromechanical switches or by power-hungry solid-state switches. MEMS offer a lightweight, low-power alternative to such switches.

MEMS switches also enable "active aperture phase array antennas." These systems consist of groups of antennas phase-shifted from each other to take advantage of constructive and destructive interference in order to achieve high directionality. If the phase separations can be actively controlled, then such systems allow for electronically steered, radiated, and received beams, which have greater agility and will not interfere with the satellite's position. An adaptive phase array can also be used to combat a jamming signal by pointing a null toward the interfering signal source. A key component in a phase array is the phase shifting element that is associated with each individual antenna in the array. Such phase shifters have been implemented with solid-state components. However, they are power-hungry, and have large insertion losses and problems with linearity. In contrast, phase shifters implemented with microelectromechanical switches have lower insertion loss and require less power, especially in the range of 8 to 120 GHz.[7] This makes MEMS an enabling technology for lightweight, low-power, electronically steerable antennas for small satellites. Rebeiz has written a thorough review of RF MEMS, which is recommended to anyone who has interest in the field.[8]

The first microfabricated relay was designed by Kurt Petersen in the late 1970s.[9] He used bulk micromachining techniques to create a switch with an actuation voltage of 70 V, 5 Ω of DC resistance in the closed state and a 10-μs switching time. The most active groups currently in the field of microwave switches are the Rockwell Science Center (RSC), Raytheon (begun at Texas Instruments),[10–12] Hughes Research Laboratories (HRL),[13–16] the University of Michigan,[17–19] Cronos (which is also associated with the Raytheon effort),[12,20] OMRON corporation,[21] and UCLA.[22,23] RSC has flown its RF switches in space on a picosatellite.[24]

8.2.1 MEMS SWITCH DESIGN AND FABRICATION

The basic MEMS switch is a suspended mechanical structure that moves when actuated to vary the electrical impedance between two electrodes. To clarify the language we will refer to two conducting plates of the switch that receive the control voltage as "electrodes"; one is stationary and the other is the moving electrode. Then there is the "conducting bar" through which the signal will travel (either to complete the path or to ground, depending on switch configuration). The contacts are the points at which the conducting bar connects to the transmission line. MEM switches can be classified by configuration, contacting mode, actuation mechanism, and switch geometry.

8.2.1.1 Switch Configuration

As is illustrated in Figure 8.1, there are two general configurations for switches: series and shunt. In a series configuration, the conducting bar sits along the signal path. The on state is when the conducting bar is brought down, completing the path. In the shunt

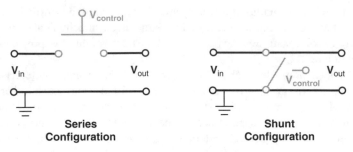

FIGURE 8.1 Different configurations for microwave switches.

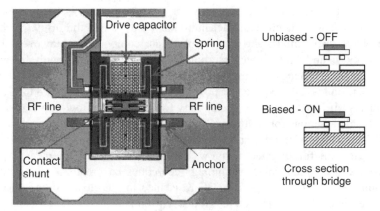

FIGURE 8.2 Structure and operation of a MEM series switch developed by the Rockwell Science Center. (Courtesy of the Rockwell Science Center and from Mihailovich, R. E., et al.)

configuration, the conducting bar sits between the signal line and ground. The on state is when the conducting bar is up, so that the signal can pass unimpeded.

Researchers have pursued switches in series configurations[15,16,26–30] and shunt configurations.[10,11,17,18] In series-configured switches, the insertion loss is determined by the impedance of the switch in its closed state, which in turn depends on the intimacy of the contact achieved by the switch. The isolation is set by the capacitance between the conducting bar and the signal line in the off state. Series switches can be implemented with both microstrip and coplanar waveguide transmission lines.[15,31–33] Figure 8.2 shows a series switch developed at RSC.

In a shunt switch, the insertion loss is the result of any impedance mismatch that occurs because of the unactuated mechanical structure (with careful calculations, the unactuated switch can be sized to match the characteristic impedance of the line), and the isolation depends on ratio between the capacitance in the "down" state and the capacitance in the "up" state. Shunt switches are only easily implemented with coplanar waveguide transmission lines.[10,17] Figure 8.3 shows a scanning electron micrograph of a shunt switch.

The impedance of a capacitor decreases with frequency. Therefore, the isolation of a series switch diminishes with frequency, while in a shunt switch that relies on a

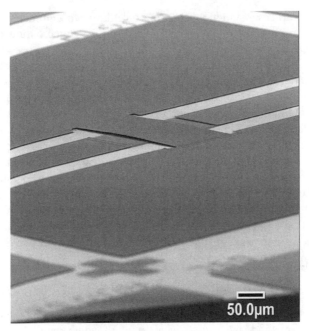

FIGURE 8.3 Scanning electron micrograph of a MEM shunt switch developed at the Johns Hopkins University Applied Physics Laboratory. (Courtesy of JHU/APL.)

capacitive contact, the isolation increases with frequency (until the capacitive reactance is comparable with the resistance of the shunt bar). Therefore, if one wishes to operate the switch at either extreme of the frequency range, the choice of switch configuration is clear. There is overlap in the frequency range of the two configurations; both switch configurations have been developed to cover the range between 10 and 40 GHz.

8.2.1.2   Contacting Modes

MEM switches are either metal contacting, in which closing the switch results in a direct electrical (preferably ohmic) contact between conductors, or capacitive coupling in which there is a thin dielectric film separating the conducting electrodes when the switch is closed. Metal-contacting switches are most often used for series switches,[25] while capacitive-coupling contacts are most often used for shunt switches.[10,11,28,29] However, there are reports of all switch and configuration combinations (although some care must be required for the control electronics in shunt, metal-contact switches).[11,28,29] Metal contacting is the natural choice for series switches because it allows for operation in lower frequencies where the series configuration is preferred, and capacitive-coupling switches are better suited to the higher frequency range of shunt switches.

In metal-contacting switches, the electrodes are typically made of gold, which has low resistivity and good chemical inertness. The advantage of the metal contact is its low resistance over a broad frequency range. Its disadvantage is that on the microscale, the forces of stiction and microwelding are commensurate with the mechanical restoring force of the switch,[11] resulting in device failure after repeated cycling (particularly "hot" cycling).

Capacitive-coupled contacts are less prone to contact failure but are not suitable to lower frequencies where the capacitive impedance even in the down state is too high to make good electrical contact.[26] In capacitively coupled contacts, some care is required to avoid dielectric charging effects.[11] The high electric field that exists in the dielectric layer when the switch is closed can cause charges to tunnel into the dielectric layer and become trapped. The charges then screen the applied electric field causing switches to require higher or lower actuation voltages and sometimes cause a stiction-like phenomena. Therefore, a simple unipolar DC control signal is often inadequate unless the charging effects can be better controlled.[11] Some groups have explored structures with both active pull-up and active pull-down in order to overcome stiction and charging forces.[33,34] In both types of contacts, the intimacy of the contact is important to the performance of the switch, requiring smooth surfaces and large contacting forces.

8.2.1.3   Actuation Mechanism

By far the most common actuation mechanism for microwave switches is electrostatic.[26] In this method, the switch is a set of movable parallel plates. When a voltage is applied between the plates it creates an electrostatic force that draws the plates together. Most of these switches are on–off devices that rely on a phenomenon colloquially known to the MEMS community as "pull-in." The balance between the force on the electrode produced by the electric field and the mechanical restoring force of the material determines the position of the movable electrode. The force of the electric field for a voltage-controlled capacitor, however, is inversely proportional to the square of the electrode separation. A force balance can only exist for small amounts of deflection. At greater levels of deflection, the electrostatic force exceeds the restoring force, resulting in a sharp instability that causes the structure to snap closed. In microwave switch design, the voltage at which this phenomenon occurs sets the actuation voltage for the switch.[17] Electrostatic actuation allows for low actuation power consumption (no steady state current required) and easy integration capability, which are two of the advantages that led researchers to investigate MEM switches as an alternative to solid-state devices.[26] In addition, electrostatically actuated switches have a relatively high speed when compared to mechanical switches that employ other actuation methods. One disadvantage of electrostatic actuation is the inherent trade-off between the gap height, which must be large for good isolation in the switch, and actuation voltage, which increases with gap height. As a result, electrostatically actuated switches generally require a large actuation voltage, which can complicate control electronics.

Thermal actuation has been explored as an alternative.[14,20,35,36] This technique takes advantage of thermal expansion. Local heating results in strain that can be used to close or open the switch. Some thermal actuators use a bimorph structure to further exaggerate the effect. The advantage of thermal actuation is that it requires a much lower actuation voltage than electrostatic actuation for the same gap height. Also, thermal actuation has a higher work force density than electrostatic, allowing for firmer contacts.[20] The disadvantage of thermal actuation is that it is generally slower and consumes considerably more power than electrostatic actuation. Since power concerns are part of the drive behind the investigation of MEM switches, this is a serious drawback of thermal actuation methods.

Some work has also been done with magnetostatic actuation, in which the moving plate of the switch is fabricated from a magnetic material, and then a miniature (but not microfabricated) electromagnet is packaged with the device.[37] The advantage of this actuation method is that like the thermal actuation method it does not require high voltages. However, the total switch is quite large, due to the external electromagnet, and the fabrication requires the processing of a magnetic material such as permalloy, which makes the process more difficult to integrate with microelectronics or other microfabricated devices.

8.2.1.4   Geometric Design

There are two issues with switch geometry, the first is the choice of lateral or vertical motion, and the second is the choice of shape for the moving electrode. Most MEM switches are "vertically contacting," with motion perpendicular to the surface; however, a few groups have explored "laterally contacting devices."[12,20,36,38] An advantage of such systems is that the actuator, contacts, conductor path, and support structure can all be defined simultaneously. Also, larger separations can generally be achieved in these structures. However, the contacts of lateral motion devices are generally worse, because the contact surfaces are determined by etching and are rough.[26] Also, it is difficult to get a large contact area with surface micromachining techniques. Vertical motion switches are more easily integrated with monolithic microwave integrated circuits (MMICs) and provide better contacts.

The moving electrode shape can be characterized as cantilever, bridge, or membrane. They all have similar mechanical behavior, with the actuation voltage for a given electrode length being lowest for a cantilever and highest for a membrane. Often bridge structures are used for shunt switches because if both anchors are connected to the ground line then the bridge structure provides a double path to ground, which increases the isolation (for shunt configured switches). A number of different anchor designs and bridge and cantilever variations have been attempted in order to minimize the actuation voltage required for a given separation. For example, some groups use serpentine springs for action of long bridge with relatively little area.[34] Also, curling or "zipping" structures have been developed.[39,40]

A few novel switches have also been developed, including a rotational switch,[14] and a "mercury microdrop" switch that employed bubble actuation to move a drop of mercury in and out of the signal path.[23]

8.2.1.5   Fabrication Methods and Materials

Most microwave MEM switches are constructed using surface micromachining techniques. The advantage of these methods is that they can be integrated relatively easily with conventional MMICs by adding the MEM devices in postprocessing steps. Microwave transmission lines are also lossy on standard undoped silicon wafers, so high-resistivity silicon, silicon-on-sapphire, or GaAs substrates are preferred.[29]

Gold is usually preferred for contact metallizations because of its noble nature, superior conductivity, and compatibility with MMICs,[26,41] although some work indicates that rhodium may be preferable to gold, because gold has a high adherence.[42,43] Hyman and Mehregany have studied gold contacts extensively and have made several observations.[16] For example, thin gold films are in general harder than bulk gold, with higher hardnesses resulting from aggressive deposition and patterning methods such as sputtering or physical deformation. Electroplated gold is three times softer than sputtered gold, and gold films that are subjected to temperatures greater than their deposition temperatures will change dramatically in cooling due to the closure of grain voids.

For capacitive switches, the closer the contact to the dielectric, the higher the capacitance and therefore the isolation. In general, the surfaces of the contacts should be as smooth as possible.[11] The developers of the Raytheon shunt switch found that hillocking of the bottom aluminum electrode greatly inhibited contact in their switches, causing them to change to tungsten electrodes. They also found that they had to be careful to avoid "wings" on metal pattern edges, which can be a problem with lift-off deposition techniques.[30] Also, some groups encountered problems with tenacious polymer residues, which caused stiction failures.[11]

The mechanical properties of the switch structural material are critical to the operation of the device. This requires strict process control of the deposited thin films. In general, low tensile stress materials are most desirable. Compressive membranes could exhibit buckling (although some groups take advantage of buckling phenomena to enhance the contact force),[21] and highly tensile membranes require too high an actuation voltage.

8.2.2 RF MEMS Switch Performance and Reliability

8.2.2.1 Figures of Merit

The figures of merit for switches are isolation, insertion loss, return loss, transition time, switching speed, control voltage, control power, maximum power capability, the IP3 point or intermodulation product (characterizing linearity), cut-off frequency, and lifetime. Isolation, insertion loss, and return loss are all quoted in decibels (dB). Isolation characterizes the difference between the input and output signal when the switch is in its blocking state. Its value is the scattering matrix coefficient S_{21} measured when the switch is open. This coefficient characterizes the amplitude of the transmitted wave at the output over that of the incident wave at the input, so when transmission is blocked this a very small quantity, or a large negative number in terms of decibels.[22] Therefore, it is desirable for the magnitude of the isolation to be large.

Insertion loss characterizes the attenuation of the signal when the switch is in its passing state, given by S_{21} when the switch is closed. The magnitude (in dB) of the insertion loss of a switch should be small. Insertion loss can be due both to impedance mismatch, which results in reflection, and resistive losses. The return loss, which is not always quoted, is the attenuation of the signal reflected back to the input.

Transition time is the time required for the signal voltage to go from 10 to 90% (for on-time) or from 90 to 10% (for off-time) of its full value. The switching time includes the transition time as well as delays in the control system.[13] The control (or actuation) voltage is the voltage required to open or close the switch. In some switches, a control current might also be specified. The control power characterizes the power required to operate the switch. This should not be confused with the power handling capability of the switch, which is how much signal power the switch is capable of routing.

Linearity with respect to power is of great importance in microwave switches, particularly for solid-state switches, which can be highly nonlinear. As power levels increase, energy will generally shift from the first-order harmonic of the signal to higher order harmonics. The IP3 point is the power at which the third-order harmonic intersects with the first-order signal. Ideally the transmission should be independent of signal level resulting in a very large IP3 point magnitude.

The cut-off frequency is often specified at a figure of merit. The cut-off frequency is the frequency at which the ratio of the off-impedance to on-impedance degrades to unity. The cut-off frequency theoretically sets the upper limit for switch function, although it neglects the effects of inductance that can become significant at high frequencies.[17]

Lifetime is usually measured in switching cycles. The switch lifetime depends on the microwave signal, and so measurement conditions must be specified if one wishes to compare different devices. "Cold-switching" refers to a measurement without any microwave signal present, and measures just the mechanical lifetime of the device. Since the predominant failure mode is degradation of the electrical contacts or dielectric charging, the cold-switching lifetime will usually be much greater than the operation lifetime. The lifetime for a signal-carrying switch is referred to as the "hot" lifetime. A long lifetime is desirable.

8.2.2.2 Example Performance

As an example, the RF switch performance goals given by the Air Force Research Laboratory (AFRL) in Rome are as follows: insertion loss < 0.1 dB from 0 to 4 GHz, isolation > 50 dB at 2 GHz, switching time < 10 µs, CMOS-compatible control voltage levels (generally 0 to 5 V), power handling capability > 30 dBm, IP3 > 55 dBm, and hot lifetime $> 10^9$ cycles.[30] It would be useful to extend these performance levels up to 40 GHz.

8.2.2.3 Failure Modes

For capacitive switches, the two dominant failure mechanisms that limit power handling are RF latching and RF self-actuation. RF self-actuation occurs when the root-mean-square (rms) signal voltage becomes large enough to close the switch with no assistance from the DC bias.[22,44] Typical MEMS switches can handle 2 to 4 W before self-actuation becomes a problem.[34,44] RF latching occurs when the switch has been actuated, and the rms signal voltage is greater than the hold voltage for the switch (this is typically much less than the actuation voltage). Latching occurs at powers as low as 0.5 W. These power levels are significantly less than what is desired, and power handling continues to be an area of MEMS switches that requires improvement. However, it should be noted that RF latching and RF self-actuation are not destructive; once the RF power is reduced the switches return to normal function.

Contact failure is the predominant failure mechanism for series switches. Lifetime depends on the signal levels and on the thermal behavior of the device.[16] Dielectric charging can limit lifetime in capacitive switches. When large voltages are applied across a dielectric, imperfections in the dielectric can lead to charge storage. This stored surface charge can have very low mobility, resulting in charge build up over several actuation cycles.[45] Over time, this will cause drift in the actuation voltage and can result in device failure. Dielectric charging can be mitigated by using alternating polarity pulses for actuation and by using a shaped signal, with a high-voltage pulse for actuation followed by a lower voltage for holding the switch in the down position.

8.3 MEMS RF PHASE SHIFTERS

Phased array antennas consist of multiple antennas where the transmission from each antenna is phase-shifted from the others to take advantage of constructive and destructive interference in order to achieve high directionality. A key component in a phase array is the phase shifting element that is associated with each individual antenna in the array.

In a comparison of MEMS phase shifters against ferrite, PIN diode, and GaAs phase shifters, it has been determined that MEMS phase shifters are particularly applicable to space-based radar because they are relatively small, lightweight, and inexpensive.[32] There are three common approaches to active phase shifters: switched line, loaded line, and reflection. MEMS phase shifters have been developed for a number of frequency ranges and applications. They have been shown to have a much lower insertion loss than current phase shifters, but they also tend to have a higher actuation voltage. They are also broadband, and are usually targeted toward military communications systems. However, examples of phase shifters specifically targeted toward satellite applications, including stub-loaded line phase shifters exist.[46]

8.3.1 Switched-Line Phase Shifters

Figure 8.4 shows a schematic of a switched line (or time-delay) phase shifter. In these systems transmission lines of different lengths are switched into the signal path to change the signal path length. These types of phase shifters are particularly good for broadband, because if the transmission lines are TEM, the phase shift is a linear function of frequency, which minimizes distortion.

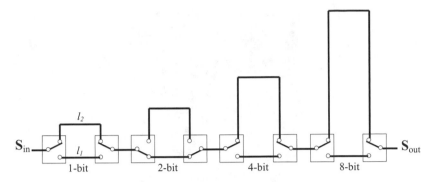

FIGURE 8.4 Schematic of a 4-bit switched-line phase shifter.

Switched-line phase shifters are usually designed for a large range of phase shifts, and by using a binary sequence of $\Delta\phi = 180°, 90°, 45°$, etc., they lend themselves to digitization. A 4-bit time-delay shifter like the circuit shown in Figure 8.4 is capable of producing 16 shift levels. Conventional switched line phase shifters are generally implemented with PIN diodes. However, the insertion loss for multiple solid-state switches is quite high, and the PIN diode switches consume significant power in operation. This has made MMIC switched-line phase shifters impractical for small and low-power applications.

In a MEMS implementation of a switched-line phase shifter, the switch is replaced by a MEMS switch such as was described in the previous section. Such devices use a microfabricated transmission line such as microstrip. Such phase shifters have been developed for a number of frequency ranges including DC-40 GHz,[47] X-band,[48] Ka-band,[49,50] Figure 8.5 shows a 2-bit phase shifter developed by the University of Michigan and Rockwell Scientific.

8.3.2   LOADED-LINE PHASE SHIFTERS

In loaded line systems, the capacitance of the line is varied to produce the desired phase shift. These systems are usually designed for 45° or less of phase shift.[51] An illustration of this type of phase shifter is shown in Figure 8.6.

One possible MEMS implementation of a loaded-line phase shifter is to use a suspended MEMS shunt switch over a coplanar waveguide to create the variable capacitive load.[19] Such shifters have been constructed for X-band and Ka-band, and have demonstrated phase shifts up to 270° with an insertion loss of less than 1.5 dB.[52] Several other groups have also demonstrated loaded-line shifters at a number of frequency ranges including U- and W-Band.[53–55] Another possible implementation is to use switches to switch in and out stubs that vary the line capacitance.[46]

8.3.3   REFLECTION PHASE SHIFTERS

A reflection phase shifter is illustrated in Figure 8.7. It makes use of a quadrature hybrid combined with a matched pair of switches. The quadrature hybrid is an

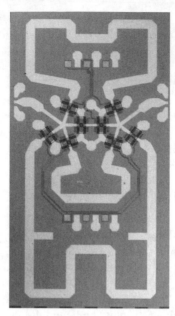

FIGURE 8.5 Photograph of a 2-bit switched-line phase shifter developed by the University of Michigan and Rockwell Scientific. (Courtesy of Rockwell Scientific Company.)

element, easily implemented in microstrip, which separates the input signal into two signals that are 90° out of phase. The two switches are tied together. If the switches are closed, the signal is reflected back into the quadrature hybrid, where the two reflected waves will add constructively at one port and destructively at another port. If the switches are open, a total phase shift of $\Delta\phi$ will be added to the signal. If the switches are perfectly matched and lossless, and the quadrature hybrid is lossless, these phase shifters should have little insertion loss. Like the switched-line phase shifter, several bits with a binary sequence of phase delays can be combined for digital phase control.

In a MEMS implementation of a reflection phase shifter, MEMS switches control the reflection stub length. There are fewer MEM reflection phase shifters

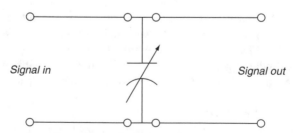

FIGURE 8.6 Schematic of a loaded-line phase shifter. Varying the capacitance alters the phase shift between the input and output.

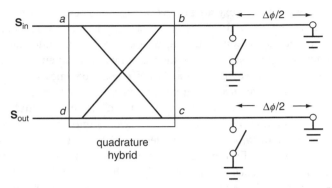

FIGURE 8.7 A schematic illustration of a reflection phase shifter.

in development than the switched-line and loaded-line types, but such phase shifters have been demonstrated for 50 to 70 GHz.[56,57] Malczewski et al.[58] have also demonstrated reflection-based X-band phase shifters based on Lange couplers.

8.4 OTHER RF MEMS DEVICES

MEMS technology is also used to create mechanical filters, variable capacitors and inductors, all of which can be used in microwave and RF filter circuits. Micromachining provides distinct advantages for all three types of components. Microelectromechanical filters based on coupled microresonators have been demonstrated for frequencies in the range of tens of kHz to tens of MHz.[59,60] MEMS filters are much smaller than SAW-based and crystal resonators, and are also more easily integrated with other microwave systems.

In monolithic microwave integrated circuits (MMICs), inductors are implemented as planar spirals. Such devices require relatively large area and also suffer from parasitic capacitances. Micromachining can be used to lift the structure off the substrate in order to reduce parasitic capacitances, as well as to increase the coupling surfaces in order to reduce overall dimensions.[61,62] Micromachining has also been used to create tunable capacitors.[63,64] Such capacitors usually involve a capacitor with a movable electrode that is positioned by electrostatic or electrothermal forces to achieve the desired capacitance level.

8.5 RF MEMS IN ANTENNA DESIGNS

8.5.1 Electrically Steered Antennas

Being able to switch different antenna sections with a given phase shift has two major applications in antenna design. In phased array antennas, multiple smaller antennas are connected in a way such that the transmission from each smaller antenna is phase-shifted from the others to take advantage of constructive and destructive interference, thus controlling the radiation pattern of the antenna.

These systems can achieve very high directionality for sending as well as receiving. A major advantage in spacecraft of this approach is that these systems do not require any attitude adjustment, either to compensate for the antenna motion or to direct the antenna. The second application is a switched antenna, where antenna arrays are connected in a way that it matches different frequency bands. This allows for rapid alteration between a low transmission rate to a higher frequency with higher transmission rate. Transmission to Earth and between satellites, commonly in different bands, could therefore be done using only one antenna array.

Phased-array antennas have been implemented for large systems, such as the AN/SPY-1 radar system (Figure 8.8), which is part of the Aegis Combat System that has been used on the U.S. Navy warships. This system does not use MEMS-based phased shifting elements and consumes large amounts of power, which is readily available on their particular deployment platform. The use of MEMS-based phase shifters could significantly reduce the power demands of such systems, making them suitable for space applications.

8.5.2 FRACTAL ANTENNAS

Switches are also the key element in reconfigurable fractal antennas.[65,66] Fractal antennas combine electromagnetic theory with fractal geometry — which describes a family of complex patterns that are self-similar or repetitive over many size

FIGURE 8.8 Photograph of a SPY-1 radar array, which is an example of a electrically steered antenna that relies on an array of smaller antennas combined with phase shifters. (Photo by F.H. Sanders. Courtesy of the Institute for Telecommunication Sciences.)

FIGURE 8.9 Illustration of Sierpinski gasket fractal geometry.

scales.[67] An example of a particular fractal geometry, called the Sierpinski gasket, is illustrated in Figure 8.9.

In a fractal antenna, the antenna elements are shaped into a fractal geometry. This creates antennas that are multiband and compact in size.[67] RF MEMS switches have been used to interconnect portions of the fractal geometry to create reconfigurable antennas, which allow for electronic steering of the radiation pattern.[65,66]

8.6 MEMS MIRRORS FOR FREE-SPACE OPTICAL COMMUNICATION

Optical communication hardware, developed in the telecom boom in recent years, is well suited to small satellites. The flight mass of an optical communications subsystem is typically 55 to 65% of that of a conventional microwave subsystem.[3] This derives from the use of low-mass detectors and semiconductor laser diodes, and fiber amplifier or fiber lasers, many of which were developed for the terrestrial fiber optics communications market.[4] In recent years, there has been a boom in MEMS applications in fiber optic communications, particularly in the area of optical interconnects formed by arrays of micromechanical mirrors.[68–72] The inter-satellite laser link application has more stringent pointing accuracy, stabilization, and vibration isolation requirements than fiber optic switching arrays; however, scanning MEMS mirrors have been demonstrated for fine tracking control with microradian resolution over a range of ± 3 mrad.[73] An example of a commercially available micromirror and a 4×1 array of such mirrors is shown in Figure 8.10.[74]

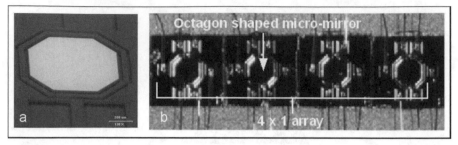

FIGURE 8.10 (a) Commercially available single micromirror and (b) a 4 × 1 MEMS micromirror array from available from MEMS Optical, Inc.

Arrays of controllable micromirrors can be used to create electrically steerable beams for optical signals,[75,76] like the phased array antennas for RF and microwave signals described in previous sections. Such systems have significant speed advantages over macro-scale, mechanically steered beams. In August 2000, DARPA initiated the STeered Agile Beams (STAB) program, which is "developing small, lightweight laser beam scanning technologies for the replacement of large, heavy gimbaled mirror systems."[77] A number of MEMS-based approaches are being developed as part of this project, including work at University of California (UC), Berkeley on "Smart Dust."[78] Other major centers of activity are at the University of California at Los Angeles and the University of Colorado, Boulder.[79]

8.6.1 FABRICATION ISSUES

An enormous amount of research and development has been conducted over the last 15 to 20 years addressing optical MEMS device fabrication[80–93] and switching applications,[94–101] leading up to the present state of knowledge. More recently there has been a surge of interest in applications of MEMS to truly free-space communications between mobile platforms.[102–106] Below, we briefly summarize the key fabrication issues.

If a silicon surface is treated properly it can provide an optical surface of extremely high quality (i.e., flat and scatter-free). Along with excellent optical surface qualities, MEMS fabrication techniques enable the construction of devices with very small high-precision displacements (on the order of a wavelength or less) required in many micro-optical applications. Additional optical components such as gratings, lenses, fibers, detectors, and laser diodes may be integrated with the MEMS devices in small-scale packages. Silicon is also totally transparent at optical communication wavelengths, another useful property for some applications.

In fact, the earliest applications of micromachined silicon enabled the fabrication of V-grooves for multiple fiber alignment and fiber switching mirrors.[94] Both bulk and surface micromachining techniques (the latter of which adds additional layers to the surface of the silicon) are used for fabrication. Small optical switches are fabricated using surface micromachining, whereas large-scale switches are made by bulk micromachining. Surface micromachining often involves selective deposition by low-pressure chemical vapor deposition (LPCVD), followed by patterning and etching, to create the desired structures on the silicon substrate. Significant progress has been made in manufacturing commercial-quality mirrors using these methods.

Stress-free optical thin film surfaces are critical for optical networking as well as free-space beamsteering applications, but film stress is difficult to control in the fabrication process. It can vary dramatically with a relatively small change in the number of atoms, and hence, the film's chemical composition. As a consequence, it is difficult to make polysilicon mirrors very flat, particularly if they need to be relatively large (~few millimeters). After a surface is initially deposited and all the supporting layers are removed, it may not remain flat. Even thin gold over-coatings can cause substantial deformation of an uncoated plate.

Bulk micromachining is used to form MEMS microstructures by either wet or dry anisotropic etching. In this case silicon on insulator (SOI) wafers are useful, especially in separating moving parts from the bulk silicon structure, and this was determined early. When a plate-type structure is freed in the fabrication process, a mirror can be produced on either side, with that surface in contact with the oxide often being superior in terms of scattering properties. The availability of both sides allows the deposition of perfectly stress-balanced gold reflection layers for enhanced reflectivity, which makes manufacturing easier and more predictable.

Leading candidates for optical switches and cross-connects are free-space micromirror switch arrays, and a scheme to do this using conventional scanning mirrors was first proposed as early as 1982.[80] Arrays of collimators are positioned such that light from each collimator is directed toward a dual-gimbaled mirror. The first mirror reflects the beam toward a corresponding mirror in the opposing array. The latter mirrors adjust their angles to send their respective beams to each receiving fiber. Light from each fiber can only be directed toward its corresponding mirror at a given instant. Likewise, the receiving mirror can only send light to its associated fiber, but both mirror arrays can be virtually infinitesimally adjusted, so that any mirror that receives a beam can send it to any of the opposing mirrors, thereby making fully free connections. The supporting parts of each mirror, such as the hinges and drive structures, are kept small to maximize mirror area fill factors.

For low-loss transmission the mirrors must be very flat, with flatness better than one fifth the operating wavelength. Mirrors with gold coatings can have reflectivities over 98%, and mirror arrays can be several square millimeters in size, with square or rectangular aspect ratios. Fiber-to-fiber losses through the cross-connect can be as low as 0.7 dB, and mirrors have been exercised over 60 billion cycles without any failures. Cross-coupling between the various channels also turns out to be negligible because even a small amount of angular offset between the input and output mirrors will cause a significant displacement of the inappropriate beam at a given output fiber entrance. Small-scale cross-connects with fewer optical switches have switching times as low as 50 μs or less, although larger $N \times N$ switches, configured into 2-D crossbar arrays, have switching times on the order of 500 μs.

Either electrostatic or electromagnetic drive mechanisms can be used to move the mirrors, but electrostatic is preferred, since it takes up little room and needs relatively low power. Large switch sizes, using relatively large mirrors (approximately few millimeters) with long focal lengths (tens of centimeters) are desired to allow the use of larger light beams, which have less beam divergence and greater useful relay distances. Larger angular deflections are also desirable.

8.6.2 PERFORMANCE REQUIREMENTS

Recent collaborative work between MEMX Corporation and the Johns Hopkins University Applied Physics Laboratory (JHU/APL)[107] has focused on developing MEMS micromirror technology for free-space multiaccess optical communications between spacecraft. Key performance issues addressed in this effort for space-based optical communications include micromirror heating due to input laser power, achievable degree of mechanical damping at ambient and partial air pressures, micromirror flatness, element size, angular field-of-regard (FOR), control-loop bandwidth, and open-loop transfer function shape. For some parameters, these devices already meet the "desired" application requirements, and for all cases, experimental tests indicate that the application requirements can be met with some redesign of existing devices. For example, the MEMX devices measured angular field-of-regard (FOR) was approximately $\pm7.9°$ optical, but would need to be $\pm12°$ for a projected redesign for GEO-to-ground links, which is quite feasible with slight micromirror redesign. Measured angular resolution was less than 360 μrad (desired greater than 1000 μrad); bandwidth was approximately 1 kHz (desired 100 to 1000 Hz); and mirror radius of curvature was approximately 0.4 m (nominally 0.5 m approximately). These and other key device parameters (and their desired range of values) are given in Table 8.1.

TABLE 8.1
Device Parameters of MEMX Micromirrors

Parameter	Nominal Value
Angular field-of-regard	$\pm12°$ (±210 Vmrad)[a]
Angular resolution	1 mrad[a]
Closed-loop bandwidth	100 to 1000 Hz
Number of elements	4×4 (minimum)
Element size	0.5 mm
Element pitch	~2 mm
Element radius of curvature	~0.5 m
Angle or voltage scale factor	10 μrad/mV

[a]Before beam expansion.

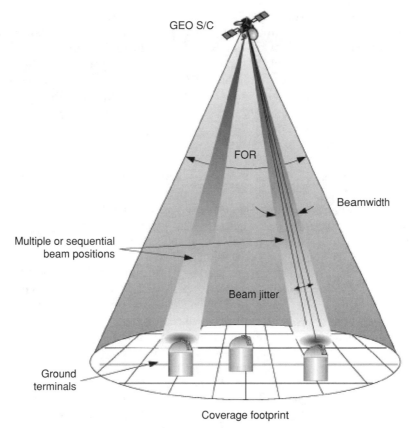

GEO S/C

FOR

Beamwidth

Multiple or sequential
beam positions

Beam jitter

Ground
terminals

Coverage footprint

FIGURE 8.11 GEO-to-ground scenario for applicability of MEMS micromirrors to multi-channel optical communications. The same terminal could support intersatellite links.

The two most basic requirements, FOR and angular accuracy, depend upon the required link range and terminal separation on the ground, as illustrated in Figure 8.11. For instance, for optical communication terminals down-linking to earth from GEO, beam widths on the order of 5 to 10 μrad are desired to support the link with reasonable laser transmitter powers (at hundreds of milliwatts), but their steered angular coverage will be limited to angles set by the dynamic limits of the MEMS mirrors and the optical transmitter beam expander design (assuming coarse steering via spacecraft attitude control). The laser beam reflecting from a given micromirror, however, must be significantly expanded to set the desired output (diffraction-limited) beamwidth to meet link margin requirements through the optical "antenna gain." The mirrors need to be physically steered to a greater angle than the output optical beam, given by the beam expansion ratio. For example, a beam expansion ratio of 250 increases the transmitter beam waist (which is assumed to be 0.5 mm at the micromirror) up to 12.5 cm, which yields a diffraction-limited beamwidth of approximately 8 μrad. Assuming that the micromirrors peak steering range is 420 mrad ($\pm$ 12°) before beam expansion, then the peak-to-peak output optical beam steering range would be approximately

1.7 mrad after beam expansion. An 8-μrad beamwidth produces a patch on the ground approximately 300 m across from GEO, and the maximum steering angle will cover a distance of approximately 60 km, corresponding to 200 beam widths. The MEMS mirror angular accuracy should be approximately 2.7 μrad (approximately 1/3 of the beamwidth) after beam expansion and 0.675 mrad before (corresponding to an angular dynamic range of 28 dB). The element pitch of such a MEMS mirror array should be adjusted in the plane of the array to enable adjacent mirrors to address adjacent areas on the earth separated by approximately 1.7 mrad. A 4 × 4 array would thus cover a square area of 240 km on a side, which is sufficient to reach terminal locations on the ground that would likely have decorrelated weather conditions, because weather cells are nominally approximately 250 km across. This is important for achieving site diversity to mitigate cloud cover.[108]

The closed-loop bandwidth requirement indicated in Table 8.1 is primarily set by the expected platform vibration environment, which can be present up to 1 kHz but is usually significant only up to approximately 100 Hz for most spacecraft. This parameter must be considered in establishing closed-loop control.[109]

A further trade-off between the transmitter power required to support the link margin and the degree of laser heat load experienced by the array elements must also be determined. The transmitter modulation waveform, such as pulse position modulation (PPM) with a variable M-ary value, is an additional degree of freedom in this trade. Under these circumstances preliminary link analyses indicate that the required average laser transmitter power should not exceed a few hundred milliwatts. Prior tests have suggested that the MEMX micromirrors can tolerate up to approximately 300 mW incident laser power. However, in the MEMS design the most efficient heat conduction path should be used, which is conduction through air or a similar gas. Additionally, the degree of micromirror curvature under steady-state conditions must be defined and maintained, and this is made easier at high partial pressures. This is the principal concern for beamwidth control.

8.6.3 PERFORMANCE TESTING FOR OPTICAL BEAMSTEERING

The particular MEMS micromirror used for recent tests at JHU/APL is shown in Figure 8.12. The diameter of this element is 1 mm, and it is supported by three legs

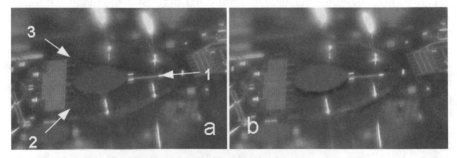

FIGURE 8.12 Close-up photographs of a specific test mirror, showing it in the quiescent state in (a) and in a nominal common-mode actuated state in (b). Note the shadow beneath the lifted mirror in (b).

disposed 120° apart around the periphery. Each leg is 0.9 mm long and is actuated by a linear electrostatic comb drive. To elevate the mirror in a piston motion to a nominal elevation of 50 μm for subsequent tip or tilt actuation, a voltage of 30 V must be applied to each leg. To cause a tip deflection, legs 2 and 3 can be held stationary while leg 1 is actuated with an appropriate (e.g., sine wave) drive signal with nominal peak amplitude of 50 V. To cause a tilt deflection, legs 2 and 3 can be driven 180° out-of-phase with each other while leg 1 is held stationary. Greater dynamic range can be achieved for the tip-case by actuating opposing legs as in the tilt case. Several test articles of this type were evaluated.

The MEMX micromirrors assessed were manufactured using polysilicon surface machining technology developed at Sandia National Laboratory; however, this technology has since been transferred to Fairchild Semiconductor in Portland, Maine.

8.7   APPLICATIONS OF MEMS TO SPACECRAFT OPTICAL COMMUNICATIONS

Optical communication links offer many advantages over microwave links. In particular, free-space laser systems can provide narrow beam widths and high gains with much smaller hardware. High gains allow for much higher data rates, on the order of Gbps for sufficiently close link ranges, for example, near terrestrial space.[4]

The Jet Propulsion Laboratory (JPL) in association with NASA is building an Optical Communications Telescope Laboratory (OCTL) transceiver station at its Table Mountain Facility, and they have explored laser communications links for deep space communications (the Galileo Optical Experiment) and near-terrestrial communications (Ground Orbit Lasercom Demonstration).[3] More recently the Mars Laser Communications Demonstrator (MLCD) program has begun to develop an optical telecomm terminal for the Mars Telecommunications Orbiter (MTO), scheduled for launch in 2009.[110] Data rates ranging from 1 to 2.5 Gbps are planned for future near-terrestrial space demonstrations and up to 30 Mbps for deep space links such as MLCD.[3] Laser downlinks have also been explored for communication with submarines via satellite.[111]

8.7.1   OPTICAL BEAM STEERING

Recent collaborative work between MEMX Corporation and JHU/APL[107] was based on previously developed MEMX optical switches. These special test units were evaluated for applications in laboratory tests as beamsteerers using a digital pointing and tracking system. Highly accurate and stabilized body-mounted tracking systems are essential to the implementation of long-haul optical communication channels and could be operated potentially from geosynchronous earth orbit (GEO) to ground-based or air-platform optical receiver terminals. For such spacecraft applications, moderate to high-powered laser diodes are likely to be required. Coupled with their potential operation at partial atmospheric pressures, MEMS mirror-shape stability and fabrication tolerances are of key concern to a system designer. To this end preliminary MEMX devices were evaluated in terms of angular jitter, focal spot stability, and open and closed-loop response versus laser transmitter power at both ambient air and lower partial pressures. The applicability and scalability of this technology to multiaccess terminals was also considered and appears to be readily transferable to a space-qualified design. For most spacecraft platforms micromirrors should be compatible with direct body-mounting because of their high intrinsic bandwidth and controllable damping. (Being able to body-mount these devices is highly desirable to take advantage of their low mass, which implies spacecraft attitude control would be used for overall coarse pointing.) Importantly, these optical beamsteerers are highly miniaturized, very lightweight, require very little prime electrical low power, and are scalable to 2-D multichannel (point-to-multi-point) links.

Initially a key concern about the MEMS micromirror performance in a space environment was the effect of partial vacuum on heat dissipation from the transmitting laser beam and on the degree of mechanical damping of the mirror. It is important that the beamsteering controller be critically damped under suitable partial or full atmospheric vapor pressure. In addition, a trade-off between the optical power required to support the link and the degree of thermal heat loading experienced by the mirror elements under pulsed laser light must also be determined. Furthermore, any micromirror curvature change induced by laser heating must be avoided. To this end preliminary optical, dynamic, and thermal measurements of the MEMX micromirrors were made using the optical test bed shown in Figure 8.13.

Using experimental measurements, physical optics modeling, and computer-based ray tracing, the laser beam quality reflected off a micromirror was evaluated. This included observing the beam waist, beam shape, and beam jitter. A quad cell detector and CCD focal plane array were used as diagnostic sensors in conjunction with the setup described in Figure 8.13, which included a vacuum chamber. The laser spot (with a minor axis of approximately 300 μm) is shown on the micromirror as well as at the CCD output focal plane in their respective insets. One concern was how much would the radius of curvature of the micromirror vary under light flux, but this was not initially evaluated because previous work had shown that a limit of about 300 mW would be sufficient to support projected link margins (even from GEO). The other concern, apart from beam jitter, is beam quality, which turned out to be poor because of an artifact of mirror fabrication, that resulted in etch pits in the mirror surface causing a diffraction pattern in the focal plane, rather than a nominal Gaussian spot, as shown in Figure 8.13 inset. This can be readily corrected in flat, smooth mirror designs specific to the application and through spatial filtering. Significant degradation, however, of the far-field beam should not be a real concern if the mirror is redesigned.

Micromirror frequency response measurements were made to establish basic dynamic performance in ambient air, angle sensitivity to deflection voltage, and dynamic response at lower pressures. The MEMX mirrors had very good frequency response, out to almost 1 kHz (or more), as indicated in Figure 8.14(a), which is

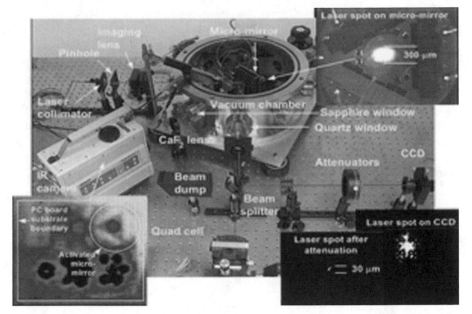

FIGURE 8.13 Optical test-bed layout to evaluate MEMX micromirror performance under partial pressure. This overall view includes sample test results, including the beam spot on the micromirror and at the CCD output focal plane, as well as the thermal camera image of the micromirror.

more than adequate to support application requirements for multiaccess free-space optical terminals on spacecraft or other moving platforms. The presence of air around the device provides viscous damping, allowing for achieving critical damping, which is best for pointing and tracking control as well as stabilizing against platform vibration. Investigation of the amplitude response versus pressure was limited to pressures well above the molecular regime, since we expect the Q would be undesirably high at lower pressures. Furthermore, at very low gas pressures, heat dissipation would be less without the conductive heat transfer effect of the air, and thus should be avoided to prevent damage and possible modification of mirror curvature. Having a controlled pressure envelope around the device also mitigates against humidity and other contamination.

Angle sensitivity was initially measured using a quad cell sensor, which for null tracking is satisfactory, as shown in Figure 8.14(b), where the quad output signal was heavily filtered to eliminate read-out noise. Without filtering the noise floor was 20 mV at the quad output, which translates into an equivalent angle noise at the mirror of 1.2 mrad. With filtering we saw much less inherent electrical noise and were unable to measure it with a digital oscilloscope, although ambient air fluctuations perturbing the micro-mirror were visibly discernable. Using a CCD array, we were able to measure low frequency (approximately 10 Hz) sine wave inputs down to 360 μrad, but this is not likely to be the actual intrinsic noise floor of the mirror.

Even at this level, however, an appropriate beam expansion factor (M) will translate this into a smaller angle (by $1/M$), which is consistent with the requirement of 1000 μrad. From the slope of the transfer characteristic in Figure 8.14(b) and assuming a perfectly linear transfer function, the maximum projected angle would be 57 mrad or approximately 3.2° (optically). Independent tests by MEMX corroborated these measurements and found a maximum envelope of ±7.9°. Future designs incorporating mirrors half of the current size should be able to achieve angular ranges on the order of ±12°. Thus, using the best measured sensitivity (360 μrad) and this projected angular range, the estimated dynamic range would be approximately 31 dB, which is very encouraging for modest field-of-regard free-space applications.

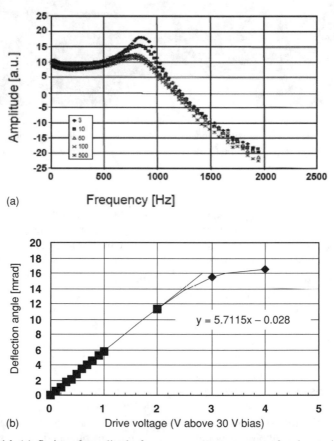

FIGURE 8.14 (a) Series of amplitude frequency response curves for decreasing ambient pressure, showing increasing Q with relatively modest decreasing ambient air pressure. The sharpest curve is at 3 Torr, followed by the 10, 50, 100, and 500 Torr curves. (b) Open-loop transfer characteristic, that is, mirror optical angular displacement versus drive voltage. The saturation effect at higher drive voltage is the result of beam vignetting on the quad photodiode detector used to make the measurement.

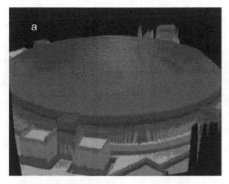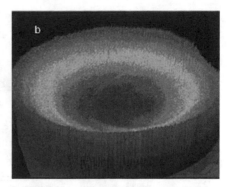

FIGURE 8.15 (a) Overall MEMX micromirror structure as viewed by an optical interferometer before curvature measurement. The textured surface appearance is due to a release-hole etch pattern; these will not be present on new mirror designs. (b) High-resolution scan by the interferometer, showing curvature of another MEMX micromirror.

Mirror curvature variation from unit-to-unit was also assessed using a commercial (Veeco) interferometer, and scans of two different mirrors are shown in Figure 8.15(a) and (b). From these measurements the radii of curvature were measured and found to vary by less than 10% (0.39 to 0.42 m), which is an acceptable degree of diopter dispersion.

An initial demonstration of image tracking for beam steering was also conducted using a commercial CMOS imager and one of the MEMS mirrors to direct a transmitting (tracking) laser beam toward a moving target laser spot actuated by a two-axis galvanometer. A simple centroiding algorithm was developed and tested using a digital control system. The transmitting laser beam was observed to track and follow a target spot as it moved across a white target plane. A block diagram of the tracking system is shown in Figure 8.16 along with a photograph of the actual tracking terminal.

A mapping between the FPA centroid position and a corresponding drive command was also measured to determine the degree of nonlinearity in the device derived from the lack of compliance of the mirror hinges at the extreme end of their angular travel. Taking the polynomial fits in two orthogonal angles, which were cross-coupled and varied with command voltages, attempts were made to linearize these and modest improvements in performance were obtained. Thus, this nonlinearity can be potentially calibrated-out and compensated-for, or, better yet, removed by redesign.

8.7.2 RECENT PROGRESS

Researchers at U.C., Berkeley, are also doing considerable work related to optical communications using MEMS devices. They are investigating distributed networks using millimeter-scale sensing elements implemented using MEMS, which are called "Smart Dust," which can be deployed either indoors or outdoors to sense and record data of interest. Each "mote" contains a power source, sensors, data

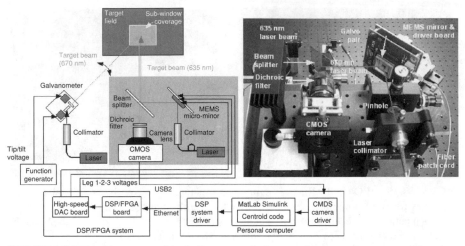

FIGURE 8.16 Tracking-system block diagram and photograph of actual test set-up, showing CMOS camera that views the target field through a dichroic filter to eliminate the tracking beam so it sees only the target beam.

storage, and a bidirectional wireless modem. A collection of such devices can be interrogated at distances up to several hundred meters by a small central transceiver. Recent efforts have been focused on implementing free-space optical communications for the interrogation of smart dust in which a novel uplink design utilizes a micro corner-cube retroreflector on each mote. A central transceiver illuminates these motes and permits transmission of information without having to radiate any power. Researchers at UC, Berkeley, as well as other institutions (Stanford, Princeton, and Sensors Unlimited) have also been funded by DARPA through the Steered Agile Beams (STAB) program to develop system architectures and novel components for high-speed, free-space optical communication between fast-moving airplanes and ground vehicles. Components under development include two-axis beam scanners fabricated using MEMS technology, as well as (1 to 5 W) InGaAsP/InP laser diodes and dual-mode (imaging and communication) InGaAs focal-plane arrays capable of operation at high bit rates (100 to 1000 Mbs). This technology may be applicable to space applications for close-range intersatellite operations.

Scaling of recent laboratory test units described in Section 8.6.1 to at least 4×4 (or larger) arrays with array pitches of 2 mm appears very achievable, which translates to chip areas not much greater than typical focal plane arrays. Multichannel DSP control hardware is needed that is well-matched to appropriate MEMS mirror designs (similar to that described above) and will need to be translated to a field programmable gate array (FPGA) chip design for spacecraft implementation to control all elements independently. Furthermore, the maturity of this technology permits prototype production of plug-in optical modules with very small form-factors that will interface to both a multifiber coupled communications bundle as well as a multi-point CCD or CMOS focal plane tracker. A 1-D concept to upscale a

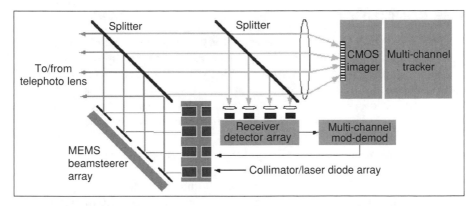

FIGURE 8.17 Conceptual 1-D MEMS-based multichannel optical communications unit.

single mirror to multiple mirrors (prior to a full 2-D design) is illustrated in
Figure 8.17 to delineate the essential elements required to implement MEMS
beam steering for optical satellite communications. A plan view of a possible 2-D
MEMX design is shown in Figure 8.18.

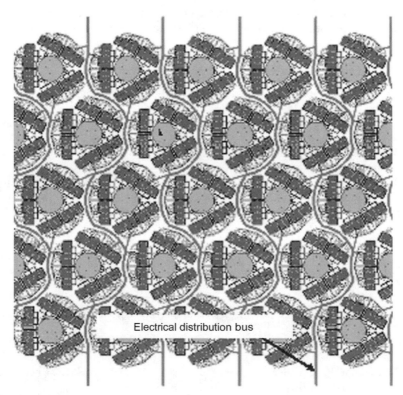

FIGURE 8.18 Plan-view of 2-D MEMS array using MEMX type micromirrors, suitable for
multichannel optical communications beam-steering.

8.8 CONCLUSION

Space communications systems are "ripe" for the insertion of MEMS-based technologies, in part due to the growth in commercial communication developments. One of the most exciting applications of MEMS for microwave communications in spacecraft concerns the implementation of "active aperture phase array antennas." These systems consist of groups of antennas phase-shifted from each other to take advantage of constructive and destructive interference in order to achieve high directionality. Such systems allow for electronically steered radiated and received beams, which have greater agility and will not interfere with the satellite's position.

Optical communications could also play an important role in low-power, low-mass, long-distance missions such as the Realistic InterStellar Explorer (RISE) mission, which seeks to send an explorer beyond the solar system, which requires traveling a distance of 200 to 1000 AU from the Sun within a timeframe of about 10 to 50 years. The primary downlink for such a satellite would need to be optical because of the distances and weight limits involved. It has been proposed that a MEMS implementation of the beam-steering mechanism may be necessary to achieve the desired directional accuracy with a sufficiently low mass.[112] MEMS in space communication may well fall under the trendy term "disruptive technology" for their potential to redefine whole systems.

REFERENCES

1. Wertz, J.R. and Larson, W.J., *Space Mission Analysis and Design*. Microcosm Press, El Segundo, California, 1999.
2. Morgan, W.L. and Gordon, G.D., *Communications Satellite Handbook*. John Wiley and Sons, New York, 1989.
3. Wilson, K. and Enoch, M., Optical communications for deep space missions, *IEEE Communications*, August 2000.
4. Begley, D.L., Laser cross-link systems and technology, *IEEE Communications Magazine*, **38** (8), 126, 2000.
5. Scott, W.B., Micromachines hold promise for aerospace, *Aviation Week and Space Technology*, March 1993.
6. Fiedziuszko, S.J., Applications of MEMS in communication satellites, *Proceedings — 13th International Conference on Microwaves, Radar and Wireless Communications*, MIKON-2000, 3, 201, 2000.
7. Rebeiz, G.M., Tan, G.-L., and Hayden, J. S., *IEEE Microwave Magazine*, June 2002, 72.
8. Rebeiz, G.M., *RF MEMS: Theory, Design, and Technology*. Wiley-Interscience, Hoboken, New Jersey, 2003.
9. Petersen, K.E., Micromechanical membrane switches on silicon, *IBM Journal of Research and Development*, **23** (4), 376, 1979.
10. Goldsmith, C.L. et al., Performance of low-loss RF MEMS capacitive switches, IEEE *Microwave Guided Wave Letters*, **8** (8), 269, 1998.
11. Yao, Z.J. et al., Micromachined low-loss microwave switches, *Journal of Microelectromechanical Systems*, **8** (2), 129, 1999.

12. Streeter, R.D. et al., VHF high-power tunable RF bandpass filter using microelectro-mechanical (MEM) microrelays, *International Journal of RF Microwave Compound Aided Engineering*, **11** (5), 261, 2001.
13. De Los Santos, H.J. et al., Microwave and mechanical considerations in the design of MEM switches for aerospace applications, *Proceedings — IEEE Aerospace Conference*, 235, 1997.
14. Larson, L.E., Microactuators for GaAs-based microwave integrated circuits, *Proceedings — Transducers '91*, 743, 1991.
15. Hyman, D. et al., GaAs-compatible surface-micromachined RF MEMS switches, *Electronic Letters*, **35** (3), 224, 1999.
16. Hyman, D. and Mehregany, M., Contact physics of gold microcontacts for MEMS switches, *IEEE Transactions on Components and Packing Technologies*, **22** (3), 357, 1999.
17. Muldavin, J.B. and Rebeiz, G.M., High-isolation CPW MEMS shunt switches — part 1: modeling, *IEEE Transactions on Microwave Theory and Techniques*, **48** (6), 1045, 2000.
18. Muldavin, J.B. and Rebeiz, G.M., High-isolation CPW MEMS shunt switches — part 2: design, *IEEE Transactions on Microwave Theory and Techniques*, **48** (6), 1053, 2000.
19. Barker, N.S. and Rebeiz, G.M., Distributed MEMS true-time delay phase shifters and wide-band switches, *IEEE Transactions on Microwave Theory and Techniques*, **46** (11), 1881, 1998.
20. Wood, R. et al., MEMS microrelays, *Mechatronics*, **8**, 535, 1998.
21. Seki, T. et al., Thermal buckling actuator for micro relays, *Proceedings — Transducers '97*, 1153, 1997.
22. Brown, E.R., RF-MEMS switches for reconfigurable integrated circuits, *IEEE Transactions on Microwave Theory and Techniques*, **46** (11), 1868, 1998.
23. Simon, J., Saffer, S., and Kim, C.-J., A liquid-filled microrelay with a moving mercury microdrop, *Journal of Microelectromechanical Systems*, **6** (3), 208, 1997.
24. Yao, J.J. et al., Microelectromechanical system radio frequency switches in a picosatellite mission, *Smart Materials and Structures*, **10**, 1196, 2001.
25. Mihailovich, R.E. et al., MEM relay for reconfigurable RF circuits, *IEEE Microwave Guided Wave Letters*, **11** (2), 53–55, 2001.
26. Yao, J.J., RF MEMS from a device perspective, *Journal of Micromechanics and Microengineering*, **10**, R9, 2000.
27. Yao, J.J. and Chang, M.F., A surface micromachined miniature switch for telecommunications applications with signal frequencies from DC up to 4 GHz, *Proceedings — Transducers '95*, 1995.
28. Goldsmith, C. et al., Micromechanical membrane switches for microwave applications, *IEEE Microwave Theory Technical Symposium Digests*, 91, 1995.
29. Goldsmith, C. et al., Characteristics of micromachined switches at microwave frequencies, *IEEE Microwave Theory Technical Symposium Digests*, 1141, 1996.
30. Randall, J.N. et al., Fabrication of micromechanical switches for routing radio frequency signals, *Journal of Vacuum Science and Technology B*, **14** (6), 3692, 1996.
31. Jones, E.J., Micro and millimeter wave MEMS for phased arrays, *Proceedings — Gomac, 98*, 1998.
32. Norvell, B.R. et al., Micro electro mechanical switch (MEMS) technology applied to electronically scanned arrays for space based radar, *Proceedings — Aerospace Conference*, **3**, 239, 1999.
33. Shen, S.-C. and Feng, M., Low actuation voltage RF MEMS switches with signal frequencies from 0.25 GHz to 40 GHz, *Proceedings — Transducers*, 1999.

34. Pacheco, S., Nguyen, C.T., and Katehi, L.P.B., Micromechanical electrostatic K-band switches, *Proceedings — IEEE MTT-S International Microwave Symposium*, 1998.

35. Zhou, S., Sun, X.-Q., and Carr, W.S., A monolithic variable inductor network using microrelays with combined thermal and electrostatic actuation, *Journal of Micromechanics and Microenginnering*, **9**, 45, 1999.

36. Kruglick, E.J.J. and Pister, K.S.J., Lateral MEMS microcontact considerations, *Journal of Microelectromechanical Systems*, **8** (3), 264, 1999.

37. Hosaka, H., Kuwano, H., and Yanagisawa, K., Electromagnetic microrelays: concepts and fundamental characteristics, *Sensors and Actuators A*, **40**, 41, 1994.

38. Kruglick, E.J.J. and Pister, K.S.J., Bistable MEMS relays and contact characterization, *Proceedings — Solid-State Sensor and Actuator Workshop*, 333, 1998.

39. Chang, C. and Chang, P., Innovative micromachined microwave switch with very low insertion loss, *Sensors and Actuators A* **79**, 71, 2000.

40. Schiele, I. et al., Surface-micromachined electrostatic microrelay, *Sensors and Actuators A*, **66**, 345, 1998.

41. Hyman, D. et al., Surface-micromachined RF MEMs switches on GaAs substrates, *International Journal of RF Microwave Computer Aided Engineering*, **9** (4), 348, 1999.

42. Schimkat, J., Contact materials for microrelays, *Proceedings — 11th Annual International Workshop on Micro Electro Mechanical Systems*, 190, 1998.

43. Schimkat, J., Contact measurements providing basic design data for microrelay actuators, *Sensors and Actuators*, **73**, 138, 1999.

44. Pillans, B., RF power handling of capacitive RF MEMS devices, *Proceedings — 2002 IEEE MTT-S International Microwave Symposium Digests*, 329, 2002.

45. Wibbeler, J., Pfeifer, G., and Hietschold, M., Parasitic charging of dielectric surfaces in capacitive microelectromechanical systems (MEMS), *Sensors and Actuators A*, **71**, 74, 1998.

46. Ko, Y.J., Park, J.Y., and Bu, J.U., Integrated 3-bit RF MEMS phase shifter with constant phase shift for active phased array antennas in satellite broadcasting systems, *Proceedings — Transducers '03*, 1788, 2003.

47. Kim, M. et al., A DC-to-40 GHz four-bit RF MEMS true-time delay network, *IEEE Microwave Wireless Component Letters*, **11** (2), 56, 2001.

48. Tan, G.-L. et al., Low-loss 2- and 4-bit TTD MEMS phase shifters based on SP4T switches, *IEEE Transactions on Microwave Theory and Techniques*, **51** (1), 297, 2003.

49. Hacker, J.B. et al., A Ka-Band 3-bit RF MEMS true-time-delay network, *IEEE Transactions on Microwave Theory and Techniques*, **51** (1), 305, 2003.

50. Pillans, B. et al., Ka-band RF MEMS phase shifters, *IEEE Microwave Guided Wave Letters*, **9** (12), 520, 1999.

51. Pozar, D.M., *Microwave Engineering*. John Wiley and Sons, New York, 1998.

52. Hayden, J.S. and Rebeiz, G.M., Very low-loss distributed X-band and Ka-band MEMS phase shifters using metal–air–metal capacitors, *IEEE Transactions on Microwave Theory and Techniques*, **51** (1), 309, 2003.

53. Ji, T.S., Vinoy, K.J., and Varadan, V.K., Distributed MEMS phase shifters by microstereolithography on silicon substrates for microwave and millimeter wave applications, *Smart Materials and Structures*, **10** (6), 1224, 2001.

54. Liu, Y. et al., K-band 3-bit low-loss distributed MEMS phase shifter, *IEEE Microwave Guided Wave Letters*, **10** (10), 415, 2000.

55. Barker, N.S. and Rebeiz, G.M., Optimization of distributed MEMS transmission-line phasie shifters — U-band and W-band designs, *IEEE Transactions on Microwave Theory and Techniques*, **48** (11), 1957, 2000.

56. Kim, H.-T. et al., A compact V-band 2-bit reflection-type MEMS phase shifter, *IEEE Microwave Guided Wave Letters*, **12** (9), 324, 2002.
57. Park, J.Y. et al., V-band reflection-typ phase shifters using micromachined CPW coupler and RF switches, *Journal of Microelectromechanical Systems*, **11** (6), 808, 2002.
58. Malczewski, A. et al., X-band RF MEMS phase shifters for phased array applications, *IEEE Microwave Guided Wave Letters*, **9** (12), 517, 1999.
59. Lin, L., Howe, R.T., and Pisano, A.P., Microelectromechanical filters for signal processing, *Journal of Microelectromechanical Systems*, **7** (3), 286, 1998.
60. Nguyen, C.T.-C., Frequency-selective MEMS for miniaturized low-power communications devices, *IEEE Transactions on Microwave Theory and Techniques*, **47** (8), 1486, 1999.
61. Lubecke, V.M., Barber, B.P., and Arney, S., Enabling MEMS technologies for communications systems, *Proceedings — Device and Process Technologies for MEMS and Microelectronics II*, 4592, 257, 2001.
62. Fan, L. et al., Universal MEMS platforms for passive RF components: suspended inductors and variable capacitors, *Proceedings — MEMS, 98, The Eleventh Annual Workshop on Micro Electro Mechanical Systems*, 29, 1998.
63. Dec, A. and Suyama, K., Micromachined electro-mechanically tunable capacitors and their applications to RF IC's, *IEEE Transactions on Microwave Theory and Techniques*, **46** (12), 2587, 1998.
64. Feng, Z. et al., Design and modeling of RF MEMS tunable capacitors using electrothermal actuators, *1999 IEEE MTT-S International Microwave Symposium Digest*, **4**, 1507, 1999.
65. Vinoy, K.J. and Varadan, V.K., Design of reconfigurable fractal antennas and RF-MEMS for space-based systems, *Smart Materials and Structures*, **10** (6), 1211, 2001.
66. Anagnostou, D. et al., Fractal antenna with RF MEMS switches for multiple frequency applications, *Proceedings — Antennas and Propagation Society International Symposium*, **2**, 22, 2002.
67. Werner, D.H. and Ganguly, S., An overview of fractal antenna engineering research, *IEEE Antennas and Propagation Magazine*, **45** (1), 38, 2003.
68. Chu, P.B., Lee, S.-S., and Park, S., MEMS: the path to large optical crossconnects, *IEEE Communications Magazine*, 80, 2002.
69. De Dobbelaere, P. et al., Digital MEMS for optical switching, *IEEE Communications Magazine*, 88, 2002.
70. Bishop, D. et al., Silicon micromachines for lightwave networks: can little machines make it big? *Annual Device Research Conference Digest. 58th Device Research Conference (58th DRC)*, Jun 19–Jun 21 2000, 7, 2000.
71. Wu, M.C., Fan, L., and Lee, S.-S., Optical MEMS: huge possibilities for lilliputian-sized devices, *Optics and Photonics News*, **9** (6), 25, 1998.
72. Bryzek, J., Petersen, K., and McCulley, W., Micromachines on the march, *IEEE Spectrum*, **31** (5), 20, 1994.
73. Suhonen, M. et al., Scanning microelectromechanical mirror for fine-pointing units of intersatellite optical links, *Smart Materials and Structures*, **10**, 1204, 2001.
74. www.memsoptical.com
75. Solgaard, O. et al., Microoptical phased arrays for spatial and spectral switching, *IEEE Communications Magazine*, **41** (3), 96, 2003.
76. Sumida, D.S. et al., All-optical, true-time-delay photonics network for high-bandwidth, free-space laser communication applications, *Proceedings — Free-Space Laser Communication Technologies XVI*, **5338**, 214, 2004.

77. www.DARPA.mil/MTO/stab
78. Warneke, B. et al., Smart dust: communicating with a cubic-millimeter computer, *Computer*, **34** (1), 44, 2001.
79. memsweb@mems.colorado.edu
80. Petersen, K., Silicon as a mechanical material, *Proceedings of IEEE*, **70** (5), 420, 1982.
81. Marxer, C. et al., Vertical mirrors fabricated by deep reactive ion etching for fiber-optic switching applications, *Journal of Microelectromechanical Systems*, **6** (3), 277, 1997.
82. Solgaard, O. et al., Optoelectronic packaging using silicon surface-micromachined alignment mirrors, *IEEE Photonics Technology Letters*, **7** (1), 41, 1995.
83. Sniegowski, J.J. and Garcia, E.J., Microfabricated actuators and their application to optics, *SPIE Proceedings*, **2383**, 46, 1995.
84. Petroz, K. et al., Integrated micro-optomechanical laser beam deflector, *Electronics Letters*, **34** (9), 881, 1998.
85. Burns, D.M. and Bright, V.M., Optical power induced damage to microelectromechanical mirrors, *Sensors and Actuators A: Physical*, **70** (1–2), 6, 1998.
86. Maeda, R., Lee, C., and Schroth, A., Development of a micromirror using piezoelectric excited and actuated structures, *Proceedings — Proceedings of the 1996 MRS Fall Meeting*, 233, 1997.
87. Sankur, H.O. et al., Fabrication of refractive microlens arrays, *SPIE Proceedings*, **2383**, 179, 1995.
88. Wilkinson, S.T. et al., Integration of thin film optoelectronic devices onto micromachined movable platforms, *IEEE Photonics Technology Letters*, **6** (9), 1115, 1994.
89. Zhu, X., Hsu, V.S., and Kahn, J.M., Optical modeling of MEMS corner cube retro-reflectors with misalignment and nonflatness, *IEEE Journal on Selected Topics in Quantum Electronics*, **8** (1), 26, 2002.
90. Kurzweg, T.P. and Morris III, A.S., Macro-modeling of systems including free-space optical MEMS, *Proceedings — 2000 International Conference on Modeling and Simulation of Microsystems — MSM 2000*, 146, 2000.
91. Zhang, J. et al., Thermal management of micromirror arrays for high-energy applications, *Proceedings — Pacific Rim/International, Intersociety Electronic Packaging Technical/Business Conference and Exhibition 2001, Advances in Electronic Packaging*, **1**, 103, 2001.
92. Mohr, J. et al., Micro-optical and opto-mechanical systems fabricated by the LIGA technique, *SPIE Proceedings*, **3008**, 273, 1997.
93. Fujita, H., Application of micromachining technology to optical devices and systems, *SPIE Proceedings*, **2881**, 2, 1996.
94. Gustafsson, K.a.B.H., Fiberoptic Switching and Multiplexing with a Micromechanical Mirror, *Proceedings of the 4th International Conference on Sensors and Actuators*, 212, 1987.
95. Daneman, M.J. et al., Laser-to-fiber coupling module using a micromachined alignment mirror, *IEEE Photonics Technology Letters*, **8** (3), 396, 1996.
96. Tien, N.C. et al., Surface-micromachined mirrors for laser-beam positioning, *Proceedings of the 1995 8th International Conference on Solid-State Sensors and Actuators, and Eurosensors IX. Part 2 (of 2)*, 352, 1995.
97. Lin, L.Y. et al., Surface-micromachined micro-XYZ stages for free-space microoptical bench, *IEEE Photonics Technology Letters*, **9** (3), 345, 1997.

98. Michalicek, M.A., Comtois, J.H., and Schriner, H.K., Geometry versus optical performance of micromirrors and arrays, *SPIE Proceedings*, **3440**, 140, 1998.
99. Hashimoto, E. et al., Micro-optical gate for fiber optic communication, *Proceedings — Proceedings of the 1997 International Conference on Solid-State Sensors and Actuators*. Part 1 (of 2), 331, 1997.
100. Field, L.A. et al., Micromachined 1 × 2 optical-fiber switch, *Sensors and Actuators A: Physical*, **53** (1–3), 311, 1996.
101. Motamedi, M.E. Hornbeck, L.J., and Pister, K.S.J, (eds), Miniaturized systems with micro-optics and micromechanics, *SPIE Proceedings*, **3008**, 378, 1997.
102. Kahn, J.M., Katz, R.H., and Pister, K.S.J., Next century challenges: mobile networking for 'Smart Dust', *Proceedings of the Annual International Conference on Mobile Computing and Networking, MOBICOM. Proceedings of the 1999 5th Annual ACM/ IEEE International Conference on Mobile Computing and Networking (MobiCom'99)*, 271, 1999.
103. Wang, J., Kahn, J.M., and Lau, K.Y., Minimization of acquisition time in short-range free-space optical communication, *Applied Optics*, **41** (36), 7592, 2002.
104. Last, M. et al., Toward a wireless optical communication link between two small unmanned aerial vehicles, *Proceedings of the 2003 IEEE International Symposium on Circuits and Systems*, **3**, 930, 2003.
105. Last, M. et al., Video semaphore decoding for free-space optical communication, *SPIE Proceedings*, **4303**, 148, 2001.
106. Leibowitz, B.S., Boser, B.E., and Pister, K.S.J., CMOS "smart pixel" for free-space optical communication, *SPIE Proceedings*, **4306**, 308, 2001.
107. Sniegowski, J.A. et al., Development, test and evaluation of MEMS micro mirrors for free-space optical communications, *SPIE Proceedings*, 1550, 2004.
108. Hahn, D.V., Edwards, C.L., and Duncan, D.D., Link availability model for optical communication through clouds, *SPIE Proceedings*, **4821**, 320, 2002.
109. Fielhauer, K.B. et al., Comparison of macro-tip/tilt and meso-scale position beam-steering transducers for free-space optical communications using a quadrant photodiode sensor, *SPIE Proceedings*, **5160**, 192, 2003.
110. Edwards, B.L. et al., Overview of the Mars Laser Communications Demonstration Project, AIAA Space 2003, September 2003.
111. Wiener, T.F. and Karp, S., The role of blue/green laser systems in strategic submarine communications, *IEEE Transactions on Communications*, **28** (9), 1602, 1980.
112. Boone, B.G. et al., Optical and microwave communications system conceptual design for a realistic interstellar explorer, *Proceedings — Free-Space Laser Communication and Laser Imaging II*, **4821**, 225, 2002.

9 Microsystems in Spacecraft Thermal Control

Theodore D. Swanson and Philip T. Chen

CONTENTS

9.1   INTRODUCTION

Thermal control systems (TCS) are an integral part of all spacecraft and instruments. Thermal engineers design TCS to allow spacecraft to function properly on-orbit.[1] In TCS design, both passive and active thermal control methods may be applied. Passive thermal control methods are commonly adopted for their relatively low cost and reliability, and are adequate for most applications. When passive thermal control methods are insufficient to meet the mission thermal requirements, active thermal control methods are warranted. Active thermal control methods may be more effective in meeting stringent thermal requirements. For example, many emerging sensor applications require very tight temperature control (to within 1 K) or isothermality across a large area; such requirements can generally be achieved only by active control techniques.

Designing effective TCS with suitable thermal control method becomes a challenging task for spacecraft thermal engineers. To develop a successful TCS, it is necessary to understand the basics of heat transfer in space, the functionality of a thermal control component, and the operation of an integrated thermal system. Miniaturization of future spacecraft results in high power densities, lower heat capacity, and reduced available power. Microelectromechanical systems (MEMS)-based solutions can provide efficient and miniaturized TCS. As the MEMS knowledge base matures, thermal controls are emerging as a viable technology for thermal engineers. These applications include specialized thermal control coatings, thermal switches, and filters for instruments. MEMS technology presents both benefits and challenges for thermal engineers. Lack of in-flight MEMS data is one of the challenges to using space-based MEMS TCS. As a consequence, in order to design a MEMS thermal control device and receive the full advantage, it is important for understanding the potential impact of the space environment on MEMS devices. The following discussion is intended to provide some insight to these issues, and it begins with a discussion of basic thermal control design consideration.

9.2   PRINCIPLES OF HEAT TRANSFER

To understand thermal control, one needs to understand the transport of heat in space. Heat transfer deals with the movement of thermal energy from one quantity of matter to the other. In the simplest terms, the discipline of heat transfer is concerned with only two things: temperature and heat flow. Temperature represents the amount of thermal energy available, whereas heat flow represents the movement of thermal energy from region to region. Heat is a form of energy transfer. It is "work" on the microscopic scale that is not accounted for at the macroscopic level. A mass of material may be considered as a thermal energy reservoir, where heat is manifested as an increase in the internal energy of the mass. A change in internal energy may be expressed as shown in the following equation:

$$\Delta E = C_p \, m \Delta T \tag{9.1}$$

where E: thermal energy change (J)
$\quad C_p$: specific heat at constant pressure (J kg^{-1} K^{-1})
$\quad M$: mass (kg)
$\quad T$: temperature change (K).

Heat transfer concerns the transport of thermal energy. There are three modes of heat transport, namely, conduction, convection, and radiation.[2] In practice, most situations involve some combination of these three modes. However, in space, all heat must ultimately be rejected by radiation.

9.2.1 CONDUCTION

Conduction is the most common mode of heat transfer. In conduction, thermal energy can be transferred through the medium from a region of high temperature to a region of low temperature. The driving force for this type of heat transfer is a temperature difference (temperature gradient), ΔT. Fourier's law of conduction is the empirical equation used to describe the conduction heat transfer. The law states that the rate of heat transfer, Q, through a homogenous solid is directly proportional to the surface area, A, (at right angles to the direction of heat flow) and to the temperature gradient, dT/dx, along the path of heat flow. For the one-dimensional plane with temperature distribution $T = f(x)$, the conduction rate equation is expressed as follows:

$$Q = -kA\frac{dT}{dx} \tag{9.2}$$

where Q: heat transfer rate (J sec^{-1} or W)
    k: thermal conductivity (W m^{-1} K^{-1})
    A: surface area (m^2)
    T: temperature (K)
    x: distance (m)

    The minus sign is a consequence of the fact that heat is transferred in the direction of decreasing temperature, that is, from the high-temperature region to low-temperature region. The material property that describes heat conduction, thermal conductivity, is typically dependent on the temperature of the material.

    In most space applications, heat conduction in a continuous medium can be properly described by Fourier's law. The same law, however, is inadequate to illustrate the heat transfer by conduction between two adjoined hardware surfaces. Thermal conduction across a physical interface is considered as a special case. At a microscopic level, such interfaces are rough and therefore significantly reduce conduction. These interfacial resistances often dominate the rate of heat flow in the process. An "interface heat conductance" is typically used to quantify this affect and is relevant to many MEMS applications. To understand the general concept of thermal conductance, C, Equation (9.2) can be rewritten for a plate of given material and thickness, λ/d as follows:

$$Q = -C \cdot \Delta T \tag{9.3}$$

$$C = \frac{kA}{d} \tag{9.4}$$

where Q: heat transfer rate (J sec^{-1} or W)
    C: thermal conductance (W K^{-1})
    k: thermal conductivity (W m^{-1} K^{-1})
    A: surface area (m^2)
    ΔT: temperature difference between the two surfaces of the material (K)
    d: thickness of the material (m).

For a simple case, the concept of conductance is equivalent and thermally describes the plate. For a complicated case, such as an interface or a complicated structure, the conductance is measured or modeled and can be used to describe the thermal transport. For example, the thermal conductance for multilayer systems is calculated with the same laws as electrical conductance. The concept of thermal conductance is important for the spacecraft thermal design with numerous structures and materials. In a model, they all can be treated as a conductance.

9.2.2 CONVECTION

Convection is heat transport in a fluid or gas by the macroscopic movement of matter. Convective heat transfer is classified as free convection or forced convection according to the nature of the flow. Forced convection employs a pressure gradient (e.g., from a fan, a mechanical pump, or a capillary wick) to drive the fluid motion, as opposed to free convection in which density gradients driven by gravity induce fluid. Free convection is of little importance for heat transport in space where as forced convection can be very effective. Important applications that make use of convective transport are heat pipes and related capillary-driven devices. These devices rely on the latent heat associated with a change of phase. When a substance changes phase — from vapor to liquid, liquid to solid, solid to liquid, or liquid to vapor — there is a significant change in the energy state of the material. Typically, this is associated with the addition or loss of thermal energy. For spacecraft, two-phase heat transfer commonly involves vapor or liquid transformations (i.e., vaporization and liquifaction); although, a few applications involve liquid to liquid or solid to solid transformations. External energy is absorbed into a two-phase device when the liquid evaporates (taking heat away from the evaporator area) and is released when it condenses. Therefore, the gas flow not only carries the heat related to its specific heat (C_p) and temperature (T), but also the "latent heat" (L), which typically is much higher. Such two-phase heat transfer is extremely efficient; several orders of magnitude more effective than normal convection and also offers the benefit of isothermality. Heat pipes play a very important role in spacecraft thermal control; however, two-phase systems tend to be more challenging to design.

9.2.3 RADIATION

The third and last form of heat transfer process is radiation. Contrary to conduction and convection, thermal radiation does not rely on any type of medium to transport the heat. Radiation heat transfer depends on the characteristics and temperature of the exposed, radiating surfaces and the effective sink temperature of their views to space. With these unique characteristics, radiation heat transfer becomes the most critical heat transfer process in space.

Any object at a temperature above absolute zero emits electromagnetic radiation. This thermal radiation is the dominant form of heat transfer in space, since the thermal radiation emitted from the outside surface of a spacecraft is the only means of losing heat.

The thermal energy per unit area (W m^{-2}) released by a body at a given temperature by radiation is termed as the surface emissive power (E). The heat flux of a radiation process is described by the Stefan–Boltzmann law as shown in the following equation:

$$E = \varepsilon \sigma T^4 \tag{9.5}$$

where E: emissive power (W m^{-2})
   ε: surface emissivity ($0 \leq \varepsilon \leq 1$)
   σ: Stefan–Boltzmann constant (5.67×10^{-8} W m^{-2} K^{-4})
   T: surface temperature (K)

In practice, radiative heat exchange occurs between real or effective surfaces; for example, between a spacecraft radiator and deep space (very cold) or between a radiator and Earth (cold, but warmer than deep space). Radiative heat transfer is calculated as a function of the difference of the surface emissivities and their respective temperature to the forth power. View factors must also be included, making the computation somewhat involved.

The surface emissivity (ε) is the ratio of the body's actual emissive power to that of an ideal black body. The emissivity depends on the surface material and finish, on the temperature (especially at cryogenic temperatures where emissivity drops off rapidly), and the wavelength. Tabulated values are available for emissivity; however, measured values are required as the actual properties of a surface can vary as "workmanship" issues impact the value. Additionally, the build-up of contamination or the effect of radiation on a surface can impact emissivity. Hence, "beginning-of-life" and "end-of-life" properties are often quoted. At cryogenic temperatures, emissivity tends to fall off rapidly. According to Kirchoff's law a surface at thermal equilibrium has the property that a given temperature and wavelength, the absorptivity equals the emissivity. By applying the conservation of energy law, get the following equation for a opaque surface:

$$1 - \varepsilon = \rho \tag{9.6}$$

where ε is the emissivity and ρ is the reflectivity of the surface. This equation measures emissivity via reflectivity which is normally simpler to measure.

Since the radiation emitted by a spacecraft falls into the infrared and far infrared regime of the electromagnetic spectrum, emissivity is normally given as an average over these wavelengths. The solar absorptivity (α) describes how much solar energy is absorbed by the material and is averaged over the solar spectrum. Surface emissivity and solar absorptivity are important parameters for spacecraft materials. Typically, a spacecraft radiator, which is used to cool the spacecraft via radiation, is built from surfaces with a high emissivity but a low solar absorptivity.

9.3 SPACECRAFT THERMAL CONTROL

The function of a TCS is to control the temperature of spacecraft components within their operational temperature ranges for all operating modes and in satellite spacecraft environments. This is a demanding requirement if the limits are tight or the environments extreme. Table 9.1 shows some of the typical temperature requirements for spacecraft components.[3]

The thermal design of a spacecraft requires accounting of all heat sources, both from within the spacecraft and imposed by the environment. Heat-producing spacecraft components include but are not limited to heaters, shunts, rocket motors, electronic devices, and batteries. Environmental heating is largely the result of solar radiation. Radiation from other heavenly bodies (such as the Earth or Moon) is typically less, but must be considered for thermal design purposes. Other spacecraft components such as a solar array and deployed devices that are in a field of view of a surface may impose a radiation heat load.

Once heat sources and environmental parameters are quantified, the thermal engineer uses analysis for the thermal design of the spacecraft. For a spacecraft, conduction (including interface conduction) and convection (if present) are considered as internal heat transfer processes. These processes affect the balance of heat energy within the spacecraft itself and may be very important. Thermal exchanges within the environment are almost completely caused by radiation exchange. The radiator area and the coating surface properties are of great importance in achieving proper thermal control. Fortunately, research has been devoted into developing coatings with specialized properties. Desired surface properties may be presented as a permanent surface coating or may be temporarily altered according to design conditions.

9.3.1 SPACECRAFT THERMAL CONTROL HARDWARE

Thermal control hardware is employed to maintain components within proper temperature ranges. Proper thermal design will maintain all components within the required operating temperature range during the entire mission. The radiator is an important element of the design. Radiators are areas on the surface of the spacecraft with high typically emissivity and low solar absorptivity and a minimum

---

TABLE 9.1
Typical Spacecraft Component Temperatures[3]

Component	Operating Temperature (°C)	Survival Temperature (°C)
Digital electronics	0 to 50	−20 to 70
Analog electronics	0 to 40	−20 to 70
Batteries	10 to 20	0 to 35
Particle detectors	−35 to 0	−35 to 35

TABLE 9.2
Passive and Active Thermal Control Hardware

Thermal Control Hardware

Passive System	Active System
Thermal surface finishes	Heaters
Multilayered insulation	Louvers
Radiators	Heat switches
Mountings and interfaces	Fluid loops
Phase change materials	Thermoelectric coolers
	Heat pipes or loops

solar exposure, that radiate excess heat into space. Other common passive thermal control elements include specialized thermal surface finishes, multilayered insulation blankets, conduction enhancing or retarding materials, phase change materials, heat pipes, and bimetallic louvers, which open and close according to the radiator temperature.

Active thermal control is required when the temperature needs to be tightly controlled, or when the thermal environment is highly variable. A summary of passive and active thermal control hardware is shown in Table 9.2. Active control provides the thermal design engineer flexibility, tighter control, and faster design turnaround. Small satellites may benefit well from this approach since they are more likely to be mass produced and need a thermal design which can meet a range of mission criteria.

9.3.2 HEAT TRANSFER IN SPACE

Radiation heat transfer is an important process between an orbiting spacecraft and its surrounding environment. This heat exchange is a final energy balance between heat absorption on spacecraft surfaces and heat rejection to space. In addition to internal heat generation, spacecraft external surfaces receive radiation from the space environment. The quantity of the radiation absorbed is related to the intensity of the external radiation, the area affected, and the solar absorptivity (α) of the surface. The quantity of heat rejected is proportional to the radiator area, temperature differential between the radiator, the "effective sink temperature" of what it is viewing, and the infrared emissivity (ε) of the surface. The ratio of solar absorptivity and infrared emissivity (α/ε) is important in determining the spacecraft surface temperature.

For passive thermal control, designing and selecting surface materials with desired α and ε is an effective way to obtain an optimal heat balance. Unfortunately, these properties are fixed once surface materials are selected. Long-term exposure to space environments degrades thermal control surfaces by increasing solar absorptivity (sometimes very significantly) with the result of increasing spacecraft surface temperature. During normal operation, spacecraft temperatures may be intermediately controlled by altering a radiation's surface solar absorptivity or infrared emissivity. Mechanical devices such as pinwheels, louvers, or shutters that can be "opened or closed" to view space may be used to achieve such effective changes in absorptivity or emissivity.

The major heat sources in the heat transfer process for a spacecraft of space include solar radiation, Earth radiation, reflected radiation (albedo), and internally generated heat. Spacecrafts reject heat by radiation to space, mainly through its designated radiator surfaces. The law for conservation of energy describes heat that is received, generated, and rejected by a spacecraft with the following equation:

$$MC_p \frac{dT}{dt} = \alpha A_p (S + E_a) + \varepsilon_E A_p E_r - \varepsilon A \sigma T^4 + Q_{int} \qquad (9.7)$$

where M: mass (kg)
C_p: heat capacity (W sec kg^{-1} K^{-1})
T: temperature (K)
t: time (sec)
α: spacecraft surface solar absorptivity
A_p: surface area for heat absorption (m^{-2})
S: solar flux (~1353 W m^{-2})
E_a: Earth albedo (~237 W m^{-2})
ε_E: Earth surface emissivity
E_r: Earth radiation (~50 W m^{-2})
ε: spacecraft surface infrared emissivity ($0 \le \varepsilon \le 1$)
A: surface area for heat radiation (m^2)
σ: Stefan–Boltzmann constant ($\sigma = 5.67 \times 10^{-8}$ W m^{-2} K^{-4})
Q_{int}: internal heat generation (W).

For a spacecraft to reach thermal equilibrium in space, the rate of energy absorption or generation and radiation must be equal. At thermal equilibrium, the spacecraft heat balance is at a steady state and the derivative term dT/dt on the left hand side of Equation (9.7) becomes zero. If one simplifies the situation and assumes that the spacecraft receives solar radiation as the only heat source, the heat balance equation (9.7) at steady state is reduced to the following equations:

$$Q = 0 = \alpha A_p S - \varepsilon A \sigma T^4 \qquad (9.8)$$

$$T = \left(\frac{A_p}{A}\right)^{1/4} \left(\frac{S}{\sigma}\right)^{1/4} \left(\frac{\alpha}{\varepsilon}\right)^{1/4} \qquad (9.9)$$

According to Equation (9.9), for a fixed spacecraft orientation and thermal exposure, surface temperature becomes a function of surface properties only. Therefore, spacecraft surface is proportional to 1/4 power of the ratio of α and ε; that is, $T = f[(\alpha/\varepsilon)^{1/4}]$. By properly selecting surface materials, spacecraft thermal control can be achieved passively (i.e., a given temperature can be achieved) by the α/ε of the surfaces. This analysis must be repeated for all conditions, as a spacecraft's thermal environment (and internal load) will typically change as it moves through its orbit.

9.4 MEMS THERMAL CONTROL APPLICATIONS

The use of nano- and picosatellites in present and future space missions require a new approach to thermal control. Small spacecraft have low thermal capacitance, making them vulnerable to rapid temperature fluctuations. At the same time, many traditional thermal control technologies, such as heat pipes, do not scale well to meet the constrained power and mass budgets of smaller satellites. MEMS are well suited for applications in small spacecraft; they are lightweight, rugged, reliable, and relatively inexpensive to fabricate.[5]

The first MEMS experiments have flown on Space Shuttle Mission STS-93 in 1999 to evaluate the effect of exposure to the space environment on the MEMS materials. During the STS-93 flight, MEMS experiments were carried in the shuttle middeck locker. These experiments examine the performance of MEMS devices under launch, microgravity, and reentry conditions. These devices included accelerometers, gyros, and environmental and chemical sensors. These MEMS experiments provide in-flight information on navigation, sensors, and thermal control necessary for future small scale spacecraft. Spacecraft MEMS thermal control applications are emerging with the Department of Defense (DoD), NASA, academia, and aerospace industry as major contributors in research and development. Most MEMS thermal control applications are developmental with technology readiness levels (TRL) up to TRL 6. Several potential applications for MEMS devices in thermal control are described below.

9.4.1 THERMAL SENSORS

At the present, many conventional MEMS thermal devices have been designed and used as thermal sensors.[6,7] MEMS thermal sensors are transducers that convert thermal energy into electrical energy. They are devices that measure a primary thermal quantity: either temperature or heat flow or thermal conductivity. One technique is to take advantage of the difference in the coefficients of thermal expansion between two joined materials. This causes a temperature-dependent deflection, creating stress on a piezoelectric material and generating an electrical signal or actuating a switch. A good example for such a MEMS thermostat or thermal switch is the Honeywell Mechanically Actuated Field Effect Transistor (MAFET)® technology.[8,9] The MAFET is a microthermal switch that is low cost, of small size (< 3.0 mm^2), and has a long operational life (1,000,000 cycles). Unlike typical thermal switches, this device uses electronic switching, thus eliminating the arcing and microwelding that occur while making or breaking metal-to-metal contact. The MAFET thermal switch uses fundamental MEMS processing technology. The heart of the thermal switch is a temperature-sensitive deflecting beam. The thermal switch has a selectable temperature action such as open or close on rise with a set point range between −65 and 175°C. According to the material's thermal expansion coefficient, heating the beam to a specific temperature causes a differential elongation of the beam. As the temperature changes from nominal, the beam deflects toward the transistor source and drain. An applied, adjustable gate voltage completes this movement and snaps the beam closed when the temperature reaches the setpoint. Contact of the beam with the substrate completes the circuit, allowing current to flow from source to drain. Honeywell's thermal switch can be used to activate an electrical signal when the switch is activated by a temperature change, much like a thermostat. Although not designed to modify conduction path, Honeywell's MEMS-based heat switch may be used to control the heater operation for spacecraft active thermal control purpose.

9.4.2 MEMS Louvers and Shutters

Mechanical thermal louvers are active thermal control devices that have been used to regulate the area of a radiator in response to its temperature. The regulation of radiator area is achieved by opening and closing of louver blades which are placed directly in front of the radiator surface. While most commonly placed over external radiators, louvers may also be used to modulate heat transfer between internal spacecraft surfaces, or from internal surfaces directly to space through the opening in the spacecraft wall.[10]

Conventional louvers have been used in different forms on many spacecraft, including Hubble Space Telescope, Magellan, Viking, and Voyager, to control the amount of cooling for a fixed size radiator. The most commonly used louver assembly is the rectangular-blade type which is spring-actuated by bimetallic metals. Hydraulically activated louvers and pinwheel louvers are used less often today than in the past. Traditional louvers typically provide closed to open effective emissivity variation of 0.1 to 0.6, are 200 to 6000 cm^2 in total area, and have a weight to area ratio of 5 to 10 kg/m^2. Disadvantages of traditional louver assemblies for small satellites are the size and weight, and the sensitivity to the solar position.

MEMS shutters and louvers have been suggested very early as a means of thermal control using MEMS for nano- and picosatellites.[11] The Johns Hopkins University Applied Physics Laboratory (JHU/APL), together with NASA Goddard Space Flight Center (NASA/GSFC), has designed, fabricated, and tested a number of louver designs using the MCNC (now MEMSCAP) MUMPs process. Figure 9.1 shows a 3 × 4 array of MEMS louvers, each 300 × 500 μm in size, and Figure 9.2 shows the infrared (IR) emissivity at 40°C at wavelengths between 8 and 12 μm of the MEMS louver array with the louvers closed, partially open, and open. The open louvers expose the high-emissivity surface. These louvers and a number of other designs such as shutters and folding structure were prototype designs of the concept to be flown on NASA/GSFC Space Technology-5 (ST5) mission as demonstration technology for variable emittance coatings (VEC). ST5 is part of a series of spacecraft in NASA's New Millennium Program (NMP) managed by Jet Propulsion Laboratory. The NMP strives to test new spacecraft

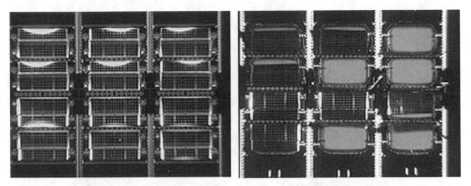

FIGURE 9.1 Microfabricated array of 300 × 500 μm louver array. The area below the louvers has been removed using deep reactive ion etch (DRIE). The right picture shows some of the louvers open, exposing the high emissivity surface below the substrate. (Courtesy: JHU/APL.)

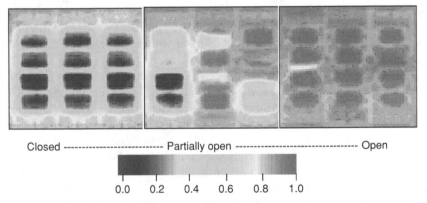

Closed ------------------------- Partially open -------------------------------- Open

0.0   0.2   0.4   0.6   0.8   1.0

FIGURE 9.2 IR emissivity of the MEMS louver array with the louvers closed, partially open, and open. (Courtesy: JHU/APL.)

technologies for future space missions whose primary objective will be to make multiple simultaneous measurements of the harsh space environment near the boundary of Earth's protective magnetic field known as the magnetosphere. The goal of NMP is to validate new technologies that will enable the reduction of weight, size, and cost for future missions. ST5, the fourth deep space mission in the NMP is designed and managed by NASA/GSFC and will validate four "enabling" technologies. Beside standard passive thermal control, these satellites will carry two VEC experiments, one of them based on a MEMS technology developed together by NASA/GSFC and JHU/APL.[12,13] These VEC experiments are technology demonstrations and are not part of the thermal control system itself, but rather independent experiments. ST5 is scheduled to launch in February of 2006. Given the limited time for prototype development, in part due to the turn-around time in MEMS fabrication, development and the need for a reliable flight

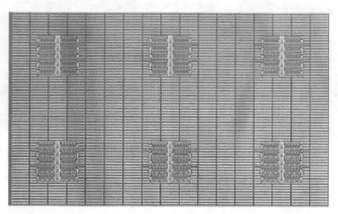

FIGURE 9.3 Shuttle arrays are on a single die, each 1.265 × 1.303 cm in size. (Courtesy: JHU/APL.)

design, JHU/APL, together with NASA/GSFC and Sandia National Laboratory (SNL), adopted a MEMS shutter design which will be flown on ST5. Fabricated with SNL's SUMMIT 5 process, six electrostatic comb drives, using SNL's high-performance design, will move an array of shutters, each 150 μm long and 6 μm wide, to either a gold surface or the silicon substrate and changing the emissivity from 0.6 (silicon) to < 0.1 (gold). A picture of such an array, 1767 × 876 μm in size, is shown in Figure 9.3. Seventy-two of these arrays are on a single die, each 1.265 × 1.303 cm in size. All arrays on a die are controlled together with a supply voltage greater than 35 V and negligible current draw. For the shutter, a single failure may cause a short and stop the entire die from working. In order to prevent such an issue, each array is connected to the supply bus via a MEMS fuse, which can be blown with a current of greater than 17 mA. Note that for normal operation, the current is minimal and the dc leakage current has been determined to be < 80 μA. A picture of the final radiator assembly is shown in Figure 9.4. Each radiator, 9 × 10 in size, contains 6 AlC substrates; which themselves contain six shutter dies each, adding up to a total of 36 dies on the radiator.

The VEC Instrument consists of two components, the previously described MEMS Shutter Array (MSA) radiator and the Electronic Control Unit (ECU). The MSA radiator is physically located on the top deck of the spin-stabilized ST5 spacecraft. The ECU is located within the spacecraft. The MSA radiator can be operated in both manual and autonomous mode, to automatically evaluate both high and low emittance states in a given test sequence as well as via ground control. A 1.5 W electrical heater is included in order to provide calibrated measurements of effective emittance changes. The radiator is located so that it receives minimal solar exposure. The MSA radiator is thermally isolated from the spacecraft, as the VEC technologies on this mission are for technology validation only. The thermal performance associated with opening and closing the shutters is measured by thermistors that are located on the underside of the MSA radiator chassis.

FIGURE 9.4 Radiator assembly. (Courtesy: JHU/APL.)

To qualify the MEMS louver, several environmental tests were conducted on the final flight articles. The device needs to pass various performance tests (burn-in, various vibration tests, and thermal vacuum) to verify its survival. In addition, life cycling tests, performance measurements (both effective IR emissivity and solar absorptivity), cycling in vacuum (over 1000 times), and exposure to a simulated space environment in solar wind and ultraviolet (UV) facilities were conducted.

9.4.3   MEMS Thermal Switch

A conventional thermal switch, sometimes referred to as heat switch, is an active thermal control device. Heat switches are devices that allow the connection or disconnection of the thermal contact between two surfaces. Thermal switches are typically installed between an insulated spacecraft structure and an external radiator or mounted between spacecraft components such as the battery on the Mars rover and the structure or the radiator. Various types of paraffin are often used in the thermal switches to create conduction paths when melted materials expand and bring components in close contact. A pedestal thermal switch designed by Starsys is 38.1×25.4 mm, weighs 100 g, and has a thermal conduction range from 1 to 100. It uses a paraffin actuator, which uses the thermal expansion of paraffin when it is melting to bring two thermally conductive surfaces into contact. The effectiveness of a thermal switch is usually characterized by a thermal conduction range which is an indicator of the improvement in effective thermal conductivity of the conduction path.

Another new variable emeltance technology, which also will be flown on the ST5 mission, uses an electrostatic thermal switch as a radiator.[4] In this design, a thin film with a high emissivity surface is suspended, thermally isolated, above the radiator. Once a voltage is applied between the radiator and the film, it is

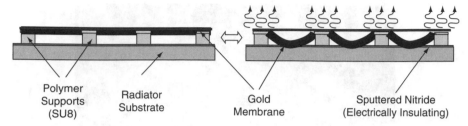

FIGURE 9.5 A schematic of SU8 fabricated device.

electrostatically attracted to the radiator and makes thermal contact, connecting the radiator to the high emissivity surface. The disadvantage of this design is the high switching voltage, typically greater than 400 V.

A similar design has been developed and fabricated in a MEMS platform by the United States Naval Academy (USNA) and JHU/APL, to be flown on the USNA Midstar satellite. For this design, the radiator consists of an array of membranes, about 400 to 500 μm wide and long, suspended a couple of microns above the surface on frames or posts at the corner. Figure 9.5 shows a schematic of the device, which is fabricated using SU8, a photosensitive epoxy with a very low thermal conductivity in the cured state, and a gold membrane. An example of a frame-supported device is shown in Figure 9.6. The devices have switching voltages between 20 and 40 V, very well within spacecraft standard voltages. While the MEMS design performs well, the thermal design needs to be improved. The thermal conductance in the off state, given by the conductivity in the support posts, is very

FIGURE 9.6 An example of a frame-supported device. (*Courtesy*: USNA.)

high due to the short length of the supports. In addition, coating the gold membrane has not yet achieved the high emissivity required.

9.4.4 MICROHEAT PIPES

To dissipate thermal energy, thermal engineers may use thermal doublers and heat pipes to spread the heat within the structure. With tight footprint restrictions, the conductive heat transfer enhancement by thermal doublers may be limited. For high heat flux dissipation, heat pipes are proven to be more effective than thermal doublers in achieving a uniform temperature distribution. Heat pipes achieve this through a capillary driven, fluid phase change process. They are sealed tubes partially filled with a working fluid, with a capillary wick acting as the pump. Large heat transfer rates can be achieved at an almost constant temperature in the system.

Conventional heat pipes were used for thermal control on spacecraft as early as 1964. A conventional heat pipe may have a capacity range from as high as a few kilowatts to as low as a few watts. Heat pipes may be classified into two main categories, constant-conductance and variable-conductance. Constant-conductance heat pipes are used for stable heat loads while variable-conductance heat pipes (VCHP) are used when the environmental sink or heat source varies, or when tighter temperature control is desired. Common heat pipes are extruded from square or finned aluminum or copper tubes ranging from about 0.5 to 2.0 cm in diameter, with lengths usually less than 2.0 m. They have the capillary grooves or a mesh wick on the inside. Common operating fluids include ammonia, water, and propylene.

MEMS-based heat exchange techniques have been investigated for cooling a central processing unit (CPU) on ground applications. The technique targeted high-performance CPUs that are used in very restricted spaces in workstations. Other versions have been planned for other types of ICs, including graphics processors and other dense ICs. In dealing with miniature or microscale heat removal within high-power density electronics, SNL has been one of the leading institutions in developing micro-machined vapor chamber heat spreaders. SNL has patented a passively "smart" heat transfer mechanism to remove heat dissipated by computer chips in the 50 W/cm^2 range. The "chip heat pipes" mechanism uses small amounts of vaporized liquid sealed in tiny flat pipes to move heat to the side edge of the computer.[14] Air fins are used to dissipate the heat into its environments.

SNL has expanded chip heat pipes into the MEMS heat pipes arena.[15] Similar to conventional heat pipes, SNL's microheat pipes contain basic components of a working fluid, a wick structure, and an envelope. As the fluid heats up and evaporates, it moves to the cooler area where it condenses. This cyclic evaporation and condensation distribute, or evaporation, and condensation distributes heat throughout the substrate. Through a capillary action, the microheat pipes are capable of removing heat from its source to a nearby heat sink passively and efficiently.

The structure of SNL's microheat pipes comprises a copper ring separating two copper plates. The advantage of SNL's microheat pipes is that they can be etched to follow curved or bent paths from a heat source to a heat sink, and go around mounting holes, screws, or standoffs. These microheat pipes are made of two pieces of silicon with microscopic wicking surfaces etched into the surfaces. The fluid path in the pipe contains finely etched lines about as deep as fingerprints and 60 μm wide. SNL's microheat pipes use methanol as cooling fluid. Methanol or other types of fluids will circulate inside the pipes and remove heat from, for example, a heat source such as a distributed battery. As is true for all heat pipes, the operation of the microheat pipes involves the phase change of the working fluid to maximize its heat transfer capability. This provides an added advantage of maintaining the hardware in a constant temperature range. Once heated, vaporized methanol flows by convection to the heat sink where it gives up its heat, condenses back to liquid phase, and returns to the hot end. The heat and cooling of the working fluid completes the evaporation and condensation cycle of the microheat pipes.

One basic issue in spacecraft thermal control is heat dissipation from densely packaged electronic parts assembled in electronic boxes mounted on a spacecraft platform.[16] In solving thermal related problems, SNL has actively engaged in the development of MEMS-based heat pipes for future space application. To further ensure MEMS advances for defense, aerospace, and commercial applications, SNL has established several cooperative research and development agreements with aerospace companies. Preliminary results demonstrated that the development of microheat pipes is steady and relatively successful. As shown in the literature, microheat pipes range in size from 1 mm in diameter and 60 mm in length to 30 μm in diameter and 10 mm in length. The methods currently used to fabricate microheat pipes with hydraulic diameters on the order of 20 to 150 μm into silicon or gallium arsenide wafers are also available.[17]

9.4.5 MEMS Pumped Liquid Cooling System

Pumped fluid loops are active thermal control systems. A simplified loop consists of a pumping device, a heat exchanger, and a space radiator. Pumped fluid loops are devices operated under the principle of forced liquid convective cooling. Cooling is accomplished when the working fluid absorbs excess heat and transports it to a heat sink. The loops can effectively maintain temperatures even when the spacecraft dissipates high power or operates under environmental extremes. For example, NASA's Mars Pathfinder used a mechanically pumped single-phase cooling loop with Freon-11 as a working fluid to achieve a cooling power capacity of 90 to 180 W. Chip level temperature control by micropumped loop device was reported by Pettigrew et al.[18] In this work, an evaporator, condenser, reservoir, and liquid and vapor lines were etched into the silicon wafer, while the glass wafer serves as a cover plate into which grooves were etched for capillary pumping. The device had a 1 × 2 mm evaporator and was capable of operating at a constant 100°C.

As spacecraft get ever smaller, many thermal control devices will be required to miniaturize. Aiming at future deep space science exploration, the NASA Jet Propulsion Laboratory (JPL) has investigated a MEMS cooling system for micro- or nanospacecraft.[19] Although the current pumped liquid cooling system is designed to transfer large amounts of thermal energy between two locations on a spacecraft, it is not capable of handling heat transfer in high power density applications. Power densities for future science instruments and engineering equipment on board a spacecraft are expected to exceed 25 W/cm^2. Some applications, such as higher power lasers, may involve fluxes in excess of 100 W/cm^2. Advanced thermal control concepts and technologies are essential to keep future payloads within allowable temperature limits and to provide accurate temperature control.

JPL's MEMS-based pumped liquid cooling is a mechanically pumped cooling system which consists of a working fluid circulated through microchannels by a micropump. Microchannel heat exchangers have been designed and fabricated in silicon. The microchannels are 50 µm deep, with widths ranging from 50 to 100 µm. In the development stage, the heat exchangers are subjected to hydraulic and thermal performance testing in simulated microspacecraft heat loads using deionized water as the working fluid. The test data will be evaluated and used for numerical thermal model validation. Optimization studies will be conducted using these numerical models on various microchannel configurations, working fluids, and micropump technologies.

The MEMS-based pumped liquid cooling is an attractive thermal control device for future missions. It may be particularly beneficial for chip level applications as The working fluid in the cooling loop provides efficient coupling to the hot surface of the electronics, and the cooling loop provides flexibility in locating the heat sink inside the spacecraft. The cooling loop provides a simple mating to semiconductor surfaces through bonding techniques. MEMS cooling system can be easily integrated with the overall spacecraft thermal control system.

Future spacecraft used for deep space science exploration are expected to reduce in size by orders of magnitude. MEMS-based pumped liquid cooling will be useful in resolving many thermally induced problems.

9.4.6 MEMS STIRLING COOLER

Stirling cooling, an active thermal control method, is theoretically able to achieve the maximum efficiency in cooling. With a minimum of moving parts, a Stirling cooler consists of a hermetically sealed capsule and a small amount of gas as its working medium. A free piston Stirling cooler has a piston to compress the internal gas and a displacer to move the gas from the cold side to the hot side, where the heat is dissipated.

Stirling coolers have been applied in several space missions. The long-life Stirling coolers, either single-stage or two-stage, are available for cooling instrument detectors. The advancement of wavelength infrared and submillimeter imaging instruments for space applications demand further improvement in areas of vibration, electromagnetic interference, and temperature stability. A two-stage linear Stirling cycle cooler has been developed for use by instruments on several Earth Observing System (EOS) spacecraft. Stirling coolers will clearly be of use to many other NASA programs in Earth science, astronomy, microgravity sciences, interplanetary sciences, and the Human Exploration Initiative. These conventional coolers are designed to have long mission life, high reliability, and low vibration, as well as being small, light weight, and efficient. A typical cooler has a weight of about 15 kg.

As the spacecraft size reduced, the scaling in heat transfer is prominent and solving high power heat transfer problems within small spacecraft becomes rather difficult. The rapidly expanding capabilities of semiconductor processing in general, and microsystems packaging in particular, present a new opportunity to extend cooling to the MEMS domain. Several MEMS-based active cooling systems to support future small spacecraft missions have been suggested.

Nakajima and his team have demonstrated a micro Stirling cycle engine having a high thermal efficiency.[20] The tiny gas filled engine can be operated with any heat source and can be driven in reverse to make an active cooling system. Thermo-mechanical actuators generally require the removal of heat energy to reestablish the previous condition. Because heat dissipation is directly related to the volume to be cooled, thermal cycling occurs much faster in microdevices than in macrodevices.

In parallel, NASA Glenn Research Center (GRC) has developed a MEMS device for active cooling and temperature control.[21] This active cooling device is aimed for future nano- or microsatellite missions with predicted efficiencies that are an order of magnitude better than current and future thermoelectric coolers.[22] GRC's MEMS-based device uses a Stirling thermodynamic cycle to provide cooling or heating directly to a thermally loaded surface. The device can be used strictly in the cooling mode or can be switched between cooling and heating modes in milliseconds for precise temperature control. Fabrication and assembly employ techniques routinely used in the semiconductor processing industry. Benefits of the MEMS cooler include scalability to fractions of a millimeter, modularity for increased capacity and staging to low temperatures, simple interfaces, limited failure modes, and minimal induced vibration.

A working model of a MEMS cooler device has been assembled and tested at the JHU/APL for MEMS regenerator performance. This 1-by-1-cm regenerator was fabricated for NASA by Polar Technologies Commercial. Piezoelectric actuators (non-MEMS) are used to drive the compression and expansion diaphragms, which are the only moving parts of the device. The diaphragms are deflected toward and away from the regenerator region in phase-shifted sinusoidal fashion to produce the Stirling cycle.

NASA GSFC is developing a small, innovative instrument, LEISA, that will incorporate a miniature cooler. To perfect the cooler technology, GSFC works with commercial cooler vendors on long life, low vibration miniature coolers. The reduction of vibration is a significant objective for cooler technology because commercial coolers presently have unacceptably large vibration which can seriously disrupt sensor readings. GSFC has specified the changes required to allow a commercial cooler to be used with the existing GSFC vibration control system. The goal at GSFC is to develop a lightweight, low-cost cooler which will meet the requirement of small satellites.

9.4.7   ISSUES WITH A MEMS THERMAL CONTROL

While controlling the temperatures of other spacecraft components, MEMS-based TCS also need to be maintained at proper temperature range. Given their small size, low heat capacity, and low thermal conductivity heat transport paths, they can exceed their survival temperatures very easily in a short time.

MEMS thermal control devices need to be properly protected for physical damages. Sensors and actuators coming into contact with the environment must be protected against adverse affects, especially if the devices are subject to long-term reliability concerns. Careful procedures need to be considered during design and handling of MEMS TCS to prevent potential threat of humidity, contamination, and charging. Ground handling is as much of a concern as in-space operations.

Similar to other components on a spacecraft, MEMS thermal control devices are susceptible to space environment induced damages. Therefore, ground-based environmental tests must be conducted on the final flight design to verify survivability of MEMS thermal control devices. Based on mission conditions, a list of environmental test requirements will typically be established for the MEMS devices.

9.5 CONCLUSION

Spacecraft TCS use both passive and active thermal control devices to maintain spacecraft systems within allowable temperature ranges. Passive thermal control is the most commonly used, while active thermal control is employed to accommodate stringent temperature control requirements, high power dissipation, and to provide design flexibility. Passive thermal control devices do not contain either moving parts or fluids. Based on these distinct characteristics, MEMS TCS are categorized as mainly an active thermal control concept.

Future space missions require complex spacecraft design, operation scenarios, and flight configurations. The push for low-cost and short-assembly schedules increases the demand for nano- or microsatellites. MEMS thermal control devices may become a critical element in such applications as they offer some unique advantages, especially for small spacecraft. Although the development of MEMS-based technology is still in its infancy, the advancement in MEMS thermal control devices is moving forward.

REFERENCES

1. Birur, G., G. Siebes, and T.D. Swanson, Spacecraft thermal control, in *Encyclopedia of Physical Science and Technology*. 2002.
2. Incropera, F.P. and D.P. DeWitt, *Introduction to Heat Transfer*. Third ed. 1996, John Wiley and Sons, Inc., New York, NY.
3. Wingate, C.A., Spacecraft thermal control, in *Fundamentals of Space Systems*, Pisacane, V.L. and R.D. Moore, Editors, 1994, Oxford University Press, New York, NY, p. 443.
4. Biter, W., S. Oh, and S. Hess, Electrostatic switched radiator for space based thermal control, in *Space Technology and Applications International Forum — STAIF 2002*. 2002, Albuquerque, NM.

5. de Aragon, A.M., et al., Future satellite services, concepts and technologies. *European Space Agency Bulletin*, 1998 (95): 99–107.

6. Maluf, N., *An Introduction to Microelectromechanical Systems Engineering*. 2000, Artech House, Inc. Boston, MA.

7. Gardner, J.W., V.K. Varadan, and O.O. Awadelkarim, *Microsensors MEMS and Smart Devices*. 2001, John Wiley and Sons, Ltd, New York, NY.

8. Honeywell, MEMS thermal switch using mechanically actuated field effect transfer (MAFET) technology, in *Preliminary Brochure*. 2000.

9. Honeywell, *New Concept in Thermal Switch Technology*. 2002, http://www.thermal-switch.com/ts-mafet.shtml

10. Gilmore, D., *Spacecraft Thermal Control Handbook*. 2002, The Aerospace Corporation, El Segundo, CA. pp. 331–352.

11. Helvajien, H., S. Janson, and E.Y. Robinson, Big benefits from tiny technologies: micro-nanotechnology applications in future space systems, in *Advancement of Photonics for Space*, Taylor, E.W., Editor. 1997, SPIE, Bellingham, WA.

12. Douglas, D., T. Michalek, and T.D. Swanson, Design of the thermal control system for the space technology 5 microsatellite, in 31st International Conference on Environmental Systems, 2001.

13. Osiander, R., et al., Microelectromechanical devices for satellite thermal control. *IEEE Sensors Journal Microsensors and Microactuators: Technology and Applications*, 2004. 4(4) 525–531.

14. German, J., As microcircuits heat up, inexpensive Sandia substrate may keep tomorrow's chip cooler, in *Sandia LabNews*. 1998.

15. Sandia, Sandia Expands Envelope of MEMS Devices, in *AW&ST*. 2000.

16. Karam, R.D., Satellite thermal control for systems engineers, in *Progress in Astronautics and Aeronautics*, series vol. 1998, American Institute of Aeronautics and Astronautics, Inc., Reston, VA.

17. Gad-el-Hak, M., Editor, *The MEMS Handbook*. 2002, CRC Press LLC, Boca Raton, FL.

18. Pettigrew, K., et al., Performance of a MEMS based micro capillary pumped loop for chip-level temperature control, in 14th IEEE International Conference on Micro Electro Mechanical Systems (MEMS 2001), January 21–25 2001. 2001, Institute of Electrical and Electronics Engineers, Inc., Interlaken.

19. Birur, G.C., et al., Micro/nano spacecraft thermal control using a MEMS-based pumped liquid cooling system, in *SPIE*. 2001.

20. Nakajima, N., K. Ogawa, and I. Fujimasa, Study on micro engines — miniaturizing Stirling engines for actuators and heatpumps, in *Micro Electro Mechanical Systems: An Investigation of Micro Structures, Sensors, Actuators, Machines and Robots*, February 20–22 1989, 1989, Salt Lake City, UT, USA: IEEE, Piscataway, NJ, USA.

21. Moran, M.E., *Multidisciplinary Analysis of a Microsystem Device for Thermal Control*. 2002.

22. Moran, M.E., *Micro-Scale Avionics Thermal Management*. 2001.

10 Microsystems in Spacecraft Guidance, Navigation, and Control

Cornelius J. Dennehy and Robert Osiander

CONTENTS

10.1   INTRODUCTION

Since the launch of Sputnik (October 4, 1957) significant resources have been invested in the design and development of guidance, navigation, and control (GN&C) systems for aerospace vehicles and platforms. As a result, the extraordinary progress in these critically important systems has been used to measure, guide, stabilize, and control the trajectory, attitude, and appendages (i.e., steerable antennas, solar arrays, robotic arms, and pointable sensors) of Earth-orbiting satellites, interplanetary spacecraft and probes, space-based robots, planetary rovers, and related platforms.

A spacecraft's GN&C system is critical to executing the typical space mission operational functions such as orbital insertion, Sun acquisition, Earth acquisition, science target acquisition, pointing and tracking, orbital or trajectory Delta-V propulsive maneuvers, as well as the articulation of multiple platform appendages. No matter what the specific mission applications are, all spacecraft GN&C systems can be deconstructed into the three basic generic functional elements of an automatic feedback control system:

- Sensors
- Processors
- Actuators

Typically, in conventional spacecraft architectures being implemented today, various individual attitude sensor units (such as star trackers, Sun sensors, Earth sensors, horizon crossing sensors, magnetometers, rate gyros, accelerometers, etc.) are physically mounted at discrete locations on the spacecraft structure and electrically harnessed to the vehicle's command and data handling system (C&DH). The attitude measurement data generated by each individual sensor are sampled, at rates ranging from 1 to 100 Hz typically, by the spacecraft's on-board digital flight processor in which attitude determination algorithms compute an updated vehicle state vector. Control law algorithms, also resident on this on-board processor, will compute the necessary attitude control torques (and/or forces) required to achieve the desired attitude, orbit, or trajectory. Command signal outputs from the processor are then directed to the appropriate attitude control actuators to generate the commanded torques or forces on the vehicle. This attitude control is cyclically repeated at rates ranging from 1 to 10 Hz, or possibly faster if the time constants of the fundamental dynamics of the vehicle to be controlled are very short and high bandwidth control is required for stabilization.

In the almost 50 years since Sputnik, the global GN&C engineering community has established and flight-proven multiple methods for determining and controlling the orientation of spacecraft.[1-3] A GN&C engineer's choice between such basic control techniques as gravity gradient stabilization, spin stabilization, and full three-axis stabilization will depend primarily on the mission-unique drivers of orbit (or trajectory), payload pointing stability and accuracy requirements, spacecraft attitude and orbital maneuvering requirements and mission life.[4] Multiple opportunities exist to infuse microelectromechanical systems (MEMS) technology in many of these attitude control and stabilization techniques, particularly in the areas of advanced attitude control system sensors and actuators. Advanced MEMS-based processors for GN&C applications are also a possibility, but that specific area of MEMS R&D will not be discussed in this chapter.

TABLE 10.1
Typical Spacecraft GN&C Attitude Sensing and Control Devices

Attitude Sensing Devices	Attitude Control Actuation Devices
Sun sensors	Thrusters
Earth sensors	Momentum wheels
Horizon sensors	Reaction wheels
Magnetometers	Control moment gyros
Gyroscopes	Magnetic torquers
Accelerometers	Antenna pointing gimbals
Fine guidance sensors	Solar array drives

While GN&C engineering and technology development efforts are primarily directed towards both controlling launch vehicle (i.e., booster) dynamics during ascent and controlling space platform dynamics in the microgravity environment of free space, they also entail the navigational aspects of maintaining precise timing (and the associated time transfer and time synchronization functions). MEMS technology can certainly be applied to the development of miniaturized spacecraft clocks and oscillators for navigational functions. Table 10.1 defines the typical set of sensing and control devices typically used to perform spacecraft GN&C functions.

10.2 MINIATURIZED MODULAR GN&C SUBSYSTEMS FOR MICROSATELLITES

Several future science and exploration mission architectures share common interests and technological requirements for microsatellites. Some envision economically mass-produced microsatellites as a means to enable new robust, flexible, and responsive space architectures for Earth (or planetary) observation and coordinated space communications and navigation functions. Others foresee clusters of microsats as affordable and reconfigurable platforms for performing new types of *in situ* or remote sensing science measurements or observations.

Consequently, many industrial and federal R&D organizations are spearheading the development of the breakthrough subsystem and component technologies needed to implement next generation microsatellites. Using data from various flight projects and cost models, some researchers have investigated the relative costs of small satellite subsystems as a way to refine the identification of technologies, which are key to reducing overall spacecraft cost. One such analysis, performed by NASA's New Millennium Program (NMP), determined that the largest cost fractions were associated with both the electrical power subsystem, 34% of total cost, and the GN&C subsystem, 27% of total cost, with the other small satellite subsystems costs being significantly less.[5] A general observation can also be made that, excluding the payload, the GN&C and the C&DH, in the range of 25 to 30% of total power, are the largest relative power consuming subsystems on small satellites. The insight gained from these types of studies is that technologies which reduce both power and mass of the GN&C subsystem will perhaps have the greatest proportional potential to lower small spacecraft costs. Applying higher risk MEMS technologies to the relatively costly and power consuming GN&C subsystems of microsatellites, and other small-scale space platforms, is a technology thrust that has potential for high payoff.

Furthermore, beyond developing technologies that simply reduce mass and power, the community must also pursue in tandem the creation of architectures that are modular and based upon commonly applied standards. When contemplating the design of microsatellites to perform future science and exploration missions, many space mission architects, space system engineers, and subsystem engineers all share a common vision in which modular, adaptive and reconfigurable system technologies enable highly integrated space platform architectures.[6] In the GN&C arena the design of modular multifunction units is being investigated and researched, by both industry and the government. Such units would effectively coalesce multiple GN&C sensing and processing functions, and in some instances communications functions, into one single highly integrated, compact, low-power, and low-cost device. Clearly MEMS technology, along with other supporting avionics systems technologies, can be exploited to enable this type of miniature GN&C hardware.

Such a unit would simultaneously provide autonomous real time on-board attitude determination solutions and navigation solutions. This "GN&C in a box" device would operate as a single self-contained multifunction unit combining the functions now typically performed by a number of hardware units on a spacecraft platform. This approach, enabled by MEMS technology and advanced electronics packaging methods, will significantly reduce the number of electrical, computer data, and mechanical interfaces for the GN&C system, relative to current engineering practice, and should therefore payoff with dramatic reductions in costly and time-consuming prelaunch integration and test activities. However, recognizing the need to satisfy a variety of future mission requirements, design provisions could be included to permit the unit to interface with externally mounted sensors and actuators, as needed, to perform all necessary GN&C functions.

The desired result is a highly versatile unit that could be configured in multiple ways to suit a realm of science and exploration mission-specific GN&C requirements. Three specific examples of modular multifunction GN&C technology developments are described in this section: the JPL MicroNavigator unit, the GSFC Microsat Attitude and Navigation Electronics (MANE), and NASA's NMP Space Technology 6 (ST6) Inertial Stellar Camera (ISC) under development at Draper Laboratory. The common design philosophy in all three cases is to merge the GN&C sensing and data processing elements into a single unit by leveraging advanced MEMS miniaturization and electronics packaging technologies. The underlying shared goal is then to be in a position to mass produce these modular GN&C units so that the overall cost of a next generation microsat is more affordable, relative to current production techniques. The evolution and eventual infusion of these innovative miniaturized modular GN&C systems will rely heavily upon continued MEMS inertial sensor (gyroscopes and accelerometers) technology maturation, not so much to further reduce device mass and power, but to significantly improve accuracy and overall sensor performance.

10.2.1 JPL MICRONAVIGATOR

Miniature high-performance, low-mass, low-power space avionics are among the high-priority technology requirements for planetary exploration missions. The spacecraft fuel and mass requirements enabling orbit insertion is the driving requirement. The MicroNavigator is an integrated hardware and software system designed to satisfy the need of a miniaturized GN&C unit for navigation, attitude determination, vehicle attitude control, pointing, and precision landing.[7]

The MicroNavigator concept depends on MEMS technology. In particular, MEMS-based gyroscope and accelerometer inertial sensors were targeted for the MicroNavigator avionics package. Miniature celestial sensors such as active pixel sensor (APS) and miniaturized GPS sensors, were also identified as key technology elements of the MicroNavigator.

The MicroNavigator has a dedicated embedded processor to perform GN&C specific computations. A state estimator hosted on this internal processor optimally filters data from the MEMS inertial sensors (as well as other sensors). A high-resolution ($0.1°$ in attitude knowledge and 10–50 m position determination accuracy) vehicle state vector is output by the MicroNavigator potentially at cycle rates of less than 1 sec. Two obvious benefits are derived here at the system-level by virtue of using the MicroNavigator: (1) the spacecraft on-board flight computer (if there is even one) is not encumbered with the task of performing the computationally intense GN&C algorithm processing and (2) the GN&C algorithms embedded within the MicroNavigator are generally applicable to a wide variety of mission applications so that new flight software design and development is not required, thus, significantly lowering the cost of implementing GN&C functionality on a given spacecraft.

Resource requirement goals for the MicroNavigator are ambitious: a mass target of less than 0.5 kg, a volume of about 8 cubic inches, and a power requirement of less than 5 W.

10.2.2 GSFC MICROSAT ATTITUDE AND NAVIGATION ELECTRONICS

In a manner very similar to the MicroNavigator the MANE represents a revolutionary leap in the design and implementation of spacecraft GN&C subsystems. MANE is a single, highly integrated, space-efficient, low-power, affordable hardware or software design concept (targeted, but not limited to, microsat applications), which autonomously provides attitude determination and navigation solutions. The MANE would obviate the need for a separate GPS receiver unit, a separate GN&C processor, a separate inertial reference unit (IRU) and a separate set of attitude-control interface electronics. An embedded (card-mounted) three-axis MEMS gyroscope sub-assembly would replace the conventional IRU which is relatively large, heavy, and power consuming.

The MANE design concept was an outgrowth of earlier work on a multifunctional GN&C System (MFGS) performed at GSFC.[8] The mass and power resource requirement goals for the MFGS were 2.5 kg and 12 W, respectively. While the MFGS design represented substantial improvement in overall GN&C subsystem resource requirements, the MANE concept was developed to drive the MFGS design to the next level of miniaturization with an ultimate, long-term, high-risk goal of developing an ultra-miniature design that captures the MFGS performance in a volume of several cubic inches employing MEMS microsystems, advanced space avionics electronics packaging or assembly technologies together with the ultra low power (ULP) electronics technology being pioneered by the University of Idaho and GSFC.[9] The very space-efficient chip-on-board (COB) technology, pioneered by the Johns Hopkins University Applied Physics Laboratory (JHU/ APL), was identified as a viable initial technique to achieve the miniaturization goal for MANE. COB achieves up to $10\times$ higher circuit density by attaching bare die directly to the underlying board.

The MANE performance capabilities will largely depend on the individual mission requirements and the available set of navigation and attitude sensor data. The MANE design utilized a single reusable GN&C software system architecture for which the performance capabilities can be tailored for individual missions obviating the need for expensive new flight software design and development. Attitude determination performance goals for the MANE ranged between 0.1 and 0.3° without the external star sensor data and 1–2 arc-seconds with the external star sensor data. The MANE design goals were to achieve power consumption of less than 3 W, a unit mass of less than 1 kg in a total volume of less than 10 cubic inches.

10.2.3 NMP ST6 INERTIAL STELLAR CAMERA

NASA's NMP is sponsoring the development of the Inertial Stellar Compass (ISC) space avionics technology that combines solid-state MEMS inertial sensors (gyroscopes) with a wide field-of-view APS star camera in a compact, multifunctional package.[10] This technology development and maturation activity is being performed by the Charles Stark Draper Laboratory (CSDL), for a Space Technology 6 (ST6) flight-validation experiment now scheduled to fly in 2005. NMP missions such as ST6 ISC are intended to validate advanced technologies that have not yet flown in space in order to reduce the risk of their infusion in future NASA missions. The ISC technology is an outgrowth of earlier CSDL research focused in the areas of MEMS inertial device development,[11] MEMS-based GN&C sensors and actuators,[12] and low-power MEMS-based space avionic systems.[13]

The ISC, shown in Figure 10.1, is a miniature, low-power, stellar inertial attitude determination system that provides an accuracy of better than 0.1° (1-Sigma) in three axes while consuming only 3.5 W and packaged in a 2.5 kg housing.[14]

The ISC MEMS gyro assembly, as shown in Figure 10.2, incorporates CSDL's tuning fork gyro (TFG) sensors and mixed signal application specific integrated

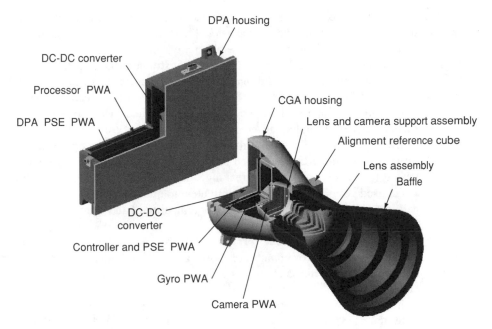

FIGURE 10.1 The NMP ST6 ISC. (*Source*: NASA JPL/CALTECH.)

circuit (ASIC) electronics designs. Inertial systems fabricated from similar MEMS gyro components have been used in PGM, autonomous vehicles, and other space-related mission applications. The silicon MEMS gyros sense angular rate by detecting the Coriolis effect on a sense mass. A sense mass is driven into oscillation by electrostatic motors. The mass oscillates in one axis and as the body is

FIGURE 10.2 NMP ST6 ISC MEMS three-axis gyro assembly. (*Source*: Charles Stark Draper Laboratory.)

rotated, Coriolis forces cause the sense mass to oscillate out of plane. This change is measured by capacitive plates and is proportional to the rotational rate of the body.

The MEMS three-axis gyro sub-assembly used in the SC is depicted in Figure 10.2. The specific MEMS inertial sensing instrument used in the ISC is the TFG14-R3, 20-μm thick gyro fabricated in a silicon-on-insulator process that incorporates novel features for high performance. Under typical operating conditions, the MEMS gyroscopes drive the ISC output attitude. The MEMS gyros sensed inertial rates are sampled, at the high sample frequency of 100 Hz, by the embedded flight processor. The raw gyro data are then processed using a Kalman filter algorithm to produce the estimated reference attitude quaternion, which is communicated to the host spacecraft in real time, at a frequency of 5 Hz. The APS star camera is used periodically (every few minutes) to obtain a camera quaternion, whose main purpose is to compensate the inherent drift of the gyros. A simple system data flow is shown in Figure 10.3.

A typical profile of attitude error, computed by simulation, is shown in Figure 10.4. The 1-sigma error bounds are shown in bold, while the actual attitude error from one simulated run is shown as a thin line. Since the error bounds are 1-sigma, the error can be expected to go outside of the bound for 32% of the time. Every 5 min, the gyros are compensated with a fresh star camera quaternion, as evidenced by the sudden narrowing of the error bounds. Used together as a tightly integrated sensor suite, the MEMS gyros and star camera enhance each other's

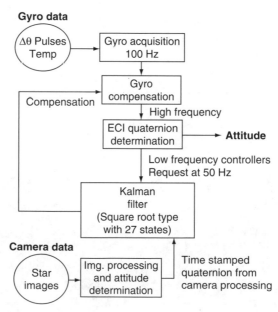

FIGURE 10.3 NMP ST6 ISC attitude determination system data flow. (*Source*: NASA CALTECH/JPL.)

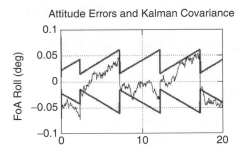

FIGURE 10.4 Typical ISC single-axis attitude error profile.

capabilities, resulting in a more robust attitude determination system than could be achieved by integrating separate star tracker and gyro units.

The ISC technology, enabled by embedded MEMS gyroscopes, is a precursor of things to come in the spacecraft avionics arena as much more highly integrated, lower power, MFGS are developed in the future. There are a wide range of science and exploration mission applications that would benefit from the infusion of the compact, low-power ISC technology. Some envisioned applications include using the ISC as a "single sensor" solution for attitude determination on medium performance spacecraft, as a "bolt on" independent safehold sensor for any spacecraft, or as an acquisition sensor for rendezvous applications. It has been estimated that approximately 1.5 kg of mass and 26 W of power can be saved by employing a single MEMS-based attitude sensor such as the ISC to replace the separate and distinct star tracker and IRUs typically used on spacecraft.[14] So in this case, MEMS is an enhancing technology that serves to free up precious spacecraft resources. For example, the mass savings afforded by using the MEMS-based ISC could be allocated for additional propellant or, likewise, the power savings could potentially be directly applied to the mission payload. Also worth noting is the fact that the significantly low ISC power consumption will have a positive secondary benefit of reducing the size and cost of the host spacecraft electrical power subsystem. These are some of the advantages afforded by using MEMS technology for GN&C applications.

10.3 MEMS ATTITUDE MEASUREMENT SENSORS

An attitude measurement is the measurement of any quantity sensitive to the attitude of the spacecraft, for example, the magnetic field vector, the direction of the Sun, a star, or some other body, the measurement of an angle such as the solar aspect or the limb of a planetary body, or the measurement of integrated angular rates. The latter is very different since it does not provide absolute attitude information. By the resolution they provide, the attitude sensors can be divided into two groups, coarse sensors such as magnetometers, sun sensors, and Earth horizon sensors, and fine sensors such as fine sun sensors and star sensors. Attitude determination using the global positioning systems (GPS) or similar systems will not be discussed here.

The MEMS technologies used for these systems are similar to those discussed in Chapter 8.

In general, there are several design considerations that must be considered in the design of spacecraft attitude sensors. Chief among these are the specific nature of the control system application. The constraints associated with predicted environmental conditions such as the prelaunch handling, launch loads (mechanical vibration or shock as well as acoustic exposure), pressure venting profiles, on-orbit operating temperatures, particle contamination, EMI or EMC effects, and radiation exposure (both the total dose and heavy ions) must be well understood and documented prior to the detailed design phase of the sensor.

Other system level, but no less important, considerations come into play with spacecraft attitude sensors such as the specific placement and orientation of the device on the spacecraft or platform structure to be controlled. Inadequate attention to these details, especially on very lightweight highly flexible structures, can lead to destabilizing (and, in extreme cases, possibly destructive) controls–structures interaction (CSI) problems for the GN&C designer.

The imminent introduction of the MEMS-based GN&C sensor technology into the spacecraft designer's inventory will herald a breakthrough in how the function of medium-to-high accuracy attitude determination will be implemented in future space missions.

10.3.1 MEMS MAGNETOMETERS

MEMS magnetometers have already been discussed in Chapter 7, Microtechnologies for Science Instrumentation Applications. A magnetometer measures the three components of the magnetic field and provides a measurement of the attitude relative to inertial coordinates. Since only the direction of the magnetic field is sensitive to the attitude, another vector measurement such as a sun sensor is required for attitude determination. For magnetometers, the largest component of the random noise for attitude determination arises not from the sensor itself, but from the magnetic field model, which, for LEO orbits, can cause an error of 0.5° at the equator, and up to 3° near the magnetic poles. Therefore, the sensitivity requirements for magnetometers as an attitude sensor are relatively weak and provide an opportunity for insertion of MEMS devices. The performance requirements for attitude determination magnetometers are a range of about ± 60 μT, with a sensitivity of ± 10 nT.

A number of miniature magnetometer developments have occurred in recent years. For the SUNSAT-1 satellite, the magnetic observatory at Hermanus manufactured a miniature fluxgate magnetometer with this performance at a size of about 130 mm $\times$ 90 mm $\times$ 36 mm and a weight of 295 g.

The University of California, Los Angeles, has developed a miniature fluxgate magnetometer for NASA's NMP ST5 small satellite mission. The magnetometer mass and power is kept low with a dual core series drive circuit. The magnetometer has two commendable ranges, 64,000 and 1000 nT. The dynamic range is changed from 64,000 to 1000 nT by altering the closed loop response from 64,000 to 5000 nT, and then amplifying the signal to get to a 1000 nT range. This method keeps the noise low in both ranges. One gain change command line switches both elements. Total mass of the ST5 magnetometer is approximately 600 g and it consumes approximately 0.55 W of power.

The magnetometers are calibrated in-orbit to correct the prelaunch gain and offset parameters. Misalignment of the orthogonal pickup coils and sensor mounting errors are also determined once in orbit. Examples of possible MEMS magnetometers are based on Lorentz force using resonating bars and membranes.[15–17]

10.3.2   MEMS SUN SENSORS

A sun sensor determines the vector direction of the Sun, and can be either a coarse or even a very fine attitude sensor. Many sun sensors rely, much like a sun dial, on the shadowing effects of some masks. When reducing the size of the masks to MEMS dimensions, problems arise due to diffraction as well as the reduced angular deflection at these small dimensions. Two categories of conventional sun sensors exist — digital and analog types. The digital sun sensors illuminate a geometric pattern on the detector plane. The presence or absence of light in these well-defined areas defines a digital signal that can be translated into the sun angle. An analog sun sensor outputs analog currents, from which the sun angles can be derived. This simple approach of the digital sun sensor, where the mask design creates a digital read out of the sun position, will no longer work. A typical approach to reduce the dimensions of a sun sensor is to use an imager and determine the centroid of a shadow pattern generated by a mask. The mask can be produced using micromachining technology, which, if it can be inserted into the same process steps, could increase the accuracy and reduce handling when producing such a sensor. One example of such a sun sensor has been produced at JPL.[18–20] This micro sun sensor is essentially a pinhole camera with multiple holes, and is comprised of a silicon wafer mask with several hundred small apertures placed on top of a charge coupled device (CCD) focal plane array at a distance of 750 μm. An image of the apertures is formed on the focal plane when the Sun illuminates this setup. Sun angles can be derived by analyzing the image. The experimental data presented indicate that this sun sensor can achieve accuracies in the order of a few arcminutes or better. It is projected that this type of sun sensor will be the size of three dimes stacked on top of each other. It will have a mass of less than 30 g and consume less than 20 mW.

10.3.3   EARTH SENSORS

Earth horizon sensors use the Earth's horizon to determine spacecraft attitude. In LEO, they concentrate on merely telling which direction is down; in geosynchronous earth orbit (GEO), they focus on the actual horizon and yield more accurate attitude measurements. Since they are typically based on an IR detector, bolometers, and uncooled imagers based on MEMS fabrication technology, as described in Chapter 7 under spacecraft instrumentation, can be effectively used. An example of such a device is the Micro Infrared Earth Sensor (MIRES) developed at LAAS-CNRS in France.[21–23] It uses an uncooled 320 × 240 infrared sensor array with a noise-equivalent temperature (NET) of 100 mK and smart processing to measure the position of the horizon.

10.3.4 STAR TRACKERS

Star sensors are very similar to sun sensors. Star cameras are star sensors that sense several stars at once. Recent developments in CCDs have reduced power requirements for these considerably, making them more practical. They are very accurate. A system involving MEMS mirrors has been developed with the Air Force Research Laboratory (AFRL) together with NASA Langley Research Center in form of the intelligent star tracker (IntelliStar).[24] It uses several novel technologies including silicon carbide housing, MEMS adaptive optics, smart active pixels, and algebraic coding theory. In addition to being lightweight, it also offers advantages of speed, size, power consumption, and radiation tolerance. The MEMS-adaptive optics, utilizing MEMS mirrors developed at AFRL, and fabricated with Sandia's SUM-MiT V process (Chapter 3), compensate for geometrical aberrations and effects, and allow the imager to match star patterns easier and faster. Research on miniature MEMS star sensors has also been performed at JPL.[25,26]

10.4 MEMS INERTIAL MEASUREMENT SENSORS

Gyroscopes (also commonly referred to as "gyros") and accelerometers are the building blocks from which most spacecraft GN&C systems are built. They are called inertial sensors since their operation takes advantage of an object's resistance to change momentum, or simply put, its inertia. Gyros have been used in space mission applications for many decades and there is a rich body of technical literature concerning the theory and practical operation of gyro instrumentation.[27]

The technology of inertial sensors, first developed in the 1920s, has continually evolved in response to the demands of the users. In the beginning the trend was to maintain the same basic designs while pushing the technology for sensor-level components (e.g., electronics, bearings, suspensions, motors, etc.) to achieve improvements in sensor performance and operational reliability. Significant increases in inertial system accuracy and reliability accomplished over this time period directly led to the successes in autonomous submarine navigation, the Apollo missions, and the ubiquitous infusion of inertial navigation on commercial aircraft. Since the 1970s or thereabout, performance plateaued and the emphasis shifted from refining the technology to achieving equivalent high performance at reduced cost. Over the past 20 years or so, MEMS technology breakthroughs have been exploited to create innovative microsystem solution for applications not previously considered feasible for inertial sensing. These emerging MEMS-based inertial sensor technologies offer little performance improvement, but provide benefits of low production and life-cycle costs, miniature size, low mass and power consumption, and are enabling for microsatellites.[28]

In certain highly maneuverable spacecraft, or propulsive upper stage applications, a three-axis gyro sensor complement for rotational sensing is combined with a three-axis set of accelerometers for translational sensing to implement a full six degree-of-freedom (6-DOF) inertial measurement unit (IMU). In navigation and flight-control systems, an IMU is used to measure angular rates and translational accelerations about three orthogonal axes of the spacecraft: the roll, pitch, and yaw. Depending on the mission applications, IMU's may have 4-for-3 gyro and accelerometer redundancy. In other attitude control systems a three-axis gyro sensor configuration alone, forming an inertial reference unit (IRU), is employed on spacecraft.

The technologies commonly used in today's IRUs include high-performance mechanical (spinning mass) gyros such as those used on the Hubble Space Telescope, Ring Laser Gyros (RLGs), Fiber Optic Gyros (FOGs) and HRGs. Three-axis IRU packages based upon these gyro technologies are the mainstay of spacecraft GN&C systems. One such IRU that has been used on a wide range of LEO, GEO and deep-space mission applications has a mass of approximately 4.5 kg and typically requires over 20 W of power to operate. Another representative IRU used on a large space platform had mass of 5 kg and consumed about 18 W of power. Consequently, these types of conventional IRU will not be amenable to microsatellite (and other mission) applications where mass and power are at a premium.

MEMS inertial sensors are therefore an attractive technology option to pursue for future microsatellite missions, and other science or exploration applications such as probes, rovers, robots and the like, where available mass and power resources are severely constrained. Microsatellite designers and developers can leverage the considerable R&D funding that has already been invested by the Department of Defense (DoD) in the development of MEMS inertial sensor technologies. The primary mission applications of these investments have been precision-guided munitions (PGMs) and unmanned robotic vehicles. In both these military applications, the MEMS-based IMUs have supplanted competing technologies (e.g., RLGs or FOGs) by virtue of their miniature size, cost, and mechanical robustness. In the case of the extended range guided munition (ERGM) the MEMS-based IMU is coupled with a GPS receiver to create a highly compact, very accurate, and jamming-resistant GPS/INS navigation system for a 5-in. artillery shell.[29]

As mentioned above, the overwhelming majority of the technology investment to date has been focused on consumer class and tactical class MEMS gyros, not MEMS gyros intended for space applications. This legacy of nonspace MEMS gyro R&D work has been extensively reviewed and reported on in the literature and will not be discussed in detail here.[30,31] While there has been a considerable R&D investment in MEMS gyros for military and commercial applications since the 1980s, it is only recently that the development of navigation class MEMS gyros (with bias stability performance in the range of 0.002 to 0.01°/h) specifically designed for space mission applications has grown at a number of R&D organizations.

Several MEMS inertial sensor technology developments specifically targeted for space mission are underway at multiple organizations. MEMS sensors in the space environment have also undergone limited testing and evaluation.[32] These developments obviously build upon the solid MEMS technology foundation already formed within industry for defense and commercial applications. MEMS microsystems are currently at a point where their inherent robustness, miniature size, and low-power and low-mass attributes make them extremely attractive to spacecraft GN&C designers. Several key issues, however, remain to be resolved before MEMS inertial sensors will displace the current family of flight-proven gyro and accelerometer technologies. When one considers the demanding GN&C requirements for most space missions it becomes apparent that a MEMS gyro, with performance and reliability characteristics suitable for guiding the relatively short-duration flight of a PGM, may not be a realistic alternative. In general, significant improvements in the standard performance metrics (drift, scale factor, etc.) of the current generation of MEMS inertial sensors must be accomplished in tandem with the ability to rigorously demonstrate the reliability specifications for space flight.

It is encouraging to observe that the majority of industrial inertial system vendors are either currently offering or actively developing MEMS-based IMUs.[33] Based upon this trend, and if current R&D investment remains stable or increases, robust and reliable higher performing space qualified MEMS-based inertial systems will be commonly available as COTS products within the next 5 to 10 years.

10.4.1 MEMS GYROSCOPES

Gyro inertial sensors are perhaps the most fundamental component of a spacecraft GN&C system. Gyroscopes or angular rate sensors are used to measure the rotation angles and rates between the axis system of a moving-body and a fixed body. Gyroscopes are stabilized by their spin and resultant angular momentum. If applied torque is zero, then angular momentum is conserved. This means that an undisturbed gyro will point in the same direction in inertial space. Hence, a stable platform is available to reference attitude. It is rare to see a spacecraft GN&C system that does not include some form of gyro instrument used to provide attitude and rate measurements for vehicle stabilization and orientation.

MEMS inertial sensors have certainly found a niche in the commercial sector; solid state silicon gyros are currently being incorporated into automotive antirollover and side airbag deployment systems, used for low-cost attitude heading reference system (AHRS) avionics for general aviation airplanes, and used for the stabilization of such platforms as the Segway® Human Transporter (HT) (Segway LLC, Bedford, NH). All signs point to a continued growth in the innovative application of MEMS inertial sensors for these nonspace product lines.

Inertial sensors have traditionally been classified or grouped as a function of their performance metrics. The accuracy of a gyro is largely determined by its bias stability or drift rate, its angle random walk (ARW), and its scale factor stability.

Other performance parameters such as angular rate sensing range and dynamic bandwidth also are used to characterize and classify gyros. There are typically four classes of gyros, in order of decreasing accuracy; precision (or strategic) class, navigation class, tactical class, and consumer class. The overwhelming majority of MEMS gyro R&D activities to date have been focused on gyros in either the tactical performance class having bias stabilities in the range of 1 to $10°/h$ or in the consumer class where bias stability may be in the range of 100 to $1000°/h$ or even greater.

With the goal of developing navigation grade MEMS gyroscopes, DARPA has invested in a number of programs, and has dramatically propelled MEMS inertial sensor technology for DoD applications. Realizing its importance for space applications, NASA, and especially JPL, has invested in the MEMS gyroscope technology for space applications.[34,35]

JPL has been developing a miniature single-axis vibratory, Coriolis force MEMS gyro, over the past several years.[36] A photograph of the JPL post resonator gyroscope (PRG) MEMS gyro can be seen in Figure 10.5. It employs a "cloverleaf" planar resonator. In this design the coupling is measured between orthogonal modes of a four-leaf clover resonator with a proof mass (the post) in the center caused by the Coriolis force.[34] The layout of the device takes the shape of a "cloverleaf" with two drive electrodes and two sense electrodes located at the quadrants (one electrode per quadrant). A relatively large post is rigidly attached to the center of the cloverleaf device formed by the four electrodes.

FIGURE 10.5 The JPL vibrating post micromachined MEMS gyro. (*Source*: NASA CALTECH/JPL.)

Excitation of the microgyro dynamics is achieved by applying a potential to the two drive electrodes. The drive electrodes and sense electrodes are suspended by silicon springs above matching electrodes on the base plate. The post adds inertia to the system which boosts the sensitivity to rotational motion. The electrical potential between the drive electrodes and their respective base plate electrodes creates an electrostatic force that, ideally, rocks the cloverleaf assembly about the y-axis. The amplitude of the rocking motion can be maximized by driving the electrodes at the natural frequency of this DoF, known as the *drive mode*. If the device is rotated about the z-axis, then the rocking about the y-axis is coupled into rocking about the x-axis via Coriolis acceleration in the x–y frame fixed to the gyro. The rocking about the x-axis is referred to as the *sense mode* and the x-axis response is related to the angular rate of rotation about z. The operating principles of the PRG microgyro, fabrication details, and preliminary performance results have been extensively documented.

Other gyroscopes, such as the ones designed by BEI Sensor and Systems Company, use tuning fork designs, where the result of the Coriolis force is a vibration mode orthogonal to the standard in plane tuning fork mode.[30,37]

Also, as previously discussed, the Draper Laboratory MEMS TFG design has been optimized for infusion in the NASA NMP ST6 ISC flight hardware. Similar to the JPL PRG, the principle of operation for the Draper Laboratory MEMS TFG is fundamentally based upon the Coriolis force. The resonant structure is composed of two proof masses which are each driven electrostatically with opposite oscillatory phases. Alternating voltages applied to the outer motor drive electrodes create electrostatic forces between the interlocking tines of the motor electrode and proof mass, which results in lateral (in the plane of the wafer) oscillatory motion. The proof masses are driven in a tuning-fork resonance mode. In response to an angular rate, Ω, being applied about the input axis, perpendicular to the velocity vector of the masses, a Coriolis acceleration is produced which forces the masses to translate in and out of the plane of oscillation. This resultant antiparallel, out-of-plane motion is measured via the capacitive pick-off, providing an output signal proportional to the rate input. Closed-loop control is employed to maintain the proof mass at constant amplitude and the rate sensing is conducted in an open loop manner.

The successful operation of this device depends on the electronics that controls the mechanism motion and senses the rate output. Each gyro axis requires an analog ASIC and a supporting field programmable gate array (FPGA), both on ball grid arrays. The gyro electronics requires only power and needs no direction from the microprocessor board except for requests for information. Gyro rate information is currently sampled at a fixed rate of 600 Hz, and the resultant information is communicated digitally through a low-voltage differential signaling (LVDS) interface to the microprocessor. The gyro electronics and the packaged gyro sensors are placed on printed wiring boards and can be assembled by standard pick and place assembly equipment.

10.4.2   A MEMS GYRO APPLICATION EXAMPLE: THE NASA/JSC AERCAM SYSTEM

One very timely application of MEMS gyro is on the miniature autonomous extravehicular robotic camera (Mini-AERCam) free-flying robotic inspection vehicle being developed by the Engineering Directorate at NASA's Johnson Space Flight Center (JSC). The Mini-AERcam system, shown in Figure 10.8, is being developed to satisfy remote viewing and inspection needs foreseen for future human space flight missions on the Space Transportation System (STS) Shuttle and the International Space Station (ISS). The Mini AERCam will provide unique free-flying video imaging of assembly, maintenance, and servicing tasks that cannot be obtained from fixed cameras, cameras on robotic manipulators, or cameras carried by EVA crewmembers. On ISS, for example, Mini-AERCam could be used for supporting robotic arm operations by supplying orthogonal views to the robot operator, for supporting crew spacewalk operations by supplying views to the ground crews monitoring the spacewalk, and for carrying out independent visual inspections of areas of interest around the ISS.[38]

Representing a significant technology breakthrough in the field of free-flying robotic space vehicles, the nanosatellite-class spherical Mini-AERCam free flyer is 7.5 in. in diameter and weighs approximately 10 lb. Advanced miniaturized avionics and instrumentation technology, together with compact mechanical packaging techniques, permit the Mini-AERCam to incorporate many additional capabilities compared to the 35 lb, 14 in. AERCam Sprint free flyer that flew as a remotely piloted shuttle flight experiment in 1997. Where the Sprint AERCam used quartz rate sensors, the Mini-AERcam GN&C hardware complement includes higher performance MEMS gyros for measuring vehicle angular rates. The Mini-AERCam attitude control system uses angular rate measurements from a three-axis Draper Laboratory developed MEMS TFG-14 gyros to estimate inertial attitude and attitude rate. The

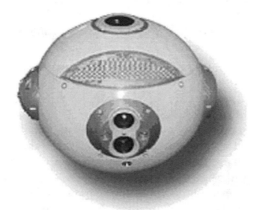

FIGURE 10.6 The NASA/JSC mini-AERCam free-flying robotic inspection vehicle. (*Source*: NASA.)

MEMS gyro package outputs digital data at a rate of 300 Hz, which is averaged down to a 25 Hz rate. The body axis roll, pitch, and yaw attitude rate measurements are converted to quaternions and then integrated to maintain an estimate of inertial attitude. The MEMS gyros are used to support the autonomous attitude determination and control functions in automatic stationkeeping, point-to-point maneuvering, and automated docking operational modes. In those modes where relative attitude estimates of the Mini-AERCam (with respect to the Shuttle, the ISS, etc.) the MEMS gyro-based inertial attitude reference is transformed to other reference frames.

10.4.3 MEMS ACCELEROMETERS

The development of MEMS accelerometers has been driven by the demand of the automobile industry for an inexpensive accelerometer as an airbag sensor. Analog devices has very successfully integrated CMOS electronics with a MEMS accelerometer in their iMEMS process design to make such devices available inexpensively.[39,40] The noise levels on the most recent models from analog devices airbag crash sensors ADI arc in the order of 100 micro/$Hz^{1/2}$ on a 2.5 mm^2 area.[41] The latest devices are built using silicon on insulator (SOI) MEMS technology, which allows for larger proof masses than surface micromachining in a single-crystal silicon layer.

In future designs, given the miniscule additional power, mass, and volume requirements imposed by a three-axis MEMS accelerometer package, it would be very reasonable to simply integrate the accelerometers, electrically and mechanically, with a three-axis gyros to create a full 6-DOF spacecraft IMU. Accelerometers could be used for navigating and perhaps also for such functions as thruster calibration, drag force measurement, monitoring launch environments g-loads, or, due to the small size, they could be placed anywhere on the spacecraft such as along a boom to measure vibrations. The Aerospace Corporation has developed a very compact triaxial accelerometer with the capability of measuring the vibration direction at specific points.[42] Similar systems have also been developed at the CSDL, with Brownian-limited noise floors at 1.0 g/pHz; orders of magnitude more sensitive than state-of-the-art surface micromachined devices such as the industry standard ADXL05.[43]

10.5 MEMS ATTITUDE CONTROL DEVICES

There are two fundamental ways to control spacecraft attitude, either by applying torques on the external via propulsion, or by changing the angular momentum with reaction wheels. Of course, any action, the spring-supported launch of a probe, or the start of a motor, will cause a change in the attitude if not compensated. Chapter 11 deals with micropropulsion, and therefore these systems will not be discussed here. It is important to know, however, that the specific impulse of some of the micropropulsion systems is enough, and especially, can be controlled to a fine enough thrust, to provide means of attitude control for even larger spacecraft.

Reaction wheels use electric motors to torque against high-inertia rotors or "wheels." When the motor exerts a torque on the wheel, an equal and opposite reaction torque is applied to the spacecraft. Reaction wheels are typically operated in a bi-directional manner to provide control torque about a single spacecraft axis. The inherently small inertia of a typical MEMS device will make them less efficient as a reaction wheel type actuator, and can only be compensated by extremely high speeds, which challenges the reliability requirements for such devices.

Microwheels for attitude control and energy storage have been suggested and designed by Honeywell.[44] They project a performance of a momentum density of 9 N m sec/kg and an energy storage of 14 W h/kg for a wheel of 100 mm diameter micromachined in a stack of silicon wafers. The advantages of microwheels increase further when the device is incorporated in the satellite's structure.

Likewise, Draper Laboratory has studied both the adaptation of a wafer spinning mass gyro and an innovative wafer-sized momentum wheel design concept (using hemispherical gas bearings) as attitude control actuators for a 1 kg nanosatellite application.[12]

A similar system, based on high-temperature superconductor (HTS) bearings, was suggested by E. Lee. It has an energy storage capacity of about 45 W h/kg, and could provide slewing rates in the order of 25°/sec for nanosatellites of 10 kg with 40 cm diameter.[45]

10.6 ADVANCED GN&C APPLICATIONS FOR MEMS TECHNOLOGY

It is fair to speculate that the success of future science and exploration missions will be critically dependent on the development, validation, and infusion of MEMS-based spacecraft GN&C avionics that are not only highly integrated, power efficient, and minimally packaged but also flexible and versatile enough to satisfy multimission requirements. Many low-TRL GN&C MEMS R&D projects are underway and others are being contemplated. In this section several ideas and concepts are presented for advanced MEMS-based GN&C R&D.

10.6.1 MEMS ATOM INTERFEROMETERS FOR INERTIAL SENSING

Atom interferometer inertial force sensors are currently being developed at several R&D organizations.[46–51] This emerging technology is based upon the manipulation of ultracold atoms of elements such as rubidium. The cold atoms (i.e., atoms which are a millionth of a degree above absolute zero) are created and trapped using a laser. These sensors use MEMS microfabricated structures to exploit the de Broglie effect. These high sensitivity sensors potentially offer unprecedented rotational or translational acceleration and gravity gradient measurement performance. Continued R&D investment to develop and test instrument prototypes to mature the

TRL of these MEMS-based atom interferometers could lead to the entirely new types of GN&C sensors.

10.6.2   MINIATURIZED GN&C SENSORS AND ACTUATORS

Generally speaking, the envisioned science and exploration mission challenges that lie ahead will drive the need for a broad array of modular building block GN&C devices. Both sensors and actuators with enhanced capabilities and performance, as well as reduced cost, mass, power, volume, and reduced complexity for all spacecraft GN&C system elements will be needed.

A great deal of R&D will be necessary to achieve significant improvements in sensor performance and operational reliability. Emphasis should be placed on moving the MEMS gyro performance beyond current tactical class towards navigation class performance. It is anticipated that some degree of performance improvements can be directly attained by simply scaling down the tactical (guided munitions) gyro angular rate range, dynamic bandwidth and operational temperature requirements to be consistent with the more modest requirements for typical spacecraft GN&C applications. For example, a typical spacecraft gyro application might only require a rate sensing range of $\pm 10°$/sec (as against a ± 1000/sec for a PGM application) and only a 10 Hz bandwidth (as opposed to a PGM bandwidth requirement of perhaps 100 Hz bandwidth). Other specific technology development thrusts for improving MEMS gyro performance could include both larger and thicker proof masses as well as enhanced low-noise digital sense and control electronics. Investigating methods and approaches for decoupling the MEMS gyro drive function from the sensing or readout function might serve to lower gyro noise.

One promising future research area could be the application of MEMS (perhaps together with emerging nanotechnology breakthroughs) to innovate nontraditional multifunctional GN&C sensors and actuators. In the latter case, the development of an array of hundreds of ultrahigh-speed (e.g., several hundred thousand revolutions per minute) miniature MEMS momentum wheels, each individually addressable, may be an attractive form of implementing nanosatellite attitude control. Building upon the initial work on the JPL MicroNavigator and the GSFC MFGS, another high-risk or high-payoff R&D area would be miniaturized into highly integrated GN&C systems that process and fuse information from multiple sensors. The combination of the continuing miniaturization of GPS receiver hardware together with MEMS-based IMU's, with other reference sensors as well, could yield low-power, low-mass, and highly autonomous systems for performing spacecraft navigation, attitude, and timing functions. Of particular interest to some mission architects is the development of novel MEMS-based techniques to autonomous sensing and navigation of multiple distributed space platforms that fly in controlled formations and rendezvous.

10.6.3   MEMS-BASED SENSITIVE SKIN FOR ROBOTIC SYSTEM CONTROL

Future robotic systems will need hardware at all points in their structure to continuously sense the situationally dynamic environment. They will use this sensed information to react appropriately to changes in their environment as they operate and maneuver in space and on lunar or planetary surfaces. Sensitive multisensor "skins" embedded with significant diagnostic resources such as pressure, stress, strain, temperature, visible or infrared imagery, and orientation sensors could be fabricated using MEMS technology for robotic control systems. A variety of sensing mechanisms reacting to temperature, force, pressure, light, etc. could be built into the outermost layer of robotically controlled arms and members. This MEMS-based sensitive skin would provide feedback to an associated data processor. The processor would in turn perform situational analyses to determine the remedial control action to be taken for survival in unstructured environments. This is one of the uses of the multisenson skin envisioned for future science and exploration missions. Modest R&D investments could be made to design and develop a working hardware robotic MEMS-based sensitive skin prototype within 5 years.

10.6.4   MODULAR MEMS-ENABLED INDEPENDENT SAFE HOLD SENSOR UNIT

Identifying and implementing simple, reliable, independent, and affordable (in terms of cost, mass, and power) methods for autonomous satellite safing and protection has long been a significant challenge for spacecraft designers. When spacecraft anomalies or emergencies occur, it is often necessary to transition the GN&C system into a safe-hold mode to simply maintain the power of the vehicle as positive and its thermally benign orientation with respect to the Sun. One potential solution that could contribute to solving this complex problem is the use of a small, low mass, low power, completely independent "bolt on" safe hold sensor unit (SHSU) that would contain a 6-DOF MEMS IMU together with MEMS sun and horizon sensors. Specific implementations would vary, but, in general, it entails one or more of the SHSUs being mounted on a one-of-a-kind observatory such as the JWST to investigate the risk of mission loss for a relatively small cost. ISC represents an enhancing technology in this application. The low mass and small volume of the SHSU precludes any major accommodation issues on a large observatory. The modest SHSU attitude determination performance requirements, which would be in the order of degrees for safe hold operation, could easily be met with current MEMS technology. The outputs of the individual SHSU sensors would be combined and filtered using an embedded processor to estimate the vehicle's attitude state. Furthermore, depending on their size and complexity it might also be possible to host the associated safe hold control laws, as well as some elements of failure detection and correction (FDC) logic, on the SHSU's internal processor. It is envisioned that such an SHSU could have very broad mission applicability across many mission types and classes, but R&D investment is required for system design and integration, MEMS sensor selection and packaging, attitude determination algorithm development, and qualification testing would require an R&D investment.

10.6.5   PRECISION TELESCOPE POINTING

Little attention has been paid to applying MEMS sensors to the problem of precision telescope stabilization and pointing. This is primarily due to the performance limitation of the majority of current MEMS inertial sensors. However as the technology pushes towards developing higher performing (navigation class) MEMS gyros, accelerometer designers could revisit the application of MEMS technology to the dynamically challenging requirements for telescope pointing control and jitter suppression. GN&C technology development investments will be required in many sub-areas to satisfy anticipated future telescope pointing needs. Over the next 5–10 years, integrated teams of GN&C engineers and MEMS technologists could evaluate, develop, and test MEMS-based approaches for fine guidance sensors, inertial sensors, fine resolution and high bandwidth actuators, image stabilization, wavefront sensing and control, and vibration or jitter sensing and control. It could be potentially very fruitful to research how MEMS technologies could be brought to bear on this class of dynamics control problem.

10.7 CONCLUSION

The use of MEMS microsystems for space mission applications has the potential to completely change the design and development of future spacecraft GN&C systems. Their low cost, mass, power, and size volume, and mass producibility make MEMS GN&C sensors ideal for science and exploration missions that place a premium on increased performance and functionality in smaller and less expensive modular building block elements.

The developers of future spacecraft GN&C systems are well poised to take advantage of the MEMS technology for such functions as navigation and attitude determination and control. Microsatellite developers clearly can leverage off the significant R&D investments in MEMS technology for defense and commercial applications, particularly in the area of gyroscope and accelerometer inertial sensors. We are poised for a GN&C system built with MEMS microsystems that potentially will have mass, power, volume, and cost benefits.

Several issues remain to be resolved to satisfy the demanding performance and environmental requirements of space missions, but it appears that the already widespread availability and accelerating proliferation of this technology will drive future GN&C developers to evaluate design options where MEMS can be effectively infused to enhance current designs or perhaps enable completely new mission opportunities. Attaining navigational class sensor performance in the harsh space radiation environment remains a challenge for MEMS inertial sensor developers. This should be a clearly identified element of well-structured technology investment portfolio and should be funded accordingly.

In the foreseeable future, MEMS technology will serve to enable fundamental GN&C capabilities without which certain mission-level objectives cannot be met. The implementation of constellations of affordable microsatellites with MEMS-enabled GN&C systems is an example of this. It is also envisioned that MEMS can be an enhancing technology for GN&C that significantly reduces cost to such a degree that they improve the overall performance, reliability, and risk posture of missions in ways that would otherwise be economically impossible. An example of this is the use of MEMS sensors for an independent safehold unit (as discussed above in Section 10.3) that has widespread mission applicability.

Future NASA Science and Exploration missions will strongly rely upon multiple GN&C technological advances. Of particular interest are highly innovative GN&C technologies that will enable scientists as well as robotic and human explorers to implement new operational concepts exploiting new vantage points; develop new types of spacecraft and platforms, observational, or sensing strategies; and implement new system-level observational concepts that promote agility, adaptability, evolvability, scalability, and affordability.

There will be many future GN&C needs for miniaturized sensors and actuators. MEMS-based microsystems can be used to meet or satisfy many, but not all, of these future challenges. Future science and exploration platforms will be resource constrained and would benefit greatly from advanced attitude determination sensors exploiting MEMS technology, APS technology, and ULP electronics technology. Much has been accomplished in this area. However, for demanding and harsh space mission applications, additional technology investments will be required to develop and mature, for example, a reliable high-performance MEMS-based IMU with low-mass, low-power, and low-volume attributes. Near-term technology investments in MEMS inertial sensors targeted for space applications should be focused upon improving sensor reliability and performance rather than attempting to further drive down the power and mass. The R&D emphasis for applying MEMS to spacecraft GN&C problems should be placed on developing designs where improved stability, accuracy, and noise performance can be demonstrated together with an ability to withstand, survive, and reliably operate in the harsh space environment.

In the near term, MEMS technology can be used to create next generation, multifunctional, highly integrated modular GN&C systems suitable for a number of mission applications and MEMS can enable new types of low-power and low-mass attitude sensors and actuators for microsatellites. In the long term, MEMS technology might very well become commonplace on space platforms in the form of low-cost, highly-reliable, miniature safe hold sensor packages and, in more specialized applications, MEMS microsystems could form the core of embedded jitter control systems and miniaturized DRS designs.

It must be pointed out that there are also three important interrelated common needs that cut across all the emerging MEMS GN&C technology areas highlighted in this chapter. These should be considered in the broad context of advanced GN&C technology development. The first common need is for advanced tools, techniques, and methods for high-fidelity dynamic modeling and simulation of MEMS GN&C sensors (and other related devices) in real attitude determination and control system applications. The second common need is for reconfigurable MEMS GN&C technology ground testbeds where system functionality can be demonstrated and exercised and performance estimates generated simultaneously. These testbed environments are needed to permit the integration of MEMS devices in a flight configuration, such as hardware-in-the-loop (HITL) fashion. The third common need is for multiple and frequent opportunities for the on-orbit demonstration and validation of emerging MEMS-based GN&C technologies. Much has been accomplished in the way of technology flight validation under the guidance and sponsorship of such programs as NASA's NMP (e.g., the ST6 ISC technology validation flight experiment) but many more such opportunities will be required to validate all the MEMS technologies needed to build new and innovative GN&C systems. The supporting dynamics models or simulations, the ground testbeds, and the flight validation missions are all essential to fully understand and to safely and effectively infuse the specific MEMS GN&C sensors (and other related devices) technologies into future missions.

REFERENCES

1. Kaplan, M.H., *Modern Spacecraft Dynamics and Control*. Wiley, New York, 1976.
2. Wertz, J.R., *Spacecraft Attitude Determination and Control*. Luwer Academic, Boston, MA, 1978.
3. Bryson, A.E., *Control of Spacecraft and Aircraft*. Princeton University Press, Princeton, NJ, 1994.
4. James, R.W. and Wiley, J.L. (eds), *Space Mission Analysis and Design, 1999, Space Technology Library*. Kluwer Academic Publishers, Dordrecht, The Netherlands, 1999.
5. Buehler, M.G., et al., Technologies for affordable SEC Missions, *Proceedings EE Big Sky Conference,* Montana, 2003.
6. Esper, J., Modular adaptive space systems, *Proceedings STAIF,* Albuquerque, NM, 2004.
7. Blaes, B.R., Chau, S.N., and Kia, T., Micro Navigator, *Proceedings — Forum on Innovative Approaches to Outer Planetary Exploration*, Houston, TX, 2001.
8. Joel G. and Neil, D., A multi-function GN&C system for future earth and space science missions, 25th Annual AAS Guidance and Control Conference, Technical Paper AAS 02–062, February 2002, 2002.
9. Maki, G.K. and Yeh, P.S., Radiation tolerant ultra low power CMOS Microelectronics: Technology Development Status, *Proceedings — Earth Science Technology Conference (ESTC)*, College Park, MD, 2003.
10. Brady, T. et al., The inertial stellar compass: a new direction in spacecraft attitude determination, *Proceedings — 16th Annual AIAA/USU Conference on Small Satellites*, 2002.
11. Connelly, J. and Kourepenis, A., Inertial MEMS Developments for Space, Draper Lab Report CSDL P-3726, 1999.
12. Connelly, J. et al., MEMS-based GN&C sensors and actuators for micro/nano satellites, *Advances in the Astronautical Sciences* 104, 561, 2000.
13. Johnson, W.M. and Phillips, R.E., Space avionics stellar-inertial subsystem, *AIAA/IEEE Digital Avionics Systems Conference — Proceedings* 2, 8, 2001.
14. Brady, T. et al., The inertial stellar compass: a multifunction, low power attitude determination technology breakthrough, *Proceedings AAS G&C Conference AAS 03–003*, 2003.
15. Wickenden, D.K. et al., MEMS-based resonating xylophone bar magnetometers, *Proceedings of SPIE* 3514, 350, 1998.
16. Kang, J.W., Guckel, H., and Ahn, Y., Amplitude detecting micromechanical resonating beam magnetometer, *Proceedings of the IEEE Micro Electro Mechanical Systems (MEMS)* 372, 1998.
17. Miller, L.M. et al., μ-Magnetometer based on electron tunneling, *Proceedings of the IEEE Micro Electro Mechanical Systems (MEMS)* 467, 1996.
18. Liebe, C.C. and Mobasser, S., MEMS based sun sensor, *IEEE Aerospace Conference Proceedings* 3, 31565, 2001.

19. Mobasser, S. and Liebe, C.C., MEMS based sun sensor on a chip, *IEEE Conference on Control Applications — Proceedings* 2, 1483, 2003.
20. Mobasser, S., Liebe, C.C., and Howard, A., Fuzzy image processing in sun sensor, *IEEE International Conference on Fuzzy Systems* 3, 1337, 2002.
21. Soto-Romero, G. et al., Micro infrared Earth sensor project: an integrated IR camera for Earth remote sensing, *Proceedings of SPIE — The International Society for Optical Engineering* 4540, 176, 2001.
22. Soto-Romero, G. et al., Uncooled micro-Earth sensor for micro-satellite attitude control, *Proceedings of SPIE — The International Society for Optical Engineering* 4030, 10, 2000.
23. Bednarek, T.J., Performance characteristics of the multi-mission Earth sensor for challenging, high-radiation environments, *Advances in the Astronautical Sciences* 111, 239, 2002.
24. Clark, N., Intelligent star tracker, *Proceedings of SPIE* 4592, 216, 2001.
25. Eisenman, A.R., Liebe, C.C., and Zhu, D., Multi-purpose active pixel sensor (APS)-based microtracker, *Proceedings of SPIE* 3498, 248, 1998.
26. Liebe, C.C. et al., Active pixel sensor (APS) based star tracker, *IEEE Aerospace Applications Conference Proceedings* 1, 119, 1998.
27. Lawrence, A., *Modern Inertial Technology*. Springer Verlag, New York, 1993.
28. Barbour, N. and Schmidt, G., Inertial sensor technology trends, *Proceedings of the 1998 Workshop on Autonomous Underwater Vehicles, 20–21 August 1998*, Cambridge, MA, 1998.
29. John, R. and Dowdle, K.W.F., A GPS/NS Guidance System for Navy 5″ Projectiles, *Proceedings — 52nd Annual Meeting, Institute of Navigation*, Cambridge, MA, June 1996.
30. Madni, A.M., Wan, L.A., and Hammons, S., Microelectromechanical quartz rotational rate sensor for inertial applications, *IEEE Aerospace Applications Conference Proceedings* 2, 315, 1996.
31. Review of MEMS Gyroscopes Technology and Commercialization Status, http://www.rgrace.com/Conferences/AnaheimExtra/paper/nasiri.doc
32. Smith, R.H., An Analysis of Shuttle-Based Performance of MEMS Sensors, AAS Technical Paper 98–143, 1998.
33. Bourne, M., Gyros to go, *Small Times* 20 February 2004.
34. Tang, T.K. et al., Packaged silicon MEMS vibratory gyroscope for microspacecraft, *Proceedings of the IEEE Micro Electro Mechanical Systems (MEMS)* 500, 1997.
35. Tang, W.C., Micromechanical devices at JPL for space exploration, *IEEE Aerospace Applications Conference Proceedings* 1, 461, 1998.
36. George, T., Overview of MEMS/NEMS technology development for space applications at NASA/JPL, *Proceedings of SPIE* 5116, 136, 2003.
37. Zaman, M., Sharma, A., Amini, B., and Ayazi, F., Towards inertial grade vibratory microgyros: a high-Q in-plane silicon-on-insulator tuning fork device, *Proceedings Solid State Sensor, Actuator, and Microsystems*, Hilton Head, 384, 2004.
38. MiniAERCam, http://aercam.nasa.gov
39. Judy, J.W. and Motta, P.S., A lecture and hands-on laboratory course: introduction to micromachining and MEMS, *Biennial University/Government/Industry Microelectronics Symposium — Proceedings* 151, 2003.
40. Lewis, S. et al., Integrated sensor and electronics processing for > 108 "iMEMS" inertial measurement unit components, technical digest — International Electron Devices Meeting 949, 2003.

41. Judy, M., Evolution of integrated inertial MEMS technology, Technical Digest of the Solid-State Sensor, Actuator and Microsystems Workshop, Hilton Head, SC, 27, 2004.

42. Smit, G.N., Potential applications of MEMS inertial measurement units, in Helvaijan, H. (ed.), *Microengineering Technology for Space Systems*, The Aerospace Press, Los Angeles, CA, 1997, 35.

43. Bernstein, J., Miller, R., Kelley, W., and Ward, P., Low-noise MEMS vibration sensor for geophysical applications, *Journal of Microelectromechanica Systems* 8 (4), 433, 1999.

44. Peczalski, A. et al., Micro-wheels for attitude control and energy storage in small satellites, *IEEE Aerospace Conference Proceedings* 5, 52483, 2001.

45. Lee, E., A micro high-temperature superconductor-magnet flywheels with dual function of energy storage and attitude control, *Proceedings of IEEE Sensors* 1, 757, 2002.

46. Durfee, D. et al., Atom interferometer inertial force sensors, Record 2000 Position, Location and and Navigation Symposium, 395, 2000.

47. Gustavson, T. et al., Atom interferometer inertial force sensors, IQEC, *Proceedings of the 1999 Quantum Electronics and Laser Science Conference (QELS '99)*, 20, 1999.

48. Kasevich, M., Atom interferometry with ultra-cold atoms, *Conference on Quantum Electronics and Laser Science (QELS) — Technical Digest Series* 74, 42, 2002.

49. McGuirk, J.M. et al., Sensitive absolute-gravity gradiometry using atom interferometry, *Physical Review A — Atomic, Molecular, and Optical Physics* 65 (3B), 033608, 2002.

50. Eriksson, S. et al., Micron-sized atom traps made from magneto-optical thin films, *Applied Physics B: Lasers and Optics* 79 (7), 811, 2004.

51. Moktadir, Z. et al., Etching techniques for realizing optical micro-cavity atom traps on silicon, *Journal of Micromechanics and Microengineering Papers from the 14th Micro-mecahnics Europe Workshop (MM'03)* 14 (9), 82, 2004.

11 Micropropulsion Technologies

Jochen Schein

CONTENTS

11.1 INTRODUCTION

Development of nanosatellites is presently a strong interest of the USAF as well as of NASA, DARPA, and MDA.[1-3] Spacecraft designs are tending towards smaller, less expensive vehicles with distributed functionality. NASA's future vision is one of reprogrammable or reconfigurable autonomous systems; small, overlapping instruments; and small, inexpensive micro-, nano- or even picosatellites. Examples include the nanosatellite program and the Orion Formation experiment. This new trend evokes the same advantages that drive computing towards distributed, parallel systems and the Internet. There are already examples of distributed satellite networks, such as the Tracking and Data Relay Satellite System (TDRSS), Intelsat, GPS, Iridium, Globalstar, and the Space-Based Infrared System (SBIRS). However, while these are groups of satellites designed to accomplish a common goal, they are nevertheless "noncooperating." The new wave of proposed constellations will be groups of vehicles that interact and cooperate to achieve mission goals. In such groups, vehicle pointing and positioning will be managed collectively. Fleets will evolve over time, extending and enhancing the overall capabilities. Also, autonomous vehicles will eliminate the need for extensive ground support. From a programmatic perspective, the concept is to replace multi-instrument observatories with low-cost, short lead-time spacecraft that would allow adaptation to changing conditions. This in turn mitigates the risk that not all formation-flying applications provide full programmatic benefits.

Tomorrow's Air Force will rely a new generation of smaller, highly capable nano and picosatellites (having masses of 10 and 1 kg respectively) that will act singly or collaboratively to accomplish various space missions. (M. Birkan, AFOSR[4])

In order to fulfill the mission requirements for the small spacecraft's new types of micro- and nanothrusters are required that offer a wide range of thrust levels from micronewton (μN) to newton levels at high overall thrust efficiencies and with very low (<1 kg) total thruster and power processing unit (PPU) mass. This chapter will try to introduce a variety of technologies that aim to satisfy these goals.

The simplest of all propulsion systems appears to be the cold gas thruster: a pressurized gas is released to produce thrust, but its exhaust velocity is so small that it would be necessary to carry a significant amount of propellant for large Δ-V missions. Systems like the so-called laser ablation thruster, where mass is energized by incident laser light to produce a higher exhaust velocity, may carry significant amounts of overhead mass. Other candidate electric propulsion engines that might be scaled down include the microcolloid thruster or the field emission electric propulsion (FEEP) thruster, which produce fairly small (μN) thrust levels and require high voltages for operation. The vacuum arc thruster as well as the micro-pulsed plasma thrusts (μPPT) have been shown to be good candidates for many missions requiring approximately μN-s to mN-s impulse bits; however, these devices are pulsed, and shot-to-shot variation can sometimes be significant.

Besides performance, another significant parameter is the system mass. Some of these technologies can benefit from the use of MEMS, which enables reduction of the mass of the thruster itself. Nevertheless, the thruster itself is only one part of a complete propulsion system, and in many cases, a small thruster requires additional overhead mass like PPU, tanks, valves, etc. to function properly. This prompts the question: How good is a MEMS thruster with a total mass of a few grams, when the PPU mass cannot be accommodated within the spacecraft budget?

Also consider that the mass of a propulsion system consists of the dry mass and the amount of propellant that needs to be carried. Mission parameters that define the requirements for propulsion systems include total Δ-V, required payload or structure of the spacecraft, and time allocated for the mission.

The amount of propellant needed depends on the Δ-V requirements and the exhaust velocity of the propulsion system, which has been expressed by Tsiolkovsky in the famous rocket equation as shown in Equation (11.1):[5]

$$\Delta V = v_e \ln\left(\frac{M_0}{M_0 - M_P}\right) \qquad (11.1)$$

with M_0 and M_P being the initial mass of the spacecraft and the amount of propellant needed, respectively, and v_e describing the exit velocity. From this equation it is obvious that for a given Δ-V and spacecraft mass, the amount of propellant required depends on the propellant velocity. The higher the velocity, the less the propellant needed. Electric propulsion (EP) systems have been shown to provide high exit velocities ranging from 10,000 up to 100,000 m/sec, whereas chemical propulsion systems are usually limited to exhaust velocities between 500 and 3000 m/sec. Therefore, at first glance, the choice seems obvious.

Apart from the propellant, both classes systems include additional mass overhead. In the case of chemical systems, this will include tanks and valves. In the case of EP systems a PPU is needed. The mass of a PPU has been shown to be a function of the average power they can handle, thereby defining a specific mass α, which commonly scales as 30 g/W. With EP thrust-to-power ratios averaging approximately 10 μN/W, the importance of taking the PPU mass into account becomes obvious. Looking at an example it can be shown how a chemical system can be more advantageous than an EP system despite its much lower exhaust velocity.

Assuming a total spacecraft mass of 5 kg, the amount of propellant needed for a ΔV of 300 m/sec can be calculated to be 15 g for a v_e of 100,000 m/sec and 696 g for a v_e of 2,000 m/sec. The average thrust T needed depends on the duration of the mission Δt, as shown in Equation (11.2).

$$T = \frac{M_P v_e}{\Delta t} \qquad (11.2)$$

For an EP system the mass of the power supply is given by Equation (11.3),

$$M_{PPU} = \frac{T\alpha}{TTP} \qquad (11.3)$$

while the overhead mass for the chemical system remains fairly constant and is assumed to be approximately 300 g.

With this information, the total mass of the propulsion system as a function of the mission duration can be estimated as shown in Figure 11.1. The faster a mission needs to be accomplished, that is, the more thrust required, the more favorable a chemical system becomes. The crossover point for this example using the parameters above is at 5×10^6 sec or approximately 58 days, which corresponds to an average thrust of approximately 300 µN.

Another way to describe the influence of exhaust velocity is by simply looking at the formula for thrust. Thrust can be described with Equation (11.4):

$$T = \frac{2P_{in}\eta}{v} \qquad (11.4)$$

which implies that for a given input power P_{in}, and a given system efficiency η, thrust is inversely proportional to exhaust velocity, which for the same conditions leads to Equation (11.5):

$$\frac{\Delta V}{\Delta t} \propto \frac{1}{v} \qquad (11.5)$$

However, using chemical thrusters of such a small size will lead to another problem. Currently, many micropropulsion devices that rely on nozzle flow have low efficiencies in terms of directed kinetic energy versus potential energy (thermal, chemical,

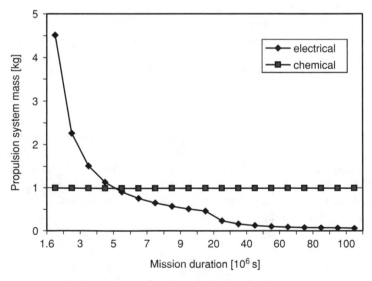

FIGURE 11.1 System dry mass as a function of mission duration.

and electrical) due to a lack of understanding of the flows in such devices. This is due to the fact that the continuum assumption commonly used in gas and plasma dynamics is no longer valid at smaller densities and characteristic dimensions of flow. The Knudsen number is defined as the ratio of the mean free path of gas molecules to a characteristic dimension of flow. As the Knudsen number increases, the collision rate becomes too low to maintain local thermodynamic equilibrium. Furthermore, the expansion of a propellant from chamber conditions to vacuum often involves flow regimes from continuum to transition to free molecular, though the smallest devices may not have any component in the continuum regime. Therefore, fairly complicated models are needed for proper evaluation, which goes beyond the scope of this review. More detailed descriptions of these effects can be found elsewhere.[6,7]

All these considerations demonstrate that both chemical and electrical propulsion systems need to be included in this chapter and that a decision between either system has to be made on a case-by-case basis. The emphasis will be put on MEMS and other low-mass systems (i.e., where the total system dry mass is less than 1000 g). The principle of operation will be discussed for each system, using few basic equations describing the performance. While simplistic, these basic equations will nevertheless help to understand the operating characteristics of the various micropropulsion technologies and calculate rough estimates of their performance.

After describing each system, its key parameters will be discussed and the performance for each system will be summarized in a table. Technologies discussed here include (a) chemical propulsion systems, such as hydrogen peroxide thrusters, cold gas thrusters, solid micro rockets and (b) electric propulsion systems, such as pulsed plasma thrusters, laser-driven plasma thrusters, field effect thrusters, ion engines, and resistojets. While many publications about these types of propulsion systems cite performance specifications of the propulsion device (i.e., the micro-manufactured emission array or the MEMS-valve), this chapter tries to take a look at the complete system, thereby providing information that is needed to successfully design a satellite. Improvements to existing systems and new propulsion technologies will emerge and may well be superior to those mentioned, which also implies that the numbers cited here are by no means absolute limitations. In this light, I would also like to refer to other review articles on micropropulsion, with the most important and complete one authored by Jürgen Müeller from NASA JPL.[8]

Regarding the formality of this chapter, I took the liberty of referring to most publications used in the beginning of each chapter, instead of placing the citations in the body of the text. By doing so, it became much easier to read, digest, and summarize. I hope that none of the original authors will take offense even if a certain thought in the body of the text may have come from a single paper only.

Enjoy!

11.2   ELECTRIC PROPULSION DEVICES

In this review, electric propulsion systems are defined as those where the majority of the energy needed for operation is electrical energy.

11.2.1   PULSED PLASMA THRUSTER

Conventionally scaled pulsed plasma thrusters have been used in the past successfully and are fully space qualified.[9-16] Thrust is produced by ablating and accelerating a solid insulator, such as Teflon, using a surface discharge initiated by high
voltage. Usually these systems are fairly massive ($\approx$5 kg), but recent efforts have
been made to shrink the PPT for the use in micro- and nano-spacecraft. A micro-
pulsed plasma thruster (μPPT) has been developed by AFRL and Busek using
coaxial thruster configurations. A power conversion system converts the bus voltage to approximately 1 kV levels to ignite the discharge. Specific impulse (I_{sp})
values can reach up to 1000 sec, with μN-sec impulse bits.

11.2.1.1   Principle of Operation

The micro-PPT uses a high voltage, capacitively driven arc discharge to ablate and
accelerate insulation and electrode material (typically Teflon and copper, respectively) in a small geometry. The acceleration process is a combination of plasma
heating and expansion as well as a Lorentz force that helps to further expand the
plasma front as shown in Figure 11.2. Therefore the I_{sp} depends on the current in the
plasma sheet and the duration of the acceleration, which results in a direct dependence on the energy, E, deposited into the plasma. Various studies have been made
and semiempirical relations like Equation (11.6) have been suggested.

$$I_{sp} = 317 \left(\frac{E}{A} \right)^{0.585} \tag{11.6}$$

where A is the area of the accelerated plasma sheath.

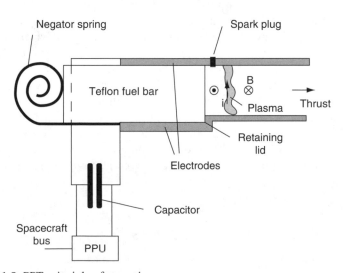

FIGURE 11.2 PPT principle of operation.

Certainly the achievable impulse bit, I, is a function of the specific impulse and the energy in the pulse. Another empirical relationship, as shown in Equation (11.7), has been formulated:

$$I \cdot I_{\text{sp}} = 1.7E^{1.65} \tag{11.7}$$

which in turn determines the achievable thrust, T, from Equation (11.8) as the product of pulse frequency and impulse bit

$$T = F \cdot I \tag{11.8}$$

By changing the energy in the plasma pulse the exhaust velocity and the mass ablated changes. This effect has shown influence on the scalability of the thruster. Thrust-to-power values decrease with decreasing energies per pulse. While thrust-to-power can reach values up to 20 μN/W for hundreds of joules, this value decreases to $\approx$10 μN/W for pulse energies of the order 5 J. As suggested by these parameters, achievable impulse bits can be large for large systems ($\approx$10 mN sec) and can approach 5 μN sec for small systems.

11.2.1.2   System Requirements

The PPT can operate in a self-triggering mode or in a controlled-pulsed mode. Both require high voltage (>1 kV) to initiate the discharge. Due to the need for high voltage and the time necessary to charge the capacitors, repetition rates may be limited to less than $\approx$10 Hz.

The driving circuit is usually fairly simple. A DC–DC converter connects the PPT to the spacecraft bus. The high-voltage output of the converter is used to charge the capacitor bank of a pulse-forming network. Once the surface breakdown voltage along the insulator occurs, a plasma sheet is produced and starts moving. Other ignition possibilities include the use of a switch to apply the high voltage or the use of a preionizer like a spark plug. Once the sheet is formed, the energy left in the capacitor drives the plasma via the Lorentz force.

A variety of different PPT configurations have been demonstrated. Many have spring-fed propellants like the side-fed ablative PPT (APPT) or the breech-fed APPT in which a block of Teflon is placed between two electrodes and pushed forward with ongoing erosion. The μPPT manufactured by Busek relies on coaxial geometries, where the propellant erosion leads to increasing recession of the plasma source with ongoing operation as shown in Figure 11.3.

The total system mass can be quite low since the power processing electronics are minimal. With its miniature electrode gaps, the micro-PPT benefits from the use of lighter, lower-voltage components. Additionally, the plasma is quasineutral, which allows for operation without an additional neutralizer. However, EMI filtering may be necessary due to the pulsed high voltages. The pulsed plasma thruster is summarized in Table 11.1 with a picture of the complete system shown in Figure 11.4.

FIGURE 11.3 Erosion pattern of coaxial μ-PPT. (*Source*: University of Illinois.)

TABLE 11.1
Performance Characteristics for Micro-PPT

I_{sp}	500 sec
I-bit (if pulsed)	15 μN sec
Rep. rate (if pulsed)	2 Hz
Power	10 W
Thrust	>30 μN
Thrust or power	10 μN/W
Impulse or prop.	5 Ns/g
Feed mechan.	No
Current system mass	600 g
(include PPU, valve, tank, etc.)	

11.2.2 VACUUM ARC THRUSTER

The need for a low-mass propulsion system motivated the development of the vacuum arc thruster.[17–21] This device is essentially a pulsed plasma thruster that uses conductive cathode materials as propellant. It requires an energy storage PPU that takes 5–24 V from the bus and converts it into an adequate power pulse for the thruster. It is a system well suited to provide small impulse bits (≈ 1 μN sec) at high specific impulse, I_{sp} (1000 to 3000 sec). Applications include positioning and drag-makeup for small spacecraft that are power- and mass-limited satellites.

FIGURE 11.4 μ-PPT (includes PPU). (*Source*: Busek Co.)

11.2.2.1 Principle of Operation

In a vacuum arc thruster, plasma is produced from the cathode material in vacuum. The plasma production takes place in tiny micron-sized emission sites, so-called arc spots. Every arc spot has a lifetime in the order of tens of nanoseconds and carries a few amperes of arc current. A highly scalable device is produced by changing the current, leading to a change of number of arc spots with the basic physics in the arc spot remaining the same. The high-density plasma created in the spots produces a very high pressure, up to 1000 atm, that accelerates the quasi-neutral plasma outward. Due to their larger mass, the ions contribute to most of the propulsion. For this design, any conducting material can be used as a propellant.

The performance of the vacuum arc thruster (VAT) is determined by the propellant mass, the degree of ionization of the plasma, the angle of expansion, the average charge state, and the ion velocity. All these parameters have been measured repeatedly in the past and verified for numerous materials and operating conditions. Typical values for the ion velocity vary between 10,000 and 30,000 m/sec. The average arc to ion current ratio has been shown to be approximately 8% and a cosine distribution has been found to emulate the plasma plume expansion very well.

With these known parameters we can predict the performance of the VAT for various materials. The ion mass flow rate $\dot{m}_{ion}$ (kg/sec) is given in Equation (11.9),

$$\dot{m}_{ion} = rI_{arc}m/(Ze) \tag{11.9}$$

where r is the ion to arc current ratio, m is the cathode material ion mass, and Z is the average ion charge state of the discharge plasma.

Therefore, the impulse size produced is given by:

$$I = \dot{m}_{ion} vt \qquad (11.10)$$

with t being the duration of the impulse. Assuming $t = 500$ μsec and $I_{arc} = 50$ A we can obtain a maximum impulse bit of the order 40 μN sec using Bi ($Z = 1.18$). This is consistent with a value of 30 μN sec measured at the JPL thrust stand for a pulse with similar current and duration for tungsten, which is slightly lighter. Thrust can be controlled by adjusting pulse power and repetition rates. Thrust-to-power values are of the order of 10 μN/W.

11.2.2.2   System Requirements

The VAT was constructed using an inductive energy storage (IES) circuit PPU and simple thruster head geometry. In the PPU, an inductor is charged through a semiconductor switch. When the switch is opened, a voltage peak, LdI/dt, is produced, which breaks down the thin metal film-coated anode cathode insulator surface at relatively low voltage levels (≈200 V). The current that was flowing in the solid-state switch (for = 1 μsec) is fully switched to the vacuum arc load. Typical currents of approximately 100 A (for ~100 to 500 μsec) are conducted with voltages of approximately 25–30 V. Consequently, most of the magnetic energy stored in the inductor is deposited into the plasma pulse. The efficiency of the PPU may thus be greater than 90%.

Based on this inductive energy storage approach, a PPU was designed to accept external TTL level signals to adjust the energy and the repetition rate of individual plasma pulses. This was accomplished by adjusting the trigger signal to the semi-conductor switch. Figure 11.5 shows an equivalent circuit diagram of the PPU. By varying the length of the trigger signal, the level of the current in the switch and thereby, the energy stored in the inductor can also be adjusted. This, in turn, changes the amount of energy transferred to the arc and the impulse bit of the individual pulse. Obviously, the repetition rate of the individual pulse can be changed by varying the input signal as well.

The mass of the PPU is small (< 300 g) resulting in a low-mass system. The plasma output is quasineutral; therefore, no additional neutralizer is needed. An EMI filter might be necessary due to the noisy characteristics of the discharge, high peak currents, and fast switching. A drawback for this technology is possible contamination from the metal propellant. A low-mass feed mechanism is available, therefore, even long missions can use this technology.

The design of the thruster head itself is very simple. A coaxial structure with a center cathode, separated from a tube-like anode by an insulator, is a possibility as well as a sandwich structure as shown in Figure 11.6.

Even smaller structures as manufactured by ChEMS are possible; however, scaling down the PPU and the thruster will lead to inefficient operation because of

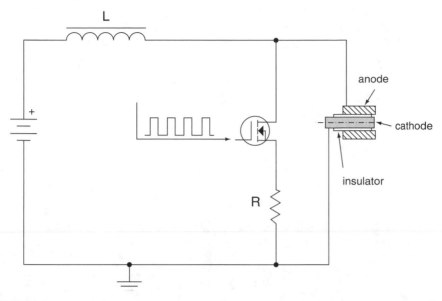

FIGURE 11.5 Inductive energy store (IES) PPU for the VAT. (*Source*: Alameda Applied Sciences Corp.)

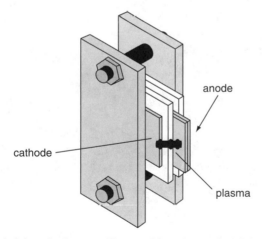

FIGURE 11.6 Sandwich-style thruster. (*Source*: Alameda Applied Sciences Corp.)

inherent losses due to the use of power semiconductors. The summary of the VAT is shown in Table 11.2 with a picture of the complete system shown in Figure 11.7.

11.2.3 FEEP

While companies in the U.S. concentrate on the fabrication and validation of colloid thrusters, the European approach has been to develop field emission or field effect electric propulsion (FEEP) systems based on the liquid metal ion source (LMIS)-

TABLE 11.2
Performance Characteristics for Vacuum Arc Thruster System

I_{sp}	1000 to 3000 sec
I-bit	10 nN to 30 μN sec
Rep. rate	Single shot 1 kHz
Power	10 W (30 W)
Thrust/Power	10 nN to 300 μN/W
Impulse/prop.	10 μN/w
	10 N sec/g
Feed mechan.	Yes
Impulse/sys.-mass	100 N sec/500 g

FIGURE 11.7 Vacuum arc thruster system (includes PPU). (*Source*: Alameda Applied Sciences Corp.)

principle.[22–26] Starting with cesium as the propellant, development of the LMIS has evolved from a single-pin emitter through linear arrays of stacked needles to the presently favored slit emitter module. Compared to other electric propulsion systems, FEEP thrusters have shown high values of thrust-to-power ratio (>100 μN/W) at high specific impulses (≈10,000 sec). FEEP thrusters appear to be well adapted to missions requiring a very fine attitude (milli arc seconds) and orbit control (relative positioning of several satellites to millimeter accuracy). This is an application domain where the FEEP system can claim several advantages compared with chemical or other electric propulsion systems, that is, continuous thrust throttling, small impulse bit, instantaneous switch-on/switch-off capability, mechanical and electrical simplicity and thruster clustering.

11.2.3.1 Principle of Operation

The basic element of a FEEP thruster is a liquid metal ion source. In a liquid metal ion source, the ions are generated directly from the surface of the liquid metal by a high electric field applied between the LMIS (anode) and an extractor geometry (cathode). When the surface of a liquid metal is subjected to a high electric field, it is distorted into a cone or a series of cones as depicted in Figure 11.8. The radius of such a cone is determined by the applied electric field E and the surface tension of the liquid metal, γ, given by:

$$r = \frac{4\gamma}{\varepsilon_0 E^2} \tag{11.11}$$

With increasing applied voltage, the radius of curvature of these cones decreases, thereby, further enhancing the local electric field. When the local field reaches values of the order 10^9 V/m, atoms of the metal tip are ionized either by field evaporation or field ionization. Subsequently, the ions are accelerated and expelled from the emitter by the same electric field that has ionized them. The charged particles leaving the liquid metal surface as an ion beam are replenished by the hydrodynamical flow of the liquid metal. The liquid metal is converted directly into an ion beam without the transitional vapor phase, which is common in the technology of other ion sources; therefore ionization operates with high power efficiency.

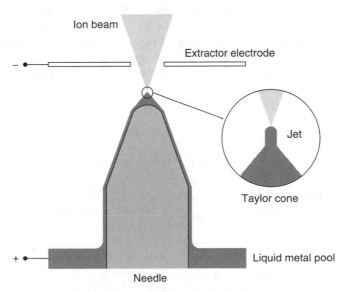

FIGURE 11.8 FEEP principle of ion current production.

With the radius of curvature of such a cone being approximately 10^{-7} m and the electrodes being 1 mm apart, voltages of the order of kilovolts are sufficient to obtain the necessary high electric fields. Applying voltages of few kilovolts (5–7 kV) results in ion currents of 10–100 µA from a single needle source with low energy spread.

The thrust can be calculated directly (for Gallium) from the following equations:

$$T = \dot{m}v \tag{11.12}$$

$$\dot{m} = I_E \frac{m_{Ga}}{e} \tag{11.13}$$

$$v = \sqrt{\frac{eV}{m_{Ga}}} \tag{11.14}$$

$$\Rightarrow T = \sqrt{\frac{m_{Ga}}{e}} I_E \sqrt{V} \tag{11.15}$$

where
T = thrust (N),
$\dot{m}$ = mass flow (kg/sec),
v = velocity (m/sec),
I_E = emission current (A),
V = extraction voltage (V),
e = 1.6×10^{-19} C, and
m_{Ga} = 1.16×10^{-25} kg

While this calculation does not account for the beam spread it will nevertheless result in a close (80%) estimate of the thrust produced. Small thrust noise and very stable operation are characteristics of this propulsion system. High mass efficiencies for FEEP thrusters can be obtained at small extraction currents on the order of a few microamperes. At higher extraction voltages, which translate into higher currents, droplets and clusters are formed, which obtain significantly smaller exhaust velocities and, therefore, lead to lower system efficiency.

11.2.3.2 System Requirements

The FEEP thruster is a very capable low-thrust, low-noise system. Reported efficiency and thrust-to-power ratios are high and the possibility of MEMS-produced ion sources exist. These MEMS structures are based on the usage of a large number of emission sites. First concepts have been evaluated using microtips immersed in liquid metal as emission sites as shown in Figure 11.9. Another approach involves the use of small tubes that lead the liquid metal with the help of capillary forces from a bulk reservoir to the emission site. Highly accurate manufacturing is necessary as small geometrical differences result in the formation of so-called hot spots, where individual pixels attract the majority of the emission current, which can lead to unwanted

FIGURE 11.9 FEEP multi-emitter design. (*Source*: Austrian Research Centre.)

heating with large droplet formation and subsequent clogging or the formation of short circuits, resulting in system failure.

When considering the use of FEEP thrusters for nano/picosatellites, a few drawbacks have to be taken into account. Due to the use of very high voltages, bulky DC/DC converters may be necessary. Additional mass has to be assigned to a neutralizer, including its power supply, because a pure metal ion beam is produced. Size reduction will most likely be limited due to possible metal droplet formation that might attach to the anode, leading to field distortions or even clogging. Contamination due to the use of metals is a problem. High-voltage wiring is necessary and an EMI filter has to be included in a final design to protect the on-board PPU from sudden high-voltage breakdowns. The summary of the FEEP thruster is shown in Table 11.3 with a picture of the complete system shown in Figure 11.10.

11.2.4 LASER ABLATION THRUSTER

Another micropropulsion alternative is the microlaser plasma thruster (μLPT).[17,27,28] The μLPT is a sub-kilogram micropropulsion option, which is intended for attitude control and station-keeping on microsatellite platforms. A lens focuses a laser diode beam on the ablation target, usually consisting of an organic material, producing a miniature jet that provides the thrust. The single impulse dynamic range has been reported to cover five orders of magnitude, and the minimum impulse bit is 1 nN sec in a 100 μsec pulse. Specific impulses of up to 1000 sec together with laser momentum coupling coefficients up to 500 μN/W have been achieved.

TABLE 11.3
Performance Characteristics for FEEP Thruster

I_{sp}	8000 to 12000 sec
I-bit (if pulsed)	DC
Rep. rate (if pulsed)	DC
Power	0.5 to 10 W (ARC FEEP 100)
Thrust	100 µN
Thrust or power	20 µN/W
Impulse or prop.	5 Nsec/g
Feed mechan.	Yes/passive
Current system mass	500 g
(includes PPU, valve, tank, etc.)	

FIGURE 11.10 FEEP thruster. (*Source*: Austrian Research Centre.)

11.2.4.1   Principle of Operation

Two modes of operation have been evaluated. In the first mode, the laser is aimed at the target at an angle to avoid deposition of the ablated material onto the sensitive optics. In this case, single layer tape can be used. More commonly, the device is operated in transmission mode ("T-mode"), as shown in Figure 11.11, to protect optics from solid contaminants produced by the ablation jet. In this mode, a lens focuses the laser diode output to a 25-µm diameter spot on the transparent side of a two-layer fuel tape. Passing through a transparent acetate substrate without damaging it, the beam heats a specially prepared absorbing coating on the opposite

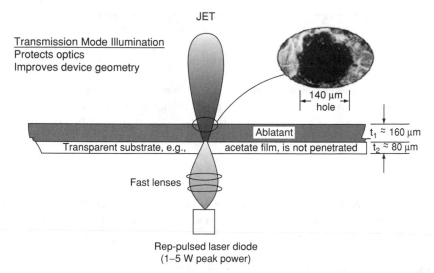

FIGURE 11.11 LAT principle of operation — transmission mode. (*Source*: Photonics Associates.)

side of the tape to high temperature, producing a miniature ablation jet. Part of the acetate substrate is also ablated. A plasma is produced and the pressure inside the plasma drives the exhaust, which produces thrust.

The μLPT can operate pulsed or CW, and power density on target is optically variable in an instant, so operating parameters can be adjusted to throttle the output of the thruster. Materials explored for the transparent substrate include cellulose acetate, PET, and Kapton™ polyimide resin. For the ablatant, over 160 materials have been studied. Many of these were so-called "designer materials" created especially for this application.

The thrust produced by this system depends on the so-called ablation efficiency, which describes the ratio of kinetic energy and laser energy.

This efficiency is defined as:

$$\eta_{AB} = C_m v_E \tag{11.16}$$

where v_E is the exhaust velocity and C_m as calculated, using the following equation, is the so-called coupling coefficient, which depends on the laser input and the material ablated through:

$$C_m = 58.3 \frac{\psi^{9/16}}{A^{1/8}(I\lambda\sqrt{\tau})^{1/4}} \left[\frac{\mu N}{W}\right] \tag{11.17}$$

$$\psi = (A/2)(Z^2(Z+1))^{1/3}$$

where A is the atomic mass number of material, Z the average charge state, I the laser intensity, λ the laser wavelength, and τ the pulse duration.

The exhaust velocity is in turn defined via the relationship seen in Equation (11.18),

$$v_E = QC_m \tag{11.18}$$

where Q is the absorbed input energy of laser light per mass unit of ablated material. It is obvious that in order to convert a certain amount of tape into a plasma, a fixed energy is necessary. This threshold intensity can be described by the empirical formula as a function of the laser pulse duration τ:

$$\Phi_{th} = (2.36 \times 10^4)(\tau)^{0.45} \tag{11.19}$$

Operation above this value leads to the production of higher charge states and increasing light absorption within the plasma, which limits the amount of energy available to ablate material.

11.2.4.2   System Requirement and Comments

While at first glance, the laser ablation thruster seems to be an extremely powerful device, it has to be noted that it includes some overheads. Apart from using a fairly powerful diode laser, a motor has to be used to move the fuel tape. Efficiency losses due to these items have to be taken into account.

The total system mass may approach 1 kg, using commercially available parts, not including the electronics that will be needed to adjust the tape speed and laser output. However, once these control mechanisms are in place, the I_{sp} can be adjusted by controlling laser output parameters like intensity and pulse lengths, which makes this thruster very versatile. The tape drive acts as a propellant feed mechanism suitable for long missions. The plasma output is quasineutral and non-metallic (although carbon might be produced with some propellants), which will minimize problems associated with contamination. Due to the lack of high voltage and the laser-produced plasma, EMI problems should be minimal.

Integration into a MEMS system might be possible with additional development. With shrinking laser size, the output power will be reduced, which in turn will reduce the spot size to submicron ranges and very small thrust levels. It is debatable if at this point the system losses like the motor drive will render this technology inefficient. The summary of the laser ablation thruster is shown in Table 11.4, with a picture of the complete system shown in Figure 11.12.

11.2.5   MICRO-ION THRUSTER

Ion thrusters are the workhorse of electrical propulsion.[29–31] Ions created in a plasma are extracted and accelerated electrostatically, thereby producing thrust. In order to achieve this, an ion thruster consists of a plasma source coupled to an extraction grid. The exhaust velocity can be adjusted by varying the extraction voltage. Although ion engines have high thrust-to-power ratios and are a well-developed and flight-proven technology, there are many difficulties

TABLE 11.4
Performance Characteristics for Laser Ablation Thruster System

I_{sp}	430 sec
I-bit (if pulsed)	0.01 µN sec
Rep. rate (if pulsed)	100 Hz
Power	8.6 W
Thrust	635 µN
Thrust or power	74 µN/W
Impulse or prop.	4.2 N sec/g
Feed mechan.	Yes
Total impulse/sys.-mass	10 Ns (450 Nsec)/750 g
Current system mass	750 g
(includes PPU, valve, tank, etc.)	

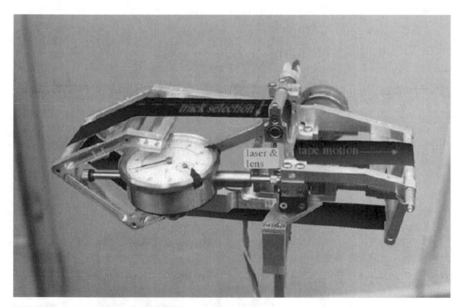

FIGURE 11.12 Laser ablation thruster. (*Source*: Photonics Associates.)

associated with a miniaturized ion thruster design. The discharge, which acts as the source for the ions, suffers from increased electron wall losses at greater surface-to-volume ratios imposed by smaller discharge chambers. Consequently, maintaining sufficient electron residence time in the chamber presents a significant challenge. Current research at NASA JPL attempts to overcome this issue and to turn the microion thruster into a competitive micropropulsion device.

11.2.5.1   Principle of Operation

The principle of an ion thruster is simple, as shown in Figure 11.13. As a first step, a plasma needs to be produced inside a discharge chamber. After that, positive ions are pulled from the plasma and accelerated using an extraction grid. Plasma production in ion engines is usually performed by ionizing a noble gas. For this purpose, an electron source is used. The produced electrons are accelerated in an electric field until they have gained sufficient energy to produce a secondary electron after colliding with an ion. Once sufficient plasma density has been achieved, a pair of electrostatic ion "acceleration" grids is needed at the exit of the thruster to properly contain the plasma and the energetic electrons, and extract and focus the ion beam.

To understand the discharge chamber performance, it is common to compare the amount of energy needed to make a single beam ion versus the propellant utilization efficiency. We define the thruster electrical efficiency, η_E, and total thruster power, PE as:

$$\eta_E = \frac{I_B V_B}{P_E}$$

$$P_E = I_B V_B + I_D V_D + P_0$$

(11.20)

where I_B, V_B, I_D, V_D, and P_0 are the beam current, beam voltage, discharge voltage, and miscellaneous power (which includes cathode operation), respectively. Combining theses relationships, we can determine an expression, as shown in Equation (11.21), for the energy per beam ion, which is inversely proportional to the engine efficiency.

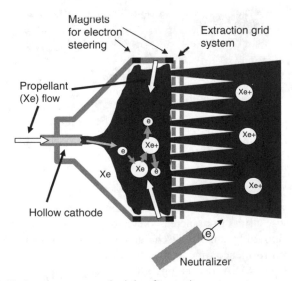

FIGURE 11.13 Xe ion thruster — principle of operation.

$$\eta_E = \frac{I_B V_B}{I_B V_B + I_D V_D + P_0} = \frac{V_B}{V_B + \varepsilon_B + \frac{P_0}{I_B}}$$

$$\varepsilon_B = \frac{I_D V_D}{I_B}$$

(11.21)

We can see from the following equation that the propellant utilization, η_u, is directly proportional to the beam current, JB, and the total propellant mass flow rate given by:

$$\eta_u = \left(\frac{I_B}{m_T}\right)\left(\frac{m_i}{e}\right)$$

(11.22)

where m_i is the mass of an ion and e is the charge of an electron. This relation does not account for the effect of multiply charged ions, which may be neglected for first order approximations and performance comparisons. An effective way of determining chamber performance is to plot the beam ion energy cost, versus the propellant utilization efficiency. To assess the relative performance of multiple thruster configurations, it is also important to compare their total efficiency values. The total efficiency of an ion thruster may be expressed as:

$$\eta_T - \frac{T^2}{2m_T P_E}$$

$$\text{where} \quad T = m_T \eta_u \sqrt{\frac{2eV_B}{m_i}}$$

(11.23)

11.2.5.2 System Requirements

While in principle, the miniaturization of the ion engine is possible, there are some problems that make the realization of a small system difficult. The chamber walls are at anode potential, which implies that electrons that are emitted from the cathode get lost to the chamber. In a very small chamber, the travel distance and thus the travel time are decreased, which limit the possibility of the electrons producing ions. One way of increasing travel time is to use magnets to insulate the anode magnetically. The magnets, however, become a mass liability. As with all ion thrusters, the positive ions that exit the thruster through the grids represent a sufficient current of positive ions to the ambient environment. This will cause the thruster and craft to quickly obtain an overall negative charge. As a result, a neutralizer cathode has to be placed near or in the beam to emit electrons into the positive ion beam. Although this will add additional mass, it is important to note that this neutralization process creates a benign, uncharged exhaust, especially in the case of noble-gas propellants such as xenon.

Care must be taken to ensure that electrical discharges do not occur across the closely spaced acceleration grids. Such discharges could seriously damage the thruster. A micro-ion thruster also requires the development of appropriately sized power conditioning units and propellant feed system. Due to the high voltages

TABLE 11.5
Performance Characteristics for Micro-Ion Engine

I_{sp}	3184s
I-bit (if pulsed)	DC
Rep. rate (if pulsed)	DC
Power	43 W
Thrust	1.5 mN
Thrust or power	35 μN/W
Impulse or prop.	32.8 g
Feed mechan.	Gas

needed and the minimum electron travel distance required for ionization, a pure MEMS manufactured ion thruster does not appear possible. A summary of the micro-ion engine is shown in Table 11.5, with a picture of the complete system shown in Figure 11.14.

11.2.6 MICRO-RESISTOJET

Resistojets have a long-standing history in space propulsion.[32–37] Because of their simplicity, they have become a workhorse of space propulsion. In principle these

FIGURE 11.14 Micro ion thruster. (*Source*: NASA.)

kinds of thrusters are based on the enthalpy increase of a propellant when it flows across an electrically heated surface, which causes an increase in pressure. The thermal energy of the propellant is subsequently transformed, in a Laval nozzle, into directed kinetic energy and thrust. Typical operational parameters are a moderate I_{sp} of the order 500 sec (this is too high to be realized in practice) with thrust levels of approximately 0.1 N at average powers of 100 W. While this technology is mature, recent efforts have been made to employ microfabrication and to reduce the size of this thruster technology to a dry mass of greater than 100 g. Therefore, it is worthwhile to take a closer look at this technology.

11.2.6.1 Principle of Operation

The principle of thrust production for this device is the same as with any chemical or electrothermal system that relies on expansion of a gas into vacuum to produce thrust. A gas under high pressure will escape into vacuum as soon as a hole is punched into the propellant tank. However, the exit velocity is very limited. Without the use of an additional nozzle, the velocity of the gas could never exceed the velocity of sound, which is a function of the gas temperature. Increasing the exit velocity would decrease the amount of fuel that needs to be carried for the missions. Therefore, a major part of resistojet development has been concentrated on designing nozzles that improve the performance of these thrusters, and has focused on the use of converging–diverging (CD) nozzles.

The exhaust velocity for a well-designed CD nozzle expanding into vacuum has been evaluated and can be described by a simple formula:

$$v = \sqrt{2T_0 \left(\frac{k}{k-1}\right) \frac{R}{M}} \tag{11.24}$$

with R being the universal gas constant, M the effective molecular weight of the propellant, $k = C_P/(C_P-R)$ is the effective ratio of specific heats (C_P = effective heat capacity of propellant) and T_0 represents the temperature of the medium in the tank.

The thrust that is produced in such a system can be calculated with the following equation as:

$$T = \dot{m}v + (p_e - p_a)A_e \tag{11.25}$$

Thus, thrust is a combination of two terms, the momentum thrust (mass flow × exit velocity) and the pressure thrust (pressure difference between nozzle exit pressure and ambient pressure × nozzle exit area).

The mass flow itself is given by:

$$\dot{m} = \frac{A_t a \rho_0}{\left(1 - \frac{k-1}{2}\right)^{\frac{1}{k-1}}} \tag{11.26}$$

where A_t represents the area of the throat of the nozzle, a is the velocity of sound, and ρ_0 is the density of the medium inside the tank. As can be seen from these equations, the nozzle design and the parameters of the tank medium, like pressure or temperature determine the performance of the thruster.

Using these equations, it is possible to calculate the performance of a sample resistojet. Assuming the gas is heated to 1200 K and the gas is purely molecular hydrogen ($k = 1.67$), the maximum exit velocity will amount to 5000 m/sec and the resulting thrust for a pressure of 2.10^6 Pa and a throat diameter of 1 mm^2 would amount to approximately 3/N.

Resistojets allow for the use of liquids — which are vaporized in the system — as a propellant. This simplifies storage and flow control compared to pure gaseous systems. Therefore, miniaturized versions of the resistojet very often use water vapor instead of hydrogen, which enables operation at lower pressures and a smaller system. Thrust values of 500 μN and exit velocities of 990 m/s are typical. The reduction in velocity is not only due to lower pressures in the system but is also affected by the decreasing influence of the nozzle with increasing Knudsen number as pointed out in the introduction.

11.2.6.2 System Requirements

Figure 11.15 shows a schematic of a resistojet, which summarizes the system requirements. Propellant tanks and valves are needed. A power supply for resistive heating of the gas has to be provided, however, no large power conversion units are necessary as the heating can be done directly from the spacecraft bus. Also, contamination is not a concern. MEMS elements like valves and nozzles can be used. Although MEMS valves have been shown to have significant problems with leakage, the leakage of these devices is not as problematic as the propellant could be in the liquid state.

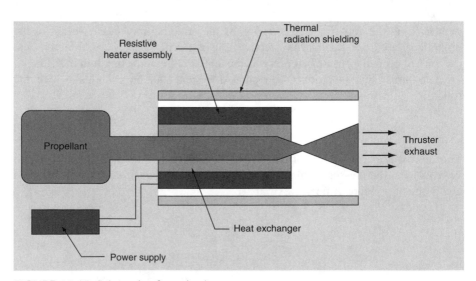

FIGURE 11.15 Schematic of a resistojet.

TABLE 11.6
Performance Characteristics for Water (Micro) Resistojet

I_{sp}	152 s (water)	100 s
Power	100 W	3 W
Thrust	45 mN	500 μN
Thrust or power	450 μN/W	≈150 μN/W
Impulse or prop.	1.5 Ns/g	≈1 Ns/g
Feed mechan.	Yes	Yes
Current system dry mass	1240 g	≈50 g
(includes PPU, valve, tank, etc.)		

Current micro-resistojets are a few centimeters in length. A pure MEMS resistojet is the vaporizing liquid microthruster, which is described in the next section, or the Free Molecular Resistojet from AFRL. A summary of the water (micro) resistojet is shown in Table 11.6.

11.2.7 Vaporizing Liquid Microthruster

One resistojet concept that is built on MEMS technology is the vaporizing liquid Microthruster (VLM) developed at the NASA JPL.[26,38,39] This microfabricated thruster device is primarily targeted for use in constellations of microspacecraft to serve as attitude control thrusters. The thruster vaporizes a suitable propellant, such as water, ammonia, or others stored compactly in its liquid phase, on demand for thrust generation. While the use of valves for gaseous propellants in MEMS devices has been problematic in the past due to unavoidable leakage of the liquid, propellant storage of the VLM reduces these concerns and, as already mentioned, reduces system mass and size requirements over high-pressure gaseous storage. The thruster chip itself is fabricated using MEMS technologies into silicon material and will ultimately be tightly integrated with a micropiezovalve to form a very compact thruster module.

11.2.7.1 Principle of Operation

The VLM is a pure resistojet, similar to the design by Surrey. Liquid propellants, like water, are pressure-fed between heater strips, vaporized, and expanded through a micronozzle, producing thrust.

To enable MEMS fabrication, innovative designs had to be employed. Due to the short distances, sufficient thermal insulation is necessary to limit power consumption to the small heating section. The current VLM concept design is T-shaped to thermally isolate the heater section from the bulk of the chip as shown in Figure 11.16. Figure 11.17 shows the laminate of three chips. The two outermost layers

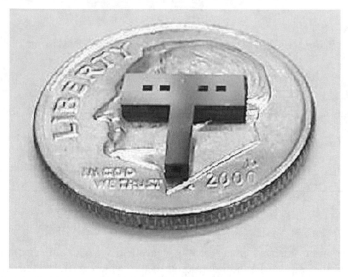

FIGURE 11.16 VLM thruster (T-shape). (*Source*: NASA.)

contain thin-film deposited gold heaters, spaced apart by a third spacer-chip. The chips are joined via a gold thermal compression bond, using a gold layer deposited in the same fabrication step as the heaters. Several chips were tested recently at JPL on a microthrust stand and yielded a thrust of 32 μN for a power level of 0.8 W, corresponding to a thrust-to-power ratio of 40 μN/W.

As indicated earlier valve fabrication is challenging. A new valve had to be developed to prevent leakage. The so-called microisolation valve (MIV) consists of two chips, one made from silicon and the other from Pyrex, anodically bonded together. The silicon chip features the flow channel, which can be blocked by a

FIGURE 11.17 Triple chip assembly. (*Source*: NASA.)

silicon barrier. An electric current is passed through the barrier, resistively heating and melting it. However, cracking of the barrier has been observed due to thermal shock, which leads to the incorporation of chip-integrated debris traps and filters.

This valve was shown to have a burst pressure of up to 3000 psi and can be opened with capacitor stored energies of 10–60 mJ using driver capacitances of 0.6–16 μF depending on barrier thickness, ranging from 25 to 50 μm tested.

11.2.7.2   System Requirements and Comments

Certainly, with a MEMS-fabricated device, thruster mass is low. However, to improve valve reliability, liquid propellants are used, which leads to a power penalty due to the need for propellant vaporization. As a benefit, the use of liquid propellants enables the use of a lower-mass and smaller propellant tank compared to an equivalent gaseous propellant storage system. Leakage concerns, often raised with the storage of high-pressure gaseous propellants, are also significantly less severe for liquid propellants, potentially increasing reliability of the system. At present, the VLM thruster uses water propellant for safety reasons and ease of handling in laboratory testing. Water is also storable at fairly high densities. In principle, any liquid propellant can be used that can be vaporized at significantly low power levels. Ammonia, for example, is another propellant candidate considered, having about half the heat of vaporization of water, which would lead to a more efficient thruster. A summary of the VLM is shown in Table 11.7, with a picture of the assembled thruster produced by NASA JPL in Figure 11.18.

11.3   CHEMICAL PROPULSION

In this chapter chemical propulsion systems are defined as those where the majority of the energy needed for operation is stored in the propellant.

TABLE 11.7
Performance Characteristics for VLM

I_{sp}	100 sec
Power	≈1.2W
Thrust	250 to 300 μN
Thrust or power	200 μN/W
Impulse or prop.	1 N sec/g
Feed mechan.	Yes
Current system dry mass (includes PPU, valve, tank, etc.)	??g

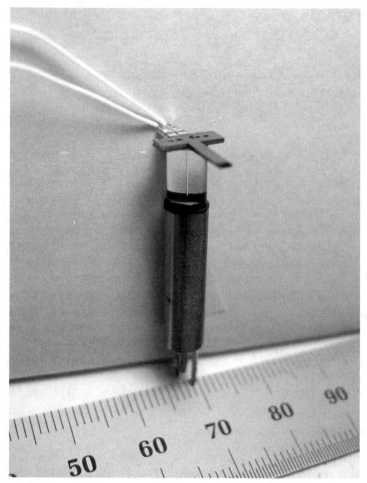

FIGURE 11.18 VLM thruster mounted to Lee Corp. solenoid valve via Pyrex thermal standoff. (*Source*: NASA.)

11.3.1 COLD GAS THRUSTER

The cold gas thruster is the simplest chemical propulsion system.[40–42] It typically consists of a pressurized gas chamber (propellant tank), a gas metering valve, a cavity chamber (gas plenum), and a converging–diverging shaped exit nozzle. By opening a valve, the pressurized gas is accelerated in the nozzle to produce thrust. Cold gas thrusters usually possess low specific impulse, and as a consequence, great care is exercised in their design to insure the efficient conversion of the pressurized fuel to thrust.

Miniaturized cold gas thrusters may become less efficient because of the increased importance of drag. Valve and nozzle design are the most important issues for this kind of thruster.

11.3.1.1 Principle of Operation

The cold gas thruster is based on gas leaving a pressurized tank into vacuum, which is accelerated in a converging–diverging nozzle as described in the resistojet section above. A valve is used to initiate and control this flow. Thrust is produced according to:

$$T \propto \dot{m} v_e \propto h_o \, \mathrm{Re} v_e \tag{11.27}$$

with $\dot{m}$ describing the mass flow, v the exit velocity, and h_o the nozzle height.

As with the resistojet, the thrust that is produced in such a system can be calculated with Equation (11.28) as:

$$T = \dot{m} \cdot v + (p_e - p_a) A_e \tag{11.28}$$

where the mass flow is given by:

$$\dot{m} = \frac{A_t a \rho_0}{\left(1 - \frac{k-1}{2}\right)^{\frac{1}{k-1}}} \tag{11.29}$$

Compared to the resistojet, it becomes obvious that the lack of additional heating forces the cold gas thruster to operate at high gas pressure. Another interesting aspect is that the colder gas leads to an increasing Reynolds number, because of the lower viscosity at lower temperature. While in principle this leads to higher thrust values, it will force a more careful production of the exhaust nozzle to keep the critical value up, which makes the production of small systems more challenging.

Looking at the cold gas thruster performance and assuming that the gas is at 300 K and purely molecular hydrogen ($k = 1.67$), the maximum exit velocity will amount to 2500 m/sec and the resulting thrust for a pressure of 2×10^6 Pa sec and a throat diameter of 1 mm^2 would amount to approximately 1.5 N.

11.3.1.2 System Requirements

The miniaturization of the cold gas thruster poses significant challenges in maintaining efficiency. As fluidic devices are miniaturized, the surface area to volume ratio increases, which can result in larger drag forces. The proper design of the exit nozzle is key to providing maximum thrust. A true 3D-axis symmetric hour-glass shape nozzle is more efficient than an extruded 2D h-glass nozzle, which is significantly easier to produce. Other issues involve leakage of gas through the closed valve, which is a common problem of MEMS devices. While MEMS-based cold gas thrusters have been developed in the past, reliability was a weak point and insufficient emphasis was put on complete system design for actual missions. Either the integrated tank was far too small, or it outsized the MEMS nozzle so much so that the advantage of using MEMS was minimal. Currently the most

TABLE 11.8
Cold Gas MiPS System Characteristics

Value	Units	Description
95	cc	Propellant volume
0.556	g/cc	Propellant density (liquid)
2028	sccm	Isobutane thruster flow rate (40 psia)
0.01	sec	Minimum pulse duration
65	sec	Specific impulse I_{sp}
55	mN	Thrust at 40 psia
1000	g	MEPSI spacecraft mass
53	g	Propellant mass
616	sec	Total thrust duration
34	N-sec	Total impulse
34	m/sec	Total delta V
0.55	mN-sec	Minimum impulse bit
61564		Max no. of minimum impulse bit firings

promising approach appears to be ChEMS™ technology, which is used for VAC-CO's cold gas systems. Using this technology eliminates tubing connections in favor of a single ChEMS manifold, so that the gas tank becomes the only "non-integrated" part. A summary of the cold gas thruster (example VACCO MIPs) is shown in Table 11.8, with a picture of the assembled thruster produced by VACCO in Figure 11.19.

FIGURE 11.19 Cold gas thruster system. (*Source*: VACCO.)

11.3.2 DIGITAL PROPULSION

Digital propulsion is a very compact and low-mass system, which relies on MEMS fabrication to provide a feasible propulsion device for small spacecraft.[43–47] A digital propulsion system (DPS) consists of a large array of sealed plenums. These plenums are filled with fuel or an inert substance in gas, liquid, or solid form. A thin diaphragm acts as the sealant. By igniting the fuel or heating the inert substance the pressure inside the plenum is increased sufficiently causing the diaphragm to rupture and release the propellant, producing an impulse. The magnitude of the impulse depends on the amount and kind of fuel stored inside the plenum.

While this kind of propulsion usually features small specific impulses, the ability to define the impulse bit by varying the fuel or plenum size and the number of plenums triggered simultaneously make this propulsion system very attractive. MEMS technology enables large number of plenums to be placed within a small area with low mass.

11.3.2.1 Principle of Operation

Typical MEMS-fabricated digital propulsion configurations consist of a three-layer sandwich. The top layer contains an array of thin diaphragms (of the order 0.5 μm thick silicon nitride). The middle layer contains an array of through-holes (often used: Schott FOTURAN® photosensitive glass, 1.5 mm thick, 300, 500, or 700 μm diameter holes), which are loaded with propellant. The bottom layer employs a matching array of polysilicon microresistors for propellant heating and fuel ignition. The bottom two layers are bonded together and then fueled. The top layer is bonded to complete the assembly as shown in Figure 11.20.

Once current is run through a microresistor underneath the plenum, heat is generated, which ignites the fuel (e.g., lead styphnate). The ignition increases the pressure in the plenum until the membrane ruptures and the gas inside is released to produce a single impulse bit. Typical pressures reach values of a few MPa. The

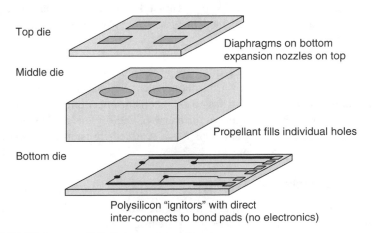

Top die

Diaphragms on bottom
expansion nozzles on top

Middle die

Propellant fills individual holes

Bottom die

Polysilicon "ignitors" with direct
inter-connects to bond pads (no electronics)

FIGURE 11.20 Layout of digital propulsion thruster.

TABLE 11.9
Digital Propulsion System Characteristics

I_{sp}	100 to 300 sec
Power	100 mJ/pulse
I-bit	≈ 100 μN sec
Thrust	100 mN
Thrust or power	1 mN/W
Impulse or prop.	≈ 0.5 N sec/g (lead styphnate)
Feed mechan.	No
Size (10,000 pixels)	10 cm × 10 cm

thrust produced is caused by the pressure difference between the plenum (P) and the vacuum, and can be described by $T = PA_E$, where A_E is the exit area. Thrust levels on the order of 10 mN can be produced.

The exit velocity depends on the mass of propellant utilized and the length of the burst. The relationship can be roughly estimated as:

$$v = \frac{\rho m}{A_E} \frac{1}{t} \tag{11.30}$$

typical exit velocities for millisecond long pulses reach 1000–3000 m/sec. The resulting impulse bits range from 1 to 100 μN sec.

The electrical power needed to ignite the fuel can be as low as 100 μJ.

11.3.2.2   System Requirements

The digital propulsion system is a very attractive system when it is based on MEMS technology. Compact arrays can be manufactured with a large number of individual pixels. Control of the amount of propellant in each pixel will enable even more flexibility by varying the impulse bit. Thrust levels can be controlled by the frequency of firing. No feed mechanism or any moving parts are needed for this system.

Problems still remaining include increasing of the pixel density while insuring the neighboring pixels are not ignited by heat transfer, enabling more efficient propellant combustion, and ensuring complete combustion of the propellant. A slight change of thrust vector has to be taken into account as well due to the changing location of thrust origin. A summary of the digital propulsion system is shown in Table 11.9, with a picture of the assembled thruster array produced by LAAS-CNRS (France) in Figure 11.21.

11.3.3   MONOPROPELLANT THRUSTER

Another chemical propulsion system employing MEMS technology is a miniaturized monopropellant thruster, such as the hydrogen peroxide microthruster.[48–50]

FIGURE 11.21 Digital propulsion array. (*Source*: LAAS, France.)

This thruster consists of a microfabricated reservoir from which the liquid propellant is injected into a catalytic chamber and due to chemical decomposition turned into the gaseous phase, which is exhausted through a converging–diverging nozzle. An I_{sp} of 130 sec is reached with this system producing thrust levels of up to 1 mN produced on an area of 2000 μm × 3000 μm.

11.3.3.1 Principle of Operation

The most important process for this thruster is the silver-catalyzed heat-assisted production of gas:

$$2H_2O_2 \text{ (l)} \rightarrow 2H_2O \text{ (l)} + O_2 + \text{heat} \tag{11.31}$$

Liquid hydrogen peroxide releases 586 cal/g of energy at 25°C. The corresponding adiabatic flame temperature is approximately 610°C. In order to achieve this process with a MEMS thruster, the liquid has to be pushed through a mesh coated with catalyst, causing the reaction to take place. The resulting gas will cause a pressure increase in the chamber and a nozzle will accelerate the flow similar to the mechanism in a resistojet.

The maximum exit velocity is therefore given as

$$v = \sqrt{2T_0 \left(\frac{k}{k-1}\right) \frac{R}{M}} \tag{11.32}$$

which in turn determines the mass flow rate for a fixed thrust level ($\approx$400 μg/sec for 1 mN).

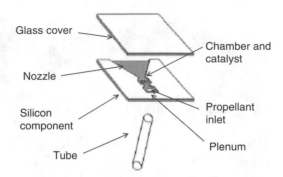

FIGURE 11.22 Principal setup of micro-hydrogen-peroxide thruster.

11.3.3.2 System Requirements

This thruster is the classical example of a downsized, well-proven macroscopic propulsion system. The thruster is produced in a three-layer step as shown in Figure 11.22. The etched features of the thruster body are connected to an inlet tube for the propellant and sealed with a Pyrex window. Great care has to be taken to ensure good coverage of silver for the catalytic chamber.

To date, complete catalytic conversion has not been obtained and a significant fraction of the propellant remains in its liquid phase. This might be due to the low Reynolds number flow inside the chamber. An SEM of the thruster is shown in Figure 11.23.

The insertion of the propellant has to be controlled with a MEMS valve. As liquid propellant is used, long-term leakage problems should be minor; however, the relatively high pressure (≈ 34 kPa) which is used might lead to problems. Storage of the hydrogen peroxide for longer periods of time might be a problem as it is known to undergo auto-decomposition under some conditions. A summary

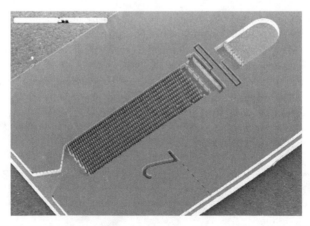

FIGURE 11.23 SEM of hydrogen peroxide thruster. (*Source*: University of Vermont.)

TABLE 11.10
Hydrogen Peroxide Thruster Characteristics

I_{sp}	130 sec
Power	<1 W
I-bit	<1 µNsec
Thrust	1 mN
Thrust or power	1 mN/W
Impulse or prop.	≈1 Nsec/g
Feed mechan.	No
Mass	<100 g

of the micro-hydrogen-peroxide system is shown in Table 11.10, with a picture of an assembled thruster shown in Figure 11.24.

11.4   RADIOISOTOPE PROPULSION

Another propulsion system worth discussing is radioisotope propulsion.[51] This kind of propulsion is one of the simplest systems that can be found and therefore, it should be introduced briefly. It is essentially a radioactive isotope emitting alpha particles. The amount of propulsion obtained can be adjusted by shielding the radioactive material or exposing it. Thrust obtainable can be adjusted from nN to mN by adjusting the kind and the mass of propellant used.

FIGURE 11.24 Hydrogen peroxide thruster (*Source*: University of Vermont.)

11.4.1 PRINCIPLE OF OPERATION

The microsatellite obtains a thrust as the radioisotope emits the alpha or beta, or both particles.

The thrust produced can be calculated as:

$$T = \dot{m}v = (A(3.7 \times 10^{10})m_P)\sqrt{\frac{2E}{m_P}} \qquad (11.33)$$

with A representing the radioisotope activity in C_i, m_P standing for the particle mass and E being the kinetic energy of the particle.

11.4.2 SYSTEM REQUIREMENTS

The system itself is extremely simple without feed mechanisms and only an optional shutter. The propellant mass can be minimized by selecting the adequate material. Alpha particles are easy to shield and therefore shielding should not require significant amounts of additional mass. A summary of the radioisotope propulsion system is shown in Table 11.11.

11.5 CONCLUSION

Ten different propulsion systems have been introduced, covering a large range of requirements for small satellite propulsion. It should be mentioned that while MEMS is an enabling technology for many of these systems, a significant amount of research needs to be done on the systems level to approach high technical readiness levels (TRL) that will lead to use of these technologies in upcoming missions. In September 2003, DARPA had initiated an effort to push MEMS technology into a different venue — that of high power electric propulsion. Field effect thrusters like colloids and FEEP systems with power levels of up to 1 kW

TABLE 11.11
Radioisotope Propulsion Characteristics

I_{sp}	$3.9.10^6$ sec	5.67 MeV alpha particles
I-bit (if pulsed)	$1.55.10^{-4}$ μNsec	Pulse rate is limited by valve — assume 1 KHz MEMS valve $\geq$ 1 msec
Rep. rate (if pulsed)	1 kHz	1 kHz MEMS valve is an option
Power	0 W—no valve/1 mW with valve	0 W required to generate particles
Thrust	0.155 μN	100 Curie of Pu-238
Thrust or power	Infinity μN/W (0 valve), 155 μN/W (with valve)	Without valve power consumption is zero, With valve operating at 1 mW
Impulse or prop.	1.58×10^{-5} Nsec/g	
Total impulse/sys.-mass	428 Nsec/2g	
Current system mass	2 g (with valve)	

---

based on MEMS technology have been proposed. In that case, MEMS technology is not used to minimize the propulsion system, but to allow for the manufacturing of large arrays in which each individual emitter operates at a highly efficient low level of emission current.

Research on MEMS propulsion is advancing quickly, and it should not be limited to the thrusters but include PPU and interface development as well to make "intelligent dust" a possibility in the near future.

REFERENCES

1. M. Birkan, *Formation Flying and Micro-Propulsion Workshop*, Lancaster, CA, October 1998.
2. J. Dunning and J. Sankovic, in 35th Joint Propulsion Conference, Los Angeles, CA, 1999. AIAA-99-2161.
3. R.A. Spores and M. Birkan, in 35th Joint Propulsion Conference, Los Angeles, CA, 1999. AIAA paper 99–2162.
4. M. Birkan, AAAF Propulsion Symposium, paper 16–355, 2002.
5. K.E. Tsiolkovsky. Issledovanie Mirovykh Prostransty Reaktivnymi Priborami (Exploration of Space with Rocket Devices). *Naootchnoye Obozreniye (Scientific Review)*, 1903.
6. R. Raju, B.P. Pandey, and S. Roy, Finite Element Model of Fluid Flow inside a Micro Thruster, AIAA paper 2002–5733, Nanotech, 2002.
7. J. Hammel, Development of an Unstructured 3-D Direct Simulation Monte Carlo/ Partice-In-Cell Code and the Simulation of Microthruster Flows, MS thesis, Worcester Polytechnic Institute, 2002.
8. J. Mueller, in *Micropropulsion for small spacecraft*, M.M. Micci and A.D. Ketsdever, (eds) *Progress in Astronautics and Aeronautics*, vol. 187, p. 45, 2000.
9. http://www.grc.nasa.gov/WWW/RT1997/6000/6910curran.htm.
10. P. Gessini and G. Paccani, Ablative Pulsed Plasma Thruster System Optimization for Microsatellites, IEPC paper 01–182, 27th IEPC, 2001
11. M. Keidar, I. Boyd, E. Antonsen, and G. Spanjers, Progress in Development of Modeling Capabilities for a Micro-Pulsed Plasma Thruster, 39th Joint Propulsion Conference, Huntsville, AL, 2003. AIAA paper 2003–5166.
12. J. Ziemer, Laser Ablation Microthruster Technology, 33rd Plasmadynamics and Lasers Conference, Maui, HI, 2002, AIAA paper 2002–2153.
13. J. Ziemer, Performance Scaling of Gas-Fed Pulsed Plasma Thrusters, PhD thesis, Princeton University, 2001.
14. C. Zakrzwski, S. Benson, P. Sanneman, and A. Hoskins, On-Orbit Testing of the EO-1 Pulsed Plasma Thruster, AIAA paper 2002.
15. F. Rysanek and R. Burton Performance and Heat Loss of a Coaxial Teflon Pulsed Plasma Thruster, IEPC paper 01–151, 27th IEPC, 2001.
16. www.busek.com
17. J. Schein, A. Anders, R. Binder, M. Krishnan, J.E. Polk, N. Qi, and J. Ziemer, Inductive energy storage driven vacuum arc thruster, *Review of Scientific Instruments*, vol. 72, no. 3, February 2002.
18. J. Schein, N. Qi, R. Binder, M. Krishnan, J.K. Ziemer, J.E. Polk, and A. Anders, Low Mass Vacuum Arc Thruster System for Station Keeping Missions, IEPC paper 01–228, 27th IEPC, 2001.

19. J. Schein, M. Krishnan, J. Ziemer, and J. Polk, Development of a miniature vacuum arc source as a space satellite thruster, *Proceedings of XXth ISDEIV, Tours, France*, 2002, pp. 664–669.

20. J. Schein, M. Krishnan, J. Ziemer, and J. Polk, Adding a "Throttle" to a Clustered Vacuum Arc Thruster, AIAA paper 2002–5716, Nanotech, 2002.

21. J. Schein, A. Gerhan, F. Rysanek, and M. Krishnan, Vacuum Arc Thruster for Cubesat Propulsion, IEPC-0276, 28th IEPC, 2003.

22. M. Tajmar, A. Genovese, N. Buldrini, and W. Steiger, Miniaturized Indium-FEEP Multiemitter Design and Performance, AIAA paper 2002–5718, Nanotech, 2002.

23. R. Forbes, Liquid–Metal Ion Sources and Electrosprays Operating in Cone-Jet Mode: Some Theoretical Comparisons and Comments, *Journal of Aerosol Science*, vol. 31, no. 1, 2000, pp. 97–120.

24. S. Marcuccio, A. Genovese, and M. Andrenucci, Experimental performance of field emission microthrusters, *Journal of Propulsion and Power*, vol. 14, no. 5, 1998, pp. 774–781.

25. A. Genovese, N. Buldrini, M. Tajmar, and W. Steiger, 2000h Endurance Test on an Indium FEEP Cluster, International Electric Propulsion Conference, IEPC-2003-102, Toulouse, 2003.

26. J. Ziemer, Performance Measurements Using a Sub-Micronewton Resolution Thrust Stand, International Electric Propulsion Conference, IEPC-01-238, Pasadena, CA, 2001.

27. C. Phipps, J. Luke, G. McDuff, and T. Lippert, Laser Ablation Powered Mini-Thruster, in C. Phipps (ed.), *High Power Laser Ablation IV, Proceedings of SPIE*, vol. 4760, 2002, pp. 833–842.

28. C. Phipps, J. Luke, and T. Lippert, Laser Ablation of Organic Coatings as a Basis for Micropropulsion, European Materials Research Society Meeting, Session H, Strasbourg, 2003.

29. R. Wirz, J. Polk, C. Marrese, J. Mueller, J. Escobedo, and P. Sheehan, Development and Testing of a 3 cm Electron Bombardment Micro-Ion Thruster, International Electric Propulsion Conference, IEPC-01-343, Pasadena, 2001.

30. H. Kaufman, Technology of Electron Bombardment Ion Thrusters, *Advances in Electronics and Electron Physics*, vol. 36, Academic Press, New York, NY, 1974, pp. 265–373.

31. R. Wirz, J. Polk, C. Marrese, and J. Mueller, Experimental and Computational Investigation of the Performance of a Micro-Ion Thruster, 38th AIAA JPC, Indianapolis, AIAA 2002–3835, 2002.

32. R.G. Jahn, Physics of electric propulsion, *McGRAW-HILL Series in Missile and Space Technology*, 1968.

33. M. Auweter-Kurtz, *Lichtbogenantriebe für Weltraumaufgaben*, B.G. Teubner, Stuttgart, 1992.

34. M. Sweeting, T. Lawrence, M. Paul, J. Sellers, L. Cowie, and D. Shields, *Results of Low-Cost Propulsion Activities at the University of Surrey*, ESA/ESTEC, Second European Spacecraft Propulsion Conference, ESTEC, Noordwijk, NL, 27–29 May 1997.

35. V.K. Bober, A.S. Koroteev, L.A. Latyshev, G.A. Popov, Yu.P. Rylov, and V.V. Zhurin, *State of Work on Electrical Thrusters in USSR*, Paper IEPC-91-003, AIDAA/AIAA/ DGLR/JSASS, 22nd International Electric Propulsion Conference, October 14–17, 1991, Viareggio, Italy.

36. Surrey Space Center, SSTL 9003–03 data sheet, 2002.

37. D. Gibbon, A. Baker, I. Coxhill, and M. Sweeting, The Development of a Family of Resistojet Thruster Propulsion Systems for Small Spacecraft. 17th Annual AIAA/USU, Conference on Small Satellites paper, SSC03-IV-8, 2003.

38. J. Mueller, I. Chakraborty, D. Bame, and W. Tang, The vaporizing liquid micro-thruster concept: preliminary results of initial feasibility studies, *Micropropulsion for Small Spacecraft, Progress in Astronautics and Aeronautics*, vol. 187, Chapter 8, M. Micci and A. Ketsdever (eds.), AIAA, Reston, VA, 2000.

39. J. Mueller, J. Ziemer, A. Green, and D. Bame, Performance Characterization of the Vaporizing Liquid Micro-Thruster (VLM), 28th International Electric Propulsion Conference, IEPC-2003-237, Toulouse, 2003.

40. P.D. Fuqua, S.W. Janson, W.W. hansen, and H. Helvajian, Fabrication of true 3D microstructures in glass/ceramic materials by pulsed UV laser volumetric exposure techniques, *Proceedings of SPIE 1999*, 1999, p. 213.

41. J. Cardin and J. Acosta, Design and Test of an Economical Cold Gas Propulsion System, 14th Annual AIAA/USU Conference on Small Satellites paper SSC00-, 2000.

42. R. Nanson III, Navier Stokes/Direct Simulation Monte Carlo Modeling of Small Cold Gas Thruster Nozzle and Plume Flows, Phd thesis. Worcester Polytechnic Institute (2002).

43. B. Larangot, C. Rossi, T. Camps, A. Berthold, P.Q. Cham, D. Briand, N.F. deRooij, M. Puig-Vidal, P. Miribel, E. Montane, E. Lopez, and J. Samitier, Solid Propellant Micro-Rockets-towards a New Type of Power MEMS', AIAA paper 2002–xxxx, Nanotech, 2002, www.laas.fr/Micropyros/Publication/Nanotech-paper.pdf.

44. D. Lewis et al., *Digital Micro-Propulsion Sensors and Actuators A, Physical*, vol. 80, no. 2, 2000, pp. 143–154.

45. S. Tanaka et al., MEMS-based solid propellant rocket array thruster with electrical feedthroughs, *Transactions of Japanese Society Aero Space Science*, vol. 46, no. 151, 2003, pp. 47–51.

46. E. Rudnyi et al., Solid Propellant Microthrusters: Theory of Operation and Modelling Strategy, AIAA paper 2002–5755, Nanotech, 2002.

47. Takahashi et al., Design and Testing of Mega-Bit Microthruster Arrays, AIAA paper 2002–5757, Nanotech, 2002.

48. D.H. Lee, J.S. Hwang, S.-E. Park, and S. Kwon, Thermochemical Design of a Micro Liquid Monopropellant Rocket with Catalytic Reaction of Hydrogen Peroxide, ASME IMECE 39193, International Mechanical Engineers Congress and Exhibition 2002, New Orleans, LA.

49. D. Hitt, C. Zakrzwski, and M. Thomas, MEMS based satellite micropropulsion via catalyzed hydrogen peroxide decomposition, *Smart Materials and Structure*, vol. 10, 2001, pp. 1163–1175.

50. A. Varia, Thermal Model of MEMS Thruster, NASA GSFC Propulsion Branch code 597 — Report, fcmci.gsfc.nasa.gov/workshop/2003/presentations/varia/Varia-MEMS_Thruster.ppt.

51. Prof. Amit Lal, Cornell University, *lal@ece.cornell.edu* — private communications.

12 MEMS Packaging for Space Applications

R. David Gerke and Danielle M. Wesolek

CONTENTS

12.1 INTRODUCTION TO FUNCTIONS OF MEMS PACKAGES

A package serves to integrate all of the components required for a system application in a manner that minimizes size, cost, mass, and complexity. It provides the interface between the components and the overall system. The following subsections present the three main functions of the microelectromechanical systems

269

(MEMS) package: mechanical support, protection from the environment, and electrical connection to other system components.

In addition to providing mechanical support, electrical connections, and thermal management, MEMS packaging for space applications must meet operational environmental requirements such as high temperature operation, thermal cycling, humidity, vibration, shock, radiation, outgassing, and depressurization, to name a few. Radiation, for example, can impact the on-board analog and digital microelectronic components of MEMS devices, the transduction mechanism of the sensor, and mechanical components of MEMS.[1]

12.1.1 MECHANICAL SUPPORT

Due to the very nature of MEMS being mechanical, the requirement to support and protect the device from thermal and mechanical shock, vibration, high acceleration, particles, and other physical damage (possibly radiation) during storage and operation of the part becomes critical. The mechanical stress endured depends on the mission or application. For example, landing a spacecraft on a planet's surface creates greater mechanical shock than experienced by a communication satellite operating in space. There is also a difference between space and terrestrial applications.

The coefficient of thermal expansion (CTE) of the package should be equal to or slightly greater than the CTE of silicon for reliability, since thermal shock or thermal cycling may cause die-cracking and delamination if the materials are unmatched or if the silicon is subject to tensile stress. Other important parameters are thermal resistance of the carrier, the material's electrical properties, its chemical properties, and resistance to corrosion.

Once the MEMS device is supported on a carrier (chip) and the wire bonds or other electrical connections are made, the assembly must be protected from scratches, particulates, and other physical damage. This is accomplished either by adding walls and a cover to the base or by encapsulating the assembly in plastic or other material. Since the electrical connections to the package are usually made through the walls, the walls are typically made from glass or ceramic. The glass or ceramic can also be used to provide electrical insulation of the leads as they exit through a conducting package wall (metal or composite materials). Although the CTE of the package walls and lid do not have to match the CTE of silicon-based MEMS as they are not in intimate contact (unless an encapsulating material is used), it should match the CTE of the carrier or the base to which the walls are connected.

12.1.2 PROTECTION FROM ENVIRONMENT

12.1.2.1 Simple — Mechanical Only

Many MEMS devices are designed to measure something in the immediate surrounding environment. These devices range from biological "sniffers" to chemical MEMS that measure concentrations of certain types of liquids. So the traditional "hermeticity" that is generally thought of for protecting microelectronic devices may not apply to all MEMS devices. These devices might be directly mounted on a printed circuit board (PCB) or a hybrid-like ceramic substrate and have nothing but a "housing" to protect it from mechanical damage such as dropping or something as simple as damage from the operator's thumb.

12.1.2.2 Traditional — Hermetic and Non-Hermetic

Many elements in the environment can cause corrosion or physical damage to the metal lines of the MEMS as well as other components in the package. Although there is little to no moisture in space, moisture remains a concern for MEMS in space applications since it may be introduced into the package during fabrication and before sealing. The susceptibility of the MEMS to moisture damage is dependent on the materials used in its manufacture. For example, aluminum lines can corrode quickly in the presence of moisture, whereas gold lines degrade slowly, if at all, in moisture. Also, junctions of dissimilar metals can corrode in the presence of moisture. Moisture is readily absorbed by some materials used in MEMS fabrication, die attachment, or within the package; this absorption causes swelling, stress, and possibly delamination.

To minimize these failure mechanisms, MEMS packages for high-reliability applications may need to be hermetic with the base, sidewalls, and lid constructed from materials that are good barriers to liquids and gases and do not trap gases releasing them later.

12.1.2.3 Custom — Vacuum Sealed Hermetic

Some MEMS chips, such as a MEMS accelerometer or MEMS magnetometer, require vacuum conditions to operate properly. For dual use devices (Earth-ambient conditions and space conditions), the chips need to be sealed in a package that contains vacuum conditions. As opposed to a typical hermetic microcircuit package that contains an inert atmosphere such as nitrogen, gases within the package can inhibit the movement of the extremely small moveable structures by causing aerodynamic drag. Selection of packaging materials becomes even more critical (than standard hermetic packages as described earlier) because even a small amount of trapped gas can raise the package internal pressure to levels which degrade the performance of the MEMS chip. In space applications, such devices could be packaged in a housing that contains a small hole to allow the atmosphere to escape after launch.

12.1.3 ELECTRICAL CONNECTION TO OTHER SYSTEM COMPONENTS

Because the package is the primary interface between the MEMS and the system, it must be capable of transferring DC power and in some designs, RF signals. In addition, the package may be required to distribute the DC and RF power to other components inside the package. The drive to reduce costs and system size by integrating more MEMS and other components into a single package increases the electrical distribution problems as the number of interconnects within the package increases.

When designs also require high-frequency RF signals, the signals can be introduced into the package along metal lines passing through the package walls, or they may be electromagnetically coupled into the package through apertures in the package walls. Ideally, RF energy is coupled between the system and the MEMS without any loss in power, but in practice, this is not possible since perfect conductors and insulators are not available. In addition, power may be lost to radiation, by reflection from components that are not impedance matched, or from discontinuities in the transmission lines. The final connection between the MEMS and the DC and RF lines is usually made with wire bonds; although flip-chip die attachment and multilayer interconnects using thin dielectric may also be possible.

12.2   TYPES OF MEMS PACKAGES

Each MEMS application usually requires a new package design to optimize its performance or to meet the needs of the system. It is possible to loosely group packages into several categories. Four of these categories are: (1) metal packages, (2) ceramic packages, (3) thin-film multilayer packages, and (4) plastic packages are presented below.

12.2.1   METAL PACKAGES

Metal packages are often used for microwave multichip modules and hybrid circuits because they provide excellent thermal dissipation and excellent electromagnetic shielding. They can have a large internal volume while still maintaining mechanical reliability. The package can either use an integrated base and sidewalls with a lid, or it can have a separate base, sidewalls, and lid. Inside the package, ceramic substrates or chip carriers are required for use with the feed-throughs.

The selection of the proper metal can be critical. CuW (10/90), Silvar® (a Ni–Fe alloy) (Semiconductor Packaging Materials, Armonk, NY), CuMo (15/85), and CuW (15/85) all have good thermal conductivity and a higher CTE than silicon, which makes them good choices. Kovar® (ESPI, Ashland, OR), a Fe–Ni–Co alloy is also commonly used. All of these materials, in addition to Alloy-42, may be used for the sidewalls and lid. Cu, Ag, or Au plating of the packages is commonly done.

Before final assembly, a bake is usually performed to drive out any trapped gas or moisture. This reduces the onset of corrosion-related failures. During assembly, the highest temperature-curing epoxies or solders should be used first and subsequent processing temperatures should decrease until the final lid seal is done at the lowest temperature to avoid later steps from damaging earlier steps. Au–Sn is a commonly used solder that works well when the two materials to be bonded have similar CTEs. Au–Sn solder joints of materials with a large CTE mismatch are susceptible to fatigue failures after temperature cycling. The Au–Sn intermetallics that form tend to be brittle and can accommodate only low amounts of stress.

Welding (using lasers to locally heat the joint between the two parts without raising the temperature of the entire part) is a commonly used alternative to solders. Regardless of the seal technology, no voids or misalignments can be tolerated since they can compromise the package hermeticity. Hermeticity can also be affected by the feedthroughs that are required in metal packages. These feedthroughs are generally made of glass or ceramic and each method (glass seal or alumina feedthrough) has its weakness. Glass can crack during handling and thermal cycling. The conductor exiting through the ceramic feedthrough may not seal properly due to metallurgical reasons. Generally, these failures are due to processing problems as the ceramic must be metallized so that the conductor (generally metal) may be soldered (or brazed) to it.

The metallization process must allow for complete wetting of the conducting pin to the ceramic. Incomplete wetting can show up as a failure during thermal cycle testing.

12.2.2   CERAMIC PACKAGES

Ceramic packages have several features that make them especially useful for microelectronics as well as MEMS. They provide low mass, are easily mass-produced, and can be low in cost. They can be made hermetic, and can more easily integrate signal distribution lines and feedthroughs. They can be machined to perform many different functions. By incorporating multiple layers of ceramics and interconnect lines, electrical performance of the package can be tailored to meet design requirements. These types of packages are generally referred to as co-fired multilayer ceramic packages. Details of the co-fired process are outlined below. Multilayer ceramic packages also allow reduced size and cost of the total system by integrating multiple MEMS and/or other components into a single, hermetic package. These multilayer packages offer significant size and mass reduction over metal-walled packages. Most of that advantage is derived by the use of three dimensions instead of two for interconnect lines.

Co-fired ceramic packages are constructed from individual pieces of ceramic in the "green" or unfired state. These materials are thin, pliable films. During a typical process, the films are stretched across a frame in a way similar to that used by an artist to stretch a canvas across a frame. On each layer, metal lines are deposited using thick-film processing (usually screen printing), and via holes for interlayer interconnects are drilled or punched. After all of the layers have been fabricated, the unfired pieces are stacked and aligned using registration holes and laminated together. Finally, the part is fired at a high temperature. MEMS and possibly other components are then attached into place (usually organically [epoxy] or metallurgically [solders]), and wire bonds are made the same as those used for metal packages.

Several problems can affect the reliability of this package type. First, the green-state ceramic shrinks during the firing step. The amount of shrinkage is dependent on the number and position of via holes and wells cut into each layer. Therefore, different layers may shrink more than others creating stress in the final package.

Second, because ceramic-to-metal adhesion is not as strong as ceramic-to-ceramic adhesion, sufficient ceramic surface area must be available to assure a good bond between layers. This eliminates the possibility of continuous ground planes for power distribution and shielding. Instead, metal grids are used for these purposes. Third, the processing temperature and ceramic properties limit the choice of metal lines. To eliminate warping, the shrinkage rate of the metal and ceramic must be matched. Also, the metal must not react chemically with the ceramic during the firing process. The metals most frequently used are W and Mo. There is a class of low temperature co-fired ceramic (LTCC) packages. The conductors that are generally used are Ag, AgPd, Au, and AuPt. Ag migration has been reported to occur at high temperatures, high humidity, and along faults in the ceramic of LTCC.

12.2.3 THIN-FILM MULTILAYER PACKAGES

Within the broad subject of thin-film multilayer packages, two general technologies are used. One uses sheets of polyimide laminated together in a way similar to that used for the LTCC packages described above, except that a final firing is not required. Each individual sheet is typically 25 μm and is processed separately using thin-film metal processing. The second technique also uses polyimide, but each layer is spun onto and baked on the carrier or substrate to form 1 to 20 μm-thick layers. In this method, via holes are either wet etched or reactive ion etched (RIE). The polyimide for both methods has a relative permittivity of 2.8 to 3.2. Since the permittivity is low and the layers are thin, the same characteristic impedance lines can be fabricated with less line-to-line coupling; therefore, closer spacing of lines is possible. In addition, the low permittivity results in low line capacitance and therefore faster circuits.

12.2.4 PLASTIC PACKAGES

Plastic packages have been widely used by the electronics industry for many years and for almost every application because of their low manufacturing cost. High-reliability applications are an exception because serious reliability questions have been raised. Plastic packages are not hermetic, and hermetic seals are generally required for high-reliability applications. The packages are also susceptible to cracking in humid environments during temperature cycling of the surface mount assembly of the package to the motherboard. For these reasons, plastic packages have not gained wide acceptance in the field of space applications. However, there are notable semiconductor designs that are beginning to be flown in space applications. Programs such as commercial off-the-shelf (COTS), which include plastic encapsulated microelectronics (PEMs) are gaining acceptance. For example, suitable PEMs were used for the Applied Physics Laboratory Thermosphere–Ionosphere–Mesosphere Energetics and Dynamics (TIMED) program. The size, cost, and weight constraints of the TIMED mission were achieved only through the use of commercially available devices.[2]

12.3 PACKAGE-TO-MEMS ATTACHMENT

The method used to attach a MEMS device to a package is a general technology applicable to most integrated circuit (IC) devices. Generally referred to as *die attach*, the function serves several critical functions. The main function is to provide good mechanical attachment of the MEMS structure to the package base. This ensures that the MEMS chip (or die) does not move relative to the package base. It must survive hot and cold temperatures, moisture, shock, and vibration. The attachment may also be required to provide a good thermal path between the MEMS structure and the package base. Should heat be generated by the MEMS structure or by the support circuitry, the attachment material should be able to conduct the heat from the chip to the package base. The heat can be conducted away from the chip and "spread" to the package base, which is larger and has more thermal mass. This spread can keep the device operating in the desired temperature range. If the support circuitry requires good electrical contact from the silicon to the package base, the attachment material should be able to accommodate the task.

The stability and reliability of the attach material are largely dictated by the ability of the material to withstand thermomechanical stresses created by the differences in the CTE between the MEMS silicon and the package base material. These stresses are concentrated at the interface between the MEMS silicon backside and the attach material and the interface between the die-attach material and the package base as shown in Figure 12.1. Silicon has a CTE between 2 and 3 ppm/°C while most package bases have higher CTE (6 to 20 ppm/°C). An expression that relates the number of thermal cycles that a die attach can withstand before failure, $N(f)$, is based on the Coffin–Manson relationship for strain. Equation (12.1) defines the case for die attach:

$$N(f) \propto \gamma^m \left(\frac{2t}{L \Delta CTE \Delta T} \right)^m \tag{12.1}$$

where
γ = shear strain
m = material constant
L = diagonal length of the die
f = thermal cycle frequency
t = die-attach material thickness
ΔT = magnitude of the temperature change in a cycle
ΔCTE = CTE between substrate and chip

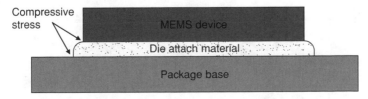

FIGURE 12.1 MEMS device in compression.

Voids in the die-attach material cause areas of localized stress concentration that can lead to premature delamination. Presently, MEMS packages use solders, adhesives, or epoxies for die attach. Each method has advantages and disadvantages that affect the overall MEMS reliability. Generally, when a solder is used, the silicon die would have a gold backing. Au–Sn (80–20) solder generally is used and forms an Au–Sn eutectic when the assembly is heated to approximately 250°C in the presence of a forming gas. When this method is applied, a single rigid assembled part with low thermal and electrical resistances between the MEMS device and the package is obtained. One problem with this attachment method is that the solder attach is rigid (and brittle) which means it is critical for the MEMS device and the package CTEs to match since the solder cannot absorb the stresses.

Adhesives and epoxies are comprised of a bonding material filled with metal flakes as shown in Figure 12.2. Typically, silver flakes are used as the metal filler since it has good electrical conductivity and has been shown not to migrate through the die-attach material.[3,4] These die-attach materials have the advantage of lower process temperatures. Generally between 100 and 200°C are required to cure the material. They also have a lower built-in stress from the assembly process as compared to solder attachment. Furthermore, since the die attach does not create a rigid assembly, shear stresses caused by thermal cycling and mechanical forces are relieved to some extent.[5,6] One particular disadvantage of the soft die-attach materials is that they have a significantly higher electrical resistivity which is 10 to 50 times greater than solder and a thermal resistivity which is 5 to 10 times greater than solder. Lastly, humidity has been shown to increase the aging process of the die-attach material.[4]

12.4   THERMAL MANAGEMENT CONSIDERATIONS

For small signal circuits, the temperature of the device junction does not increase substantially during operation, and thermal dissipation from the MEMS is not a problem.

However, with the push to increase the integration of MEMS with power from other circuits such as amplifiers perhaps even within a single package, the temperature rise in the device junctions can be substantial and cause the circuits to operate in an unsafe region. Therefore, thermal dissipation requirements for power

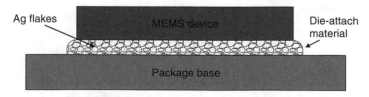

FIGURE 12.2 Schematic representation of silver filled epoxy resin.

amplifiers, other large signal circuits, and highly integrated packages can place severe design constraints on the package design. The junction temperature (T_j) of an isolated device can be determined by

$$T_j = QR + T_{case} \qquad (12.2)$$

where
  Q (W) is the heat dissipated by the junction and is dependent on the output power
    of the device and its efficiency,
  R (°C/W) is the thermal resistance between the junction and the case, and
  T_{case} (°C) is the temperature of the case.

Normally, the package designer has no control over Q and the case temperature, and therefore, it is the thermal resistance of the package that must be minimized. Figure 12.3 is a schematic representation of the thermal circuit for a typical package, where it is assumed that the package base is in contact with a heat sink or case.

It is seen that there are three thermal resistances that must be minimized: the resistance through the package substrate, the resistance through the die-attach material, and the resistance through the carrier or package base. Furthermore, the thermal resistance of each is dependent on the thermal conductance and the thickness of the material. A package base made of metal or metal composites has very low thermal resistance and therefore does not add substantially to the total resistance. When electrically insulating materials are used for bases, metal-filled via holes are routinely used, under the MEMS, to provide a thermal path to the heat sink. Although thermal resistance is a consideration in the choice of the die-attach material, adhesion and bond strength are even more important. To minimize the thermal resistance through the die-attach material, the material must be thin, there can be no voids, and the two surfaces to be bonded should be smooth.

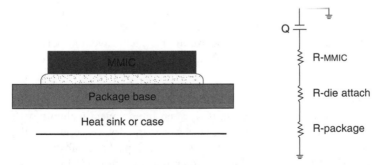

FIGURE 12.3 Cross section of MMIC attached to a package and its equivalent thermal circuit.

12.5 MULTICHIP PACKAGING

12.5.1 MCM/HDI

Multichip packaging of MEMS can be a viable means of integrating MEMS with other microelectronic technologies such as complementary metal oxide semiconductor (CMOS). One of the primary advantages of using multichip packaging, as a vehicle for MEMS and microelectronics, is the ability to efficiently host die from different or incompatible fabrication processes into a common substrate. High-performance multichip module (MCM) technology has progressed rapidly in the past decade, which makes it attractive for use with MEMS.

The chip-on-flex (COF) process has been adapted for the packaging of MEMS.[7] One of the primary areas of the work was reducing the potential for heat damage to the MEMS devices during laser ablation. Additional processing has also been added to minimize the impact of incidental residue on the die.[8]

12.5.1.1 COF/HDI Technology

COF is an extension of the high density interconnect (HDI) technology developed in the late 1980s. The standard HDI "chips first" process consists of embedding bare die in cavities milled into a ceramic substrate and then fabricating a layered thin-film interconnect structure on top of the components. Each layer in the HDI interconnect overlay is constructed by bonding a dielectric film on the substrate and forming via holes through laser ablation. The metallization is created through sputtering and photolithography.[9]

COF processing retains the interconnect overlay used in HDI, but molded plastic is used in place of the ceramic substrate. Figure 12.4 shows the COF process flow. Unlike HDI, the interconnect overlay is prefabricated before chip attachment. After the chip(s) have been bonded to the overlay, a substrate is formed around the components using a plastic mold forming process such as transfer, compression, or injection molding. Vias are then laser drilled to the component bond pads and the metallization is sputtered and patterned to form the low impedance interconnects.[10]

For MEMS packaging, the COF process is augmented by adding a processing step for laser ablating large windows in the interconnect overlay to allow physical access to the MEMS devices. Figure 12.5 depicts the additional laser ablation step for MEMS packaging. Additional plasma etching is also included after the via-drilling and large area laser ablations to minimize adhesive and polyimide residue that accumulates in the exposed windows.

12.5.2 FLIP-CHIP

Controlled collapse chip connection (C4) is an interconnect technology developed by IBM during the 1960s as an alternative to manual wire bonding. Often called "flip-chip," C4 attaches a chip top-face-down on a package substrate as shown in Figure 12.6. Electrical and mechanical interconnects are made by means of plated solder bumps between bond pads and metal pads on the package substrate.

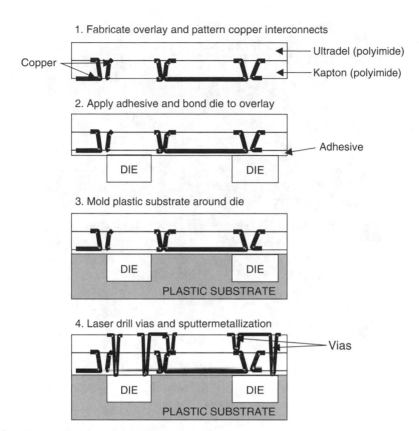

FIGURE 12.4 Chip-on-flex (COF) process flow.

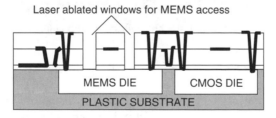

FIGURE 12.5 Large area ablation for MEMS access in COF package.

The flip-chip process is self-aligning, that is, the surface tension in the molten solder is sufficient to correct for misalignments in the positioning process. The action compensates for the slight chip-to-substrate misalignment incurred during die placement. The attachment between chip and package substrate is close in proximity, with typical spacing on the order of 50 to 200 μm.

Flip-chip allows the placement of bond pads over the entire chip, enabling increased interconnect density unlike wire bonding which requires that bond pads be

FIGURE 12.6 C4 (controlled collapse chip connection) flip-chip.

placed around the periphery of the die. The ability to closely pack a number of distinct chips on a single package makes flip-chip technology especially attractive to the MEMS industry.[11] An added feature of flip-chip is the ability to rework. Several techniques exist that allow for removal and replacement of the chip without scrapping the chip or the substrate. In fact, rework can be performed several times without degrading quality or reliability. For improved reliability, chip underfill may be injected between the joined chip and the package substrate. Care should be taken that the underfill covers the entire underside without air pockets or voids, and forms complete edge fillets around all four sides of the chip to avoid high-stress concentrations.

12.5.3 SYSTEM ON A CHIP

System on a chip (SOAC) may not necessarily be classified as a packaging technology. It is derived from the wafer fabrication process where numerous individual functions are processed on a single piece of silicon. These processes, generally CMOS technology, are compatible with the MEMS processing technology. Most SOAC chips are designed with a microprocessor of some type, some memory, some signal processing and others. It is very conceivable that a MEMS device could one day be incorporated on a SOAC.

12.6   EXAMPLE APPLICATIONS OF MEMS FOR SPACE

Many types of MEMS devices have been proposed for application to space systems, all of which serve to reduce size, weight, cost, and power consumption. Examples of common sensors and actuators that are considered for space applications include inertial sensors such as accelerometers, gyroscopes, and magnetometers; remote sensors such as spectrometers, shutters or filters, bolometers, and optical elements; and subsystems such as propulsion and active mechanical and thermal control systems. This section will focus on MEMS packaging technologies incorporated in applications of space-science instruments and subsystems.

12.6.1   VARIABLE EMITTANCE COATING INSTRUMENT FOR SPACE TECHNOLOGY 5

Novel packaging techniques that are needed to place MEMS-based thermal control devices on the skin of a satellite are addressed in the Variable Emittance Coating Instrument developed by the Johns Hopkins University Applied Physics Laboratory (JHU/APL). The instrument consists of two components: the MEMS shutter array (MSA) radiator and the electronic control unit (ECU). The MSA radiator is located on the bottom deck of the spin-stabilized Space Technology 5 (ST5) spacecraft, whereas the ECU is located within the spacecraft.

The instrument consists of an array of 36 dies, each 12.65×13.03 mm, which consists of arrays of 150-μm long and 6-μm wide shutters driven by electrostatic comb drives, mounted on a radiator. The gold-coated shutters open and close over the substrate and change the apparent emittance of the radiator. The device had to be on the exposed side of the radiator, and any cover had to be infrared transparent well into the far infrared. An additional requirement was that the substrate be thermally and electrically coupled to the radiator to allow heat transfer and prevention of electric charging effects.

In order to manage the thermal expansion mismatch between Al and Si for the survival temperature range, -45 to 65°C, an intermediate carrier made from aluminum nitrate was used. Sets of six dies, with wirebonds connecting all the common inputs, are attached to the aluminum nitride substrate, shown in Figure 12.7, with conductive epoxy, which themselves are attached to the aluminum radiator with epoxy. The radiator package contains heaters and is pigtailed to the connectors for the electronic control unit inside the spacecraft.

A photograph of the entire package is shown in Figure 12.8. In order to eliminate the concern associated with potential particulates from integration and

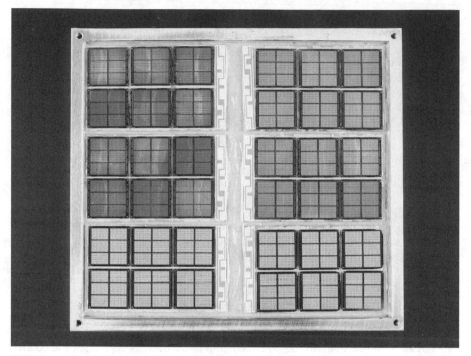

FIGURE 12.7 Sets of six VEC-MEMS shutter array die attached to the aluminum nitride substrate. (*Source*: JHU/APL.)

test or the launch environment, protection of this instrument is achieved using a polymer commercially known as CP1 that is both IR transmissive and electrically conductive. A film, less than 4-μm thick, is sandwiched in tension between two window frames and bonded in place. The CP1 film is suspended several millimeters above the shutters, providing a barrier between the MEMS die and the environment. Electrical conductivity of the film is achieved through the application of a thin coating of $InSnO_2$. This oxide coating serves to protect the CP1 from degradation in the presence of atomic oxygen.[12]

12.6.2   FLAT PLASMA SPECTROMETER FOR THE USAFA FalconSAT-3

MEMS microfabrication and packaging techniques enabled fabrication and system integration of a miniature flat plasma spectrometer (FlaPS) capable of making fine resolution measurements of the kinetic energy spectra and angular distributions of ions in a space environment. Conceived conceptually by NASA Goddard Space Flight Center (GSFC) in conjunction with the Air Force Academy, and designed, fabricated, and packaged by JHU/APL, the FlaPS reduces a plasma spectrometer for space from the size of a coffee-urn to that of a teacup. FlaPs will be placed as a payload on the USAFA FalconSAT-3 satellite, and will measure ion spectra differential in energy with a DE/E ~ 5%. The instrument includes a sensor-head array,

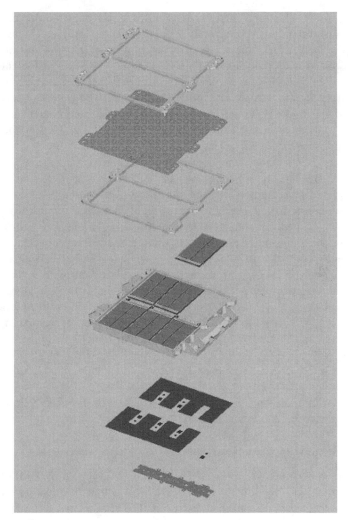

FIGURE 12.8 ST5 — VEC-MEMS shutter array (MSA) radiator. (From JHU/APL.)

printed circuit board with rad hard amplifier array electronics, power supply, and chassis and occupies a volume of approximately 200 cm^3 in a 0.5 kg, 300 mW package. The sensor head, fabricated and assembled at the wafer-level and stacked in a planar geometry, consists of an array of five identical spectrometer modules, each with a different fixed field-of-view consisting of a collimator, electrostatic analyzer, energy selector masks, microchannel plates (MCP), and anode plate for detection.[13]

The region of the sensor head comprising the collimator, electrostatic analyzer, and energy selector masks consists of three layers of silicon and two layers of beryllium copper. This total of five layers of materials aligned and stacked in an alternating fashion provided some unique assembly and packaging challenges. Each silicon layer has five dies (one per pixel) with an array of aperture slits on each die. The CuBe plates were precision machined to achieve the array of channels, each with a fixed field-of-view, with placement accuracy between arrays sufficient to allow for the integration of an array of five silicon die. The wafers were diced such that each of the five dies could be individually aligned and bonded to the CuBe plates using a flip-chip die-attach bonding technique. Low stress conductive and nonconductive epoxies were selected for bonding the five layers to each other because of high mismatches in coefficients of thermal expansion between the silicon and CuBe. The bonded components of the sensor head were packaged within the iridite-plated aluminum supporting structure via mounting brackets and aluminum rods used for maintaining a 1-mm offset to the MCP. The lower flange of the MCP was adhesively attached to the insulating mounting plate, which is attached to the housing with screws around the perimeter. The remaining items were assembled and packaged into the spacecraft mechanical interface housing using 2–56 and 4–40 screws. Spot welding a high voltage lead to the upper and lower plates of the MCP provided electrical connection from the HV power supply. A AuNi plated Kovar lead was welded to the CuBe electrostatic analyzer in order to provide an accessible site for soldering a scan voltage supply and ground wire from the PCB. An Sn63Pb37 solder was used to connect the power supply to the PCB, and from each preamplifier discriminator circuit on the PCB to the plated through vias on each anode. In addition, ensuring a conductive bleed path from every conductive surface to spacecraft ground mitigated potential charging effects. The packaging scheme of the FlaPS instrument is illustrated in Figure 12.9.

12.6.3 Micromirror Arrays for the James Webb Space Telescope

In support of the James Webb Space Telescope (JWST), equipped with the multi-object-spectrometer, individually addressable MEMS mirror arrays serving as a slit mask for the spectrometer, will selectively direct light rays from different regions of space into the spectrometer. An integrated micromirror array or CMOS driver chip was designed at NASA GSFC. System requirements posed several challenges to the packaging of the integrated MEMS chips. However, flip-chip technology to bump-bond the large chips (9 × 9 cm) onto a silicon substrate in a 2 × 2 mosaic pattern was used to eliminate the concern for global thermomechanical stresses due to mismatched coefficients of thermal expansion between the chip and substrate. Alignment of the chips forming the mosaic pattern was also a critical system specification. The relative tilt angle between the chips was held within 0.05° by making use of the restoring force of the solder bumps to self-align the chips during flip-chip solder reflow. The attached MMA or CMOS assembly was placed inside a package and fixed via peripheral pressure contacts. And, finally, input or output leads were made via tape-automated bonding from the package to the chips.[14]

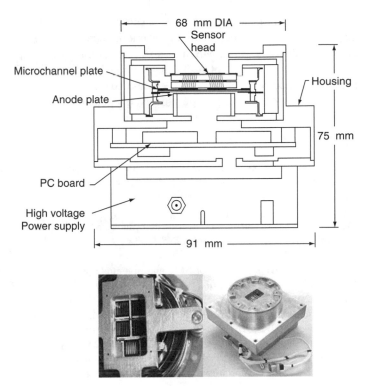

FIGURE 12.9 FlaPS for FalconSAT-3. (Top): schematic of FlaPS package. (Bottom): top view of sensor-head array (left); packaged instrument showing chassis enclosure housing amplifier array electronics, spacecraft interface bus, power supply, and sensor-head array (right). (*Source*: JHU/APL.)

12.7 CONCLUSION

We have shown a number of packaging approaches which can be, and have been, used for MEMS devices in space applications. The examples also showed that, while in the semiconductor industry, packaging is a way to protect the devices from environmental conditions including radiation, this no longer holds for packaging of MEMS devices. Many of the devices, actuators, sensors, etc. need to be exposed to the environment to perform their function. It seems like an oxymoron, a package that provides protection and allows exposure at the same time. In addition, due to the individuality of different MEMS devices, there is no general package solution; almost each device requires its own package approach.

Using MEMS devices in space applications increases the challenge even further. The package needs to protect the device in a number of changing conditions such as environmental tests, storage in humid air at prelaunch, environmental conditions during launch, and space environment with radiation, micrometeorites, UV light, vacuum, and high temperature variations. There have been very little flight opportunities so far which has allowed a good assessment of packaging approaches for MEMS in space applications. As shown, most designs and materials are still based on the experience with the semiconductor devices. In order to accelerate the introduction of MEMS into spacecraft, more flight opportunities are neseccary to allow a selection of packaging approaches, and a strong exchange of knowledge is required between the engineers and space institutions to omit error repetition. This chapter should help to get this exchange started.

REFERENCES

1. Muller, L., M.H. Hecht, et al., Packaging and qualification of MEMS-based space systems, *Proceedings of the 1995 9th Annual International Workshop on Micro Electro Mechanical Systems, February 11–15 1996, San Diego, CA, USA*, IEEE, Piscataway, NJ, USA, 1996.
2. Moor, A., et al., The case for plastic encapsulated microcircuits in space flight applications, *The Johns Hopkins University Applied Physics Lab Technical Digest*, Vol. 20, No. 1, 1999.
3. Hvims, H.L., Conductive adhesives for SMT and potential applications, *IEEE Transactions on Components, Packaging, and Manufacturing Technology — Part-B*, Vol. 18, No. 2, pp. 284–291, May 1995.
4. Rusanen, O. and J. Lenkkeri, Reliability issues of replacing solder with conductive adhesives in power modules, *IEEE Transactions on Components, Packaging, and Manufacturing Technology — Part-B*, Vol. 18, No. 2, pp. 320–325, May 1995.
5. Tuhus, T. and A. Bjomeklett, Thermal cycling reliability of die bonding adhesives, *1993 IEEE Annual International Reliability Physics Symposium Digest*, 208 pp, March 23–25, 1993.
6. Yalamanchili, P. and A. Christou, Finite element analysis of millimeter wave MMIC packages subjected to temperature cycling and constant acceleration, *1993 GaAs REL Workshop Programs and Abstracts*, October 10, 1993.
7. Butler, J., V. Bright, and J. Comtois, Advanced multichip module packaging of microelectromechanical systems, *Tech Digest of the 9th International Conference on Solid-State Sensors and Actuators (Transducers '97)*, Vol. 1, pp. 261–264, June 1997.
8. Butler, J., V. Bright, R. Saia, and J. Comtois, Extension of high density interconnect multichip module technology for MEMS packaging, *SPIE*, Vol. 3224, pp. 169–177, 1997.
9. Daum, W., W. Burdick Jr., and R. Fillion, Overlay high-density interconnect: a chips-first multichip module technology, *IEEE Computer*, Vol. 26, No. 4, pp. 23–29, April 1993.
10. Filtion, R., R. Wojnarowski, R. Saia, and D. Kuk, Demonstration of a chip scale chip-on-flex technology, *Proceedings of the 1996 International Conference on Multichip Modules, SPIE*, Vol. 2794, pp. 351–356, April 1996.
11. Maluf, N., *An Introduction to Microelectromechanical Systems Engineering*, Artech House, Inc, Boston, 2000.
12. Darrin, M.A., R. Osiander, J. Lehlonen, D. Farrar, D. Douglas, and T. Swanson, Novel micro electro mechanical systems (MEMS) packaging for the skin of the satellite, *Proceedings of 2004 IEEE Aerospace Conference 4*, pp. 2486–2494, 2004.
13. Wesolek, D.M., A. Darrin, R. Osiander, J.S. Lehtonen, R.L. Edwards, and F.A. Hererro, Micro processing a path to aggressive instrument miniaturization for micro and picosats, *2005 IEEE Aerospace Conference, March 2005*, Big Sky, Montana.

14. Lu, G.-Q., J. Calata, et al., Packaging of large-area, individually addressable, micro-mirror arrays for the next generation space telescope. *Design, Test, Integration, and Packaging of MEMS/MOEMS 2002, May 6–8 2002, Cannes, France*, 2002.

13 Handling and Contamination Control Considerations for Critical Space Applications

Philip T. Chen and R. David Gerke

CONTENTS

13.1 INTRODUCTION

No characteristic of microelectromechanical systems (MEMS) devices sets them apart from integrated circuits (ICs) more clearly than their sensitivity to surface contamination. An IC wafer leaves the foundry passivated for normal environmental exposure; a MEMS wafer does not. As a result, standard back-end processing steps (dicing, pick and place, die attach, wire bonding or bumping, and packaging) commonly used for ICs cannot be used for MEMS.

13.2 WAFER HANDLING

Particular attention and concern must be given to the handling and processing of MEMS devices to optimize yield and reduce the possibility of inducing latent defects into the product line. Handling considerations from the microelectronic world are not totally adequate and need to be supplemented in the MEMS device arena. Although MEMS wafer fabrication has been developed around the microcircuit IC type manufacturing process and equipment, the devices are different. MEMS fabrication does not identically track the CMOS IC world either. In general, the IC design rules have been the basis for the MEMS fabrication tools. These rules and practices, however, have their limitations.

MEMS designs often call for a double-sided wafer processing as opposed to the one-sided processing world of microelectronic manufacturing. Second-side processing for microcircuits is normally only a backside grind or other thinning techniques such as etching. In addition, for MEMS, there may be significant variation in the thickness of the wafer as seen in such components as accelerometers, and of course different thickness for different technology types. MEMS devices are heterogeneous as opposed to the homogeneity of ICs. Material variations in the MEMS device arena cover a broader range than in the microcircuit work where there is a more limited choice of substrate material (germanium, silicon, gallium arsenide, silicon on insulator, etc.) with the dominant material being silicon. Moreover, traditional substrate material from the microcircuit world such as glass and quartz may be used with MEMS.

At the MEMS wafer level, concern and attention must be paid to prevent damage to structures from tooling, considering both the front and back sides of the wafer. For example, with processing that involves both sides of the wafer, care must be taken to minimize particles and other contaminants at the back of the wafer. Equipment may need to be modified to work with both sides of the wafer. Wafer thickness and equipment versatility must also be considered. Since different technologies drive different thicknesses, awareness that the clean room will have various thicknesses in process at the same time adds to the complexity of tooling. It would not be unrealistic to find thicknesses of device types varying by as much as several hundred microns. Foundries are more vulnerable to the concerns of varying MEMS thickness in manufacturing than a custom MEMS house. Foundries attempt some design guidelines in this area to minimize the complexity of their processing job. Traditional processing tools often must be adapted in order to accommodate these differences.

Inspection criteria do not exist for anomalies induced by back and front side handling, and therefore extra precaution at the design stage is necessary. For example, it is essential to assure that there is an adequate, dedicated ring area around the perimeter of the wafer for handling. These areas may typically have a width of 7 mm. Tool marks and contact with chucks should be constrained to these areas, and the area of interest should be clean. However, marks or anomalies that will be removed or cleaned up at a later step should not be cause for rejection.

13.3   HANDLING DURING DIE SINGULATION, RELEASE, AND PACKAGING

Wafers will often be received, if coming from an external or an in-house foundry, in a nonreleased condition in order to protect the MEMS devices during transportation. In a typical process, the assembly house will have the responsibility to separate the wafer into smaller dies and perform the release. In most cases, the fabrication of surface micromachined devices involves layering, or intercalation steps, or both to add mechanical protection. This protective material (often SiO_2) must be etched in order to liberate or "release" the device from its carrier substrate. The drying procedure is critical to minimize stiction for many structures.

13.3.1   DIE SINGULATION

Die singulation is a process in which a wafer is sawed into many single die segments. Slicing and dicing down to the single die unit in the MEMS industry has more intricate concerns than it does in the IC industry. Once released, active movable components on the surface of the chips (front and back sides) are particularly susceptible to damage from traditional microcircuit handling and cleaning methods. The easiest way to get around this is to singulate the die before the MEMS devices are released. This way the moving structures are protected and can hardly be damaged by particulates and contamination such as saw slurry, particles generated by laser scribing and from scribe and break. In some cases, when the devices are less sensitive to particulates, it can be advantageous to release the die before singulation. It might be less labor intensive to release an entire wafer instead of hundreds of small dies. Traditional techniques such as using forced inert gas to blow particulates off the chip and other handling methods, such as vacuum pick-ups, may compromise the devices. For skilled laborers in these areas, who are used to these techniques and have used them successfully in the IC industry, retraining is required to preclude damage when handling MEMS. Unlike standard ICs, MEMS devices cannot be easily cleaned once they have been released. For this reason MEMS wafers must be singulated (cut up into individual die) and assembled using very specialized techniques.

13.3.2   HANDLING DURING RELEASE

Once the dies are singulated, they can be released. Special handling and process controls will normally be put into place to reduce the possibility of stiction in the drying process. The surface tension during drying can pull the moveable members together increasing the likelihood of stiction. Several design options are available to reduce the possibility of stiction, including the use of stand-off bumps, sacrificial polymer, and polymer columns sustaining the released structure.[1] Options for the process include special mixes of methanol, hydrofluoric acid (HF) vapor, and supercritical drying. The supercritical CO_2 drying method[2] takes advantage of the supercritical transition of a fluid, avoiding the formation of an interface between the liquid and gas. Vapor-phase HF etching[3] at an elevated temperature has often been used for dry etching of the sacrificial layer.

The typical Microelectronics Center of North Carolina (MCNC) (http://www.mcnc.org/) procedure used for multiuser MEMS processes (MUMPS) runs, which etches 2 μm thick phosphosilicate glass (PSG) layers with a minimum of 30 μm between release holes, is as follows:

1. Coat the dies with photoresist for protection in sawing, strip it by soaking in acetone for 20 min; mild agitation is helpful.
2. Etch the PSG in fully concentrated (49%) HF at room temperature for 2.5 min, with gentle agitation.
3. Quickly transfer to DI water and rinse for 10 min.
4. Quickly transfer to isopropyl alcohol for a 5-min rinse.
5. Remove from methanol and *immediately* bake for 10 to 15 min at 110°C.

The importance of safely handling chemicals associated with these process steps must be given the utmost attention. At all times the Material Safety Data Sheets (MSDS) must be carefully read and followed.

It cannot be stressed enough that handling these materials will injure individuals who diverge from the instructions of the MSDS. HF is one of the strongest and most corrosive of inorganic acids. Therefore, special safety precautions are necessary while using it. HF is used in a variety of industrial and research applications including glass etching, pickling of stainless steel, removal of sand and scale from foundry castings, and as a laboratory reagent. Exposure usually is accidental and most likely due to inadequate use of protective measures (face shields, safety goggles, acid gloves, and acid aprons). In the U.S., more than 1000 cases of HF exposure are reported annually. Actual incidence rate is unknown.

13.3.3 PACKAGING

Given the huge range of MEMS applications (accelerometers, RF switches, optical mirror arrays, etc.) contamination covers a range of issues. Consider an RF switch when the open contacts are 300 nm apart, any particle in submicrometer range (less than 1.0 μm) lodged in this space will obviously be a problem. Chemical contamination of these surfaces can alter the electrical characteristics of the switch and affect service life. Water vapor or other species with high surface tension can cause stiction effects.

For reliability, the MEMS device must be isolated in a hermetic package. Often the damage is done before packaging. Also, while hermeticity specifications are defined in terms of leakage in and out of a "sealed" cavity, the issue is far more complex. The permeation of contamination in a solution must be prevented which occurs when contaminants diffuse through the seal over time. The outgassing must also be limited where materials internal to the hermetic cavity (such as polymers or epoxies) release trace quantities of gases or vapors which contaminate active surfaces of the device.

An approach to address the four issues is delineated below:[6]

- *Direct contamination.* The active device is placed in a hermetic cavity at wafer scale as an integral part of the MEMS foundry flow before any postfoundry operations occur. Hermeticity is established at the earliest possible step of the manufacturing process.
- *Leakage.* The leakage standard set by the MIL-STD-883 fine leak testing protocol may need to be reevaluated.
- *Permeation.* No materials such as polymers or epoxies known to allow long-term permeation should be used. In fact, typically, no new materials should be added. The cap used is normally a combination of Si and SiO_2. It is applied to the MEMS wafer without adhesives or other bonding materials. Covalent bonding between prepared surfaces of conventional semiconductor process materials occurs without applied pressure, temperature, or electric field. Prepared materials are aligned, and they simply bond on contact. The menu of bondable materials is large and includes materials common to IC and MEMS processing, such as Si, SiO_2, and Si_3N_4.
- *Outgassing.* Because the process takes place at wafer scale, the cavity formed can be arranged to include only the active MEMS device. Other materials used for die attach, bump preparation, or packaging, are not included in the hermetic cavity. This is a large change from what is convention today when all of these materials are in the package. With this approach, materials known to create outgassing effects are simply excluded from the hermetic cavity.

Minimizing MEMS packaging contamination sometimes requires "thinking outside the box." Capping the devices before dicing provides advantages that could be gained in both contamination control and in cost. Ziptronix has developed a process whereby a cap-wafer is placed over the production MEMS wafer before any postfoundry operation. The cap-wafer has cavities etched on the surface to provide the headspace for the MEMS devices. The bonding operation of the cap-wafer uses no glue, solder, or elevated temperatures. But most importantly, the headspace is designed to include only the MEMS device.[7]

Because of this precapping, the postprocessing operations can proceed along more conventional lines with only minimal customized MEMS postprocessing needed. With precapping, the dicing operation results in separation of devices which are already totally sealed. If conventional wire bonding is required, the cap-wafer leaves these leads available to be bonded without exposing the MEMS device.

13.4   IN-PROCESS HANDLING AND STORAGE REQUIREMENTS

The following conditions are recommended for the proper storage of MEMS devices once released. Any deviation from the following conditions should be avoided. MEMS devices should be stored in cabinets with an atmosphere of inert gas, dry air, or dry nitrogen. The temperature range should be 18 to 24°C with a relative humidity (RH) range well below 30% and an air cleanliness of Class 1000 per FED-STD-209. All materials shall be electrostatic discharge control (ESD) protective and specifically chosen to preclude ESD damage to the devices. In addition any special instructions should be included but not limited to:

- Die attaches, material, and properties
- Bond wire size and down bonds (pad number and electrical potential)
- Bonding method (e.g., thermo compression, thermo sonic, ultrasonic, etc.)
- Descriptions of any other unique materials or exposed surfaces that may require special protection during assembly
- Suggested limitations on handling methods, or die-attach pressures
- Suggested wire bonding sequence, quantity of bond wires on power and ground pins, and stitch bond (connection) requirements between ground pins
- Suggested lid sealing material and sealing procedure
- Packaged component ESD sensitivity
- Environmental conditions necessary to ensure long-term die reliability (e.g., special scaling)
- Maximum recommended allowable peak die assembly process temperatures or times
- Dimensional data (for features of top metal and mask layers) in a backside surface roughness finish type, etc.
- Environmental conditions and storage duration prior to shipment
- Unusual die material properties (e.g., SiC backside coatings)
- Moisture resistance data (for nonhermetic applications) based upon accelerated stress studies (e.g., 85% RH — 85°C, highly accelerated stress testing, autoclave, etc.) of existing nonhermetic packaged product
- Recommended die coat material, thickness, and application process as required.

13.5   ELECTROSTATIC DISCHARGE CONTROL

If not handled properly, several elements used in MEMS can be damaged by ESD. Therefore, every process and design should be characterized to determine ESD sensitivity. Regardless of these results, all MEMS devices should be treated as sensitive to ESD damage. An ESD handling and training program is essential to maintain a low level of ESD-attributed failures.

Inspection, test, and packaging of MEMS should be carried out in a static-free environment to assure that delivered products are free of damage. Devices should be packaged in conductive carriers and delivered in static-free bags. All handling and inspection should be performed in areas meeting "Class 1" handling requirements. Both the manufacturer and the user share the responsibility of assuring that an adequate procedure is in place for protection against ESD.

In general, the following measures can help reduce or eliminate ESD problems in device manufacturing and test areas:

- Ensure that all workstations are static-free
- Handle devices only at static-free workstations
- Implement ESD training for all operators
- Control RH to within 40 to 60%
- Transport all devices in static-free containers
- Ground yourself before handling devices

Because of the catastrophic failure caused by ESD, all personnel who work with MEMS should be trained in the proper procedures for handling the devices. Furthermore, these procedures should be documented and readily available for reference. Typically, the procedures include the methods, equipment, and materials used in the handling, packaging, and testing of MEMS. Further guidance for device handling is available in the Electronics Industry Association JEDEC Publication EIA 625[8] and MIL-STD-1686.[9]

13.6   CONTAMINATION CONTROL

In aerospace applications, contaminants are commonly referred to as any undesired foreign materials which emerge at any phase of a mission. The presence of contaminants, either molecules or particles, degrades the performance of hardware to various degrees of severity. In a worst-case scenario, contaminants may render an instrument worth millions of dollars useless. As a result, maintaining hardware to its designed cleanliness conditions through all mission phases becomes a demanding task. Therefore, an effective contamination control program starts with conceptual design phase of the mission and proceeds through its on-orbit operations.

13.6.1   CONTAMINATION CONTROL PROGRAM

The effort of contamination control depends on the specific mission goals, instrument designs, and planned operating scenarios. This dependence may be simply interpreted as the "contamination sensitivity" of the mission. It is noticeable that a mission with high contamination sensitivity requires a more elaborate contamination control effort. In the cases of payloads which are not sensitive to contamination, this program may still be required due to cross-contamination potentials to other payloads or orbiter systems. The contamination control program is applicable to all payloads, subsystems, instruments, and components during all mission phases. A typical mission, small or big, consists of sequential phases from its conceptual design, fabrication, assembly, integration and test, storage, transport, launch site preparation, launch, to its on-orbit operations. In certain occasions, the last phase of contamination control is extended to handle space-returned hardware such as the investigation of the returned hardware of the Long Duration Exposure Facility (LDEF) mission. To accomplish contamination control, it is necessary for each mission to provide a contamination control plan (CCP), which defines the comprehensive contamination control program that will be implemented in the mission. Additionally, specific verification plans and requirements must be defined in the CCP.

Regardless of contamination sensitivity, the implementation of the CCP needs to be addressed in all mission phases in order to prevent any detrimental contamination damage. Among spacecraft systems, performance of optical and thermal control is most vulnerable to contamination degradation. A high contamination sensitivity mission is primarily one which relies on optical sensing, and imaging or spacecraft or both, which require very strict temperature control, while a low contamination sensitivity mission is generally one with very insensitive optics and relatively flexible thermal control requirements. Highly sensitive missions usually require design and implementation of a strict contamination control program accompanied with ongoing monitoring and cleaning procedures.

Table 13.1 describes top-level contamination requirements of high contamination-sensitive hardware in a conventional mission. As hardware dimensions decrease due to spacecraft miniaturization, surface cleanliness levels become increasingly significant. Micrometer-sized particles of lesser impact on conventional spacecraft become extremely critical for miniaturized spacecraft.

13.6.2 MEMS CONTAMINATION CONTROL

The contamination effect on MEMS devices is enhanced by the relative dimensions between contaminants and MEMS devices. Because MEMS devices may contain exposed moving parts, they do not function well in the presence of liquid, vapor, particles, or other contaminants. A contamination assessment needs to be made early in the program to determine whether the possibility exists that the MEMS devices will be unacceptably degraded by molecular or particulate contaminants, or if it will be a source of contaminants itself. This preliminary assessment can be achieved by carefully examining mission-specific environments and contamination sources are shown in Table 13.2. The assessment should take into account all the various factors during the entire mission phases including selection of materials

---

TABLE 13.1
Contamination Requirements for a High Contamination Sensitive Mission

Requirement Category	Quantitative Level
Clean room needed (when optics are exposed)	Class 100 per FED-STD-209
Clean room needed (other operations)	Class 10,000 per FED-STD-209
Optics allowable molecular (EOL)	<100 Å
Nonoptics allowable molecular (EOL)	$\leq$Level A per MIL-STD-1246
Optics allowable particulate (EOL)	$<$Level 100 per MIL-STD-1246
Nonoptics allowable particulate (EOL)	Level 200–300 per MIL-STD-1246

TABLE 13.2
Mission Specific Environments and Contamination Sources

Mission Phase	Molecular	Particulate
Design	Configuration, operation conditions, material selection	Configuration, operation conditions, material selection
Fabrication	Materials outgassing, machining oils, fingerprints, air fallout	Shedding, flaking metal chips, filings, particle fallout, personnel
Assembly	AMC, outgassing, personnel, cleaning, solvents, soldering, lubricants, bagging material	Particle fallout, personnel, soldering, drilling, bagging material, shedding, flaking
Integration and test	AMC, outgassing, personnel, test facilities, purges	Particle fallout, personnel, test facilities, purges, shedding, flaking, redistribution
Storage	Bagging material, outgassing, purges, containers	Bagging material, purges, containers, shedding, flaking
Transport	Bagging material, outgassing, purges, containers	Bagging material, purges, containers, vibration, shedding, flaking
Launch Site	Site bagging material, AMC, outgassing, personnel, purges bagging material, air fallout	Bagging material, particle fallout, personnel, shedding, flaking, checkout activities, other payload activities
Launch	Ascent outgassing, venting, engines, companion payloads separation maneuvers	Vibration and redistribution, venting, shedding, flaking
On-orbit	Outgassing, UV interactions, atomic oxygen, propulsion systems	Micrometeoroid and debris impingement, material erosion, redistribution, shedding, flaking, operational events

(quantity and location), manufacturing processes, integration and test, packing and packaging, transportation, launch, on-orbit operations, and return to Earth, if applicable. In addition, the assessment should identify the types of substances that may contaminate and cause unacceptable degradation. The assessment results serve as a general guideline to how extensive a CCP should be instituted.

Actual contamination control implementation of MEMS devices can be divided into three major levels: design, packaging, and postpackaging. In the design level, contamination control is focused in MEMS device configuration, operation conditions, and material selection with an aim to minimize the contamination generation potential. At the MEMS packaging level, adequate fabrication, assembly environments and processes are key to prevent contaminants from reaching MEMS devices. The postpackaging level includes the integration and test of MEMS devices with spacecraft and transport until their final operations on-orbit. At this final stage, contamination control is essential in reducing accumulation of contaminants and mitigating contamination impact on MEMS devices.

The goal of contamination control at the design level is to minimize contamination sources and to remove contaminants from MEMS devices whenever it is feasible on-ground or on-orbit. By eliminating contaminants before they ever have chance to generate, this design level contamination control is not only effective but also very cost-saving. Unfortunately this critical stage of contamination control is often neglected due to the lack of the involvement from a contamination engineer. Material selections for MEMS devices are critical for effective contamination control. Single-crystal Si, polysilicon, Si_3N_4, and SiO_2, and other materials are well recognized for constructing MEMS devices. In addition SiC, shape memory alloy (SMA) metals, permalloy, and high-temperature superconductive materials are potential candidates. Although these materials have certain unique properties which are attractive for certain MEMS applications, contamination issues may result from the usage of these materials. For example, silica material used in fiber optics is brittle and is prone to fracture including delayed fracture.

13.6.3 CONTAMINATION CONTROLS DURING FABRICATION

Contamination concerns start at the beginning of the MEMS fabrication life. Problem areas in the foundry can be with both inferior materials and chemicals or due to inadequate or not followed processing steps. Entire lots due to the homogeneous nature of fabrication runs may need to be destroyed due to contamination related yield losses such as streamers, corrosion, and other results from impurities or improper processing. The greater concern at the foundry level is allowing contamination to reside with a lot only to appear at a later date found through failure of the component. At the foundry level the most common source of contamination is organics that have not been adequately removed. Most foundries ship product with the photoresists still present, which protect the MEMS from damage, but are absolutely necessary to be removed prior to release. Other sources of contamination include those from humans such as finger oils, makeup, human spittle, and processing materials. Often, dicing films are special adhesives that must be properly removed. Bubbles forming during the release step can "protect" the material in the sacrificial area yielding a nonfunctioning or only partially functioning device.

The recommended solvent should be used to assure the complete removal of organics. Oxygen plasma and piranha etch are often used. Oxygen plasma is just gaseous oxygen electrically charged into plasma. Organics placed in oxygen plasma will etch quite thoroughly. Piranha etch is an etching compound formed of 70% sulfuric acid and 30% hydrogen peroxide that will consume almost all organics, but leave behind nonorganics. Piranha etch can remove some metal so it is necessary to test pieces before committing a lot to any particular solution.

13.6.4 MEMS PACKAGE CONTAMINATION CONTROL

The discussion of package level contamination control for MEMS devices for space flight use must be devoted to controlling contaminants from damaging the devices. Risk of contamination is present at the bare die level, packaged, and through on-orbit. MEMS package contamination control requires comprehensive contamination control protocols for fabrication and assembly. The contamination effort deals with both molecular and particulate contaminants resulted from facility environments and packaging procedures. It is important not to jump to the conclusion that contamination is the culprit. The types of failures associated with stiction and particulates could also be caused by design or manufacturing discrepancies such as over or under etching.[10–12]

The bulk of today's MEMS devices are manufactured in the traditional semiconductor clean room facilities with air cleanliness ranges from Class 100 to Class 10,000 per FED-STD-209. Examples of damage caused by unwanted molecular and particulate contamination suggest the deficiency of conventional facility, equipment, and process at the MEMS package level. One hard-to-detect failure in MEMS devices is particulate contamination that occurs during fabrication. The effect produced by dust adhering to the wafer in the water process differs according to the process. Particles also affect thermal management in photonic packages. A typical edge-emitting communications laser diode will have an energy flux through the facet of up to 2 million watts per square centimeter. The influence of even slight levels of impurities or contaminating particles is disastrous for thermal control. Therefore, the best contamination control approach is to not allow contaminants to generate, stay around, and finally adhere to surfaces.

Contamination-induced effects can be reduced by fabricating MEMS devices in a better clean room facility with more stringent clean room protocols. Class 100 clean room environments with localized Class 10 work areas are optimal for post-singulation processing. As a minimum, the device should be in a Class 100 clean room environment from its release point until it is safely sealed in a clean, hermetic package. Dust generated by equipment adheres directly to wafers, and thus has a large effect. Sufficient consideration should be given to dust when selecting equipment models; it is also important for device manufacturers to take steps to reduce dust generation when setting process conditions or performing maintenance during production. It is important to package MEMS devices in a controlled, hermetic, particle-free environment. Every step, from die preparation to package seal, must be performed in a Class 100 clean room environment until the device is safely sealed in a clean hermetic package. Clean room techniques normally reserved only for wafer fabrication must be extended to the probe, die-prep, and assembly areas.

Further contamination control improvement can be achieved by implementing better assembly processes for MEMS devices. Certain unwanted organic compound residues in the adhesives can lead to catastrophic optical damage (COD) of the laser die. Outgassing occurs when materials used for die attach, bump preparation, or packaging are included in the hermetic cavity. Improved processes keep these materials from being included in the package, thus eliminating potential contamination sources. Because the process takes place at wafer scale, the cavity formed can be arranged to include only the active MEMS device. With this approach, materials known to create outgassing effects are simply excluded from the hermetic cavity.

For particulate contamination, Blanton and others at CMU have developed a tool called contamination and reliability analysis of microelectromechanical layout

(CARAMEL) for analyzing the impact of particles on the structural and material properties of surface-micromachined MEMS. CARAMEL accepts as input a microelectromechanical design represented as a layout in Caltech Interchange Format (CIF), a particulate description, and a process (fabrication) recipe. It performs process simulation that includes the foreign particle and creates a three-dimensional representation of the resulting defective microelectromechanical structure. CARAMEL then extracts a mesh netlist representation of the defective structure whose form is compatible with finite-element analysis (FEA) tools. Performing FEA of the CARAMEL mesh output correlates the contamination of concern to a defective structure and a faulty behavior. CARAMEL has been used to investigate the impact of particles on electrostatic comb-drive actuated microresonator.[13] This technique is demonstrated on a resonator as shown in Figure 13.1. Interestingly enough, experiments through CARAMEL reveal that the resonator is susceptible to a variety of misbehaviors as a result of a single particle contamination. Figure 13.2 shows two representative defects caused by particles.

Protection of MEMS devices from the environment is an important concern as a hermetic package significantly increases the long-term reliability of the devices. Traditional hermetic IC packaging techniques, when applicable, offer protection from contamination; however, only a subset of devices can be packaged in this manner. This subset includes accelerometers, which may be packaged with the hermetic schemes used for ICs. Numerous devices however require interaction with the environment such as gas detectors, optical switches (requiring optical windows) and lab-on-chip systems. In this case, while functionality must be maintained, vulnerabilities must be reduced. MEMS devices, which require free space to function, may be at particular risk. There are few standardized solutions to this problem and for the low quantities required by the space industry most solutions will be customized.

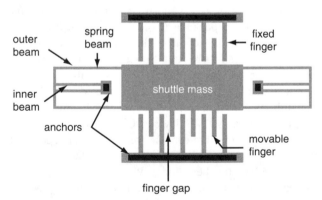

FIGURE 13.1 Top view of a surface-micromachined, electrostatic comb-drive actuated structure that is suspended over the die substrate and is anchored only at the shuttle movement to a capacitance change between the moveable and fixed potential difference between the shuttle and fixed fingers, or from an inertial force caused by external acceleration. (Courtesy: CMU S. Blanton.)

FIGURE 13.2 Example of a resonator defect due to particulate contamination: A small particle between two fingers that does not fuse the fingers and hence the inter-finger capacitance is greatly increased due to the significant gap reduction. (Courtesy: CMU S. Blanton.)

The digital micromirror device (DMD) is a microchip consisting of a superstructure array of Al micromirrors functionally located over CMOS memory cells. The mirrors are hermetically sealed beneath nonreflecting glass to prevent contamination-induced failure. For reliability, the device must be isolated in a hermetic package. The key question lies in when the device is isolated since the damage often occurs before packaging. Also, while hermeticity specifications are defined in terms of leakage in and out of a "sealed" cavity, the issue is more complex. A hermetic package prevents the diffusion of gases, moisture, and outgassed hydrocarbons through its walls. A robust contamination solution must also stop permeation, which occurs when contaminants diffuse through the seal over time, and outgassing, where materials internal to the hermetic cavity (such as polymers or epoxies) release trace quantities of gases or vapors, which contaminate active surfaces of the device.

Particular attention should be paid to the protection of devices that are not hermetically packaged such as environmental sensors; however, hermetic parts are also susceptible to contamination problems. Any contaminant once sealed in a hermetic package has a wonderful "growth medium" that has accelerators such as voltage and temperature. Of particular concern is the presence of liquid, vapor, gases, particles, or other contaminants. Controls for packaging cleanliness used in the microcircuit industry are not adequate for the MEMS world as MEMS devices are affected by particles, especially nonmetallics, which might not affect an IC. Modifications may have to be made to standard assembly equipment, assembly handling methods and tooling, and equipment environments to accommodate the intensive handling and particle control requirements for packaging microstructures.

13.6.5 MEMS POSTPACKAGE CONTAMINATION CONTROL

Postpackaged MEMS devices must be considered as contamination-sensitive flight hardware and handled accordingly. Additional contamination control precautions are needed for nonhermetically packaged MEMS devices that are more susceptible to contaminants. Measures to protect nonhermetically packaged MEMS devices, may include temperature control, humidity control, gas purging, and protective enclosures. In addition, for nonhermetically sealed MEMS devices, especially if mounted on the skin of the spacecraft, the need to identify the component and "red tag" the item for special handling is essential.

MEMS postpackage level contamination control is concentrated on maintaining proper surface cleanliness levels, that is, molecular and particulate contamination budget. Therefore, the amount of performance degradation that is allowed for MEMS contamination-sensitive surfaces needs to be established. From this degradation limit, the amount of contamination that can be tolerated, that is, the contamination allowance, can be established. This allowable degradation should also be included as a contamination budget stated in CCP.

The contamination budget describes the quantity of contaminant and the degradation that may be expected during various phases in the lifetime of a MEMS device. The established contamination budget for MEMS devices is monitored as the program progresses. When the contamination budget exceeded requirements, MEMS surfaces may be cleaned periodically to reestablish a budget baseline. In addition, contamination-preventive methods, such as clean rooms and MEMS device covers, should be included.

The integration and test (I&T) of conventional spacecraft is generally performed in clean rooms with air cleanliness classes ranges from Class 1000 to as high as Class 100,000. Integration through launch conditions may provide numerous opportunities for gaseous and particulate contaminants to be deposited on MEMS surfaces. For optical MEMS (MOEMS) gaseous contaminants can degrade performance by condensing on critical windows or alternatively by absorbing light along the line-of-sight.

There is a concern for MEMS devices when they are exposed to uncontrolled ambient humidity. During I&T, MEMS devices with sliding and rotational motion may experience wear since speeds can approach 1 million rpm in the devices. According to study results from Sandia National Laboratory, the RH is critical for proper operations of MEMS devices. Low humidity may increase resistance and wear of MEMS devices, while high humidity may cause corrosion, wear, and stiction. The ideal range appears to be somewhere between 20 and 60% for the I&T of MEMS devices. However, specific RH requirements may depend on distinct MEMS hardware design and applications.

As stated in Table 13.2, considerable amounts of contaminants may be generated during launch and on-orbit operations. Microscopic particles can dislodge or even form during these operations. To prevent contaminants, materials with a less potential of generating particles should be chosen for fabricating MEMS devices. Besides particles, material outgassing as a major contamination source is also a well-recognized fact. Outgassed contaminants are greatly promoted by the space environments of high vacuum and elevated temperatures. On-orbit degradation due to contamination can truncate the mission lifetime and degrade data quality. These degradations may include long-term changes in the optical surfaces or changes in absorptivity of a thermal control surface, which will eventually reduce its effectiveness and cause loss of performance. It is necessary to minimize contribution to spacecraft contamination through outgassing product in modern MEMS packaging materials. All nonmetallic materials should be selected for low outgassing characteristics and baked out in meeting their outgassing requirements. The thermal vacuum bake is an effective method to assure that outgassed materials have been removed. Generally, the hotter and longer the item can be baked, the better the chance that the item will not contaminate the chamber or test article. Space flight hardware are typically baked at 50°C or higher, under 5×10^{-6} torr vacuum environment for at least 48 h unless otherwise noted. Visible degradation of the material during bakeout will obviously result in the rejection of the material. Some materials must be qualified for use by monitoring the outgassing levels during the bakeout. The use of MSFC-SPEC-1238[14] is recommended for critical optical applications. Bakeouts of MEMS devices are required unless it can be satisfactorily demonstrated that the contamination allowance can be met without bakeouts.

MEMS devices operated on-orbit require proper protection from various contamination sources. Plume impingement poses a great threat to MEMS devices with both thermal heating and contamination degradation effects. Propulsion systems and attitude control systems are major contributors to plume contamination. Plumes contain particulates that may be impinged on the exposed surfaces. For example, solid rocket motors emit Al_2O_3 and gaseous HCl, H_2O, CO, CO_2, N_2, and H_2. The shuttle Orbiter and International Space Station may also release water vapor and ice particles along with gases leaking from the pressurized cabins.[15] To warrant proper on-orbit operations, it is necessary to protect MEMS devices from plume impingement. The protection is attained by a combination of mitigation methods including placing plume shields, optimizing thruster operations, or installing active decontamination devices.

13.6.6   CONTAMINATION CONTROL ON SPACE TECHNOLOGY 5

The Space Technology 5 (ST5) mission, as part of NASA's New Millennium Program (NMP), is a technology demonstration mission designed and managed by NASA Goddard Space Flight Center (GSFC) that consists of three nanosatellites flying in Earth's magnetosphere. A thermal management method developed by NASA and JHU/APL as one of the demonstration techniques of variable emittance surfaces is a MEMS-based device that regulates the heat rejection of the small satellite.[16] This system consists of MEMS arrays of gold-coated sliding shutters, fabricated with the Sandia ultraplanar, multilevel MEMS technology fabrication process, which utilizes multilayer polycrystalline silicon surface micromachining. The shutters can be operated independently to allow digital control of the effective emissivity.

For variable emissivity radiators the concerns of contamination and handling drove the packaging design. The shutters open only 6 μm by 105 μm with a concern that a small particle can lodge in the devices within the hinges of the MEMS shutters and prohibit movement. Placing a protective window over the

MEMS shutter array (MSA) was the obvious solution, but even the protective window must meet the NASA GSFC material requirements. In this application the external surface of the window must be electrically conductive, and if made of an organic material, must be resistant to the attack by atomic oxygen in space. In addition, for the shutter application, high infrared transparency was required.

The protective windows used are a fluorinated polyimide material developed by NASA Langley Research Center (LaRC) located in Newport News, Virginia. LaRC-CP1® polyimide is a high-performance material with a wide variety of uses in space structures, thermal insulation, electrical insulators, industrial tapes, and advanced composites. This polyimide material may be dissolved readily in a number of solvents for use in various applications such as castings and coatings. CP1 was selected for the ST5 application for its infrared transparency and space environment survivability for a 10-year life in geosynchronous earth orbit (GEO). CP1 is colorless and offers better space UV-radiation resistance than most known polymer materials (including other polyimides, polyesters, Teflon, Teflon-based materials, and others). The MEMS dies are fabricated in wafer format using Sandia's processing as described in Chapter 3. The wafers go through a standard backside grind process and then are released, diced, tested, gold coated, and functionally tested again, in preparation for final attach. The individual dies are bonded to aluminum nitride (AlN) carriers that are subsequently bonded to the MSA chassis. This design allows for optimum rework or replacement of each MEMS shutter die (MSD) as necessary.

Of most significance is the window assembly. As stated previously, the micro-machined comb drives are sensitive to the abundant contamination in space. The CP1 fluorinated polyimide material was selected for the fabrication of MEMS device. A CP1 film, less than 4 mils thick, is sandwiched in tension between two window frames and bonded in place, as shown in Figure 13.3. CP1 in its relaxed

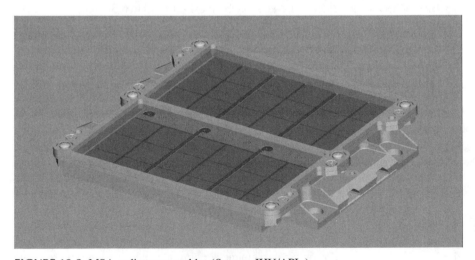

FIGURE 13.3 MSA radiator assembly. (*Source*: JHU/APL.)

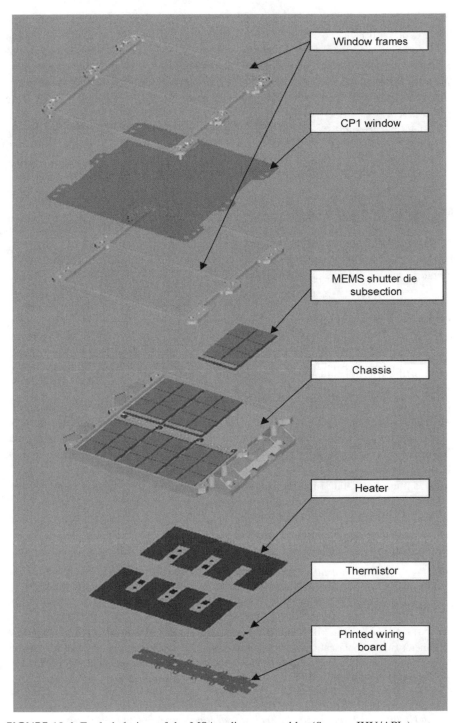

FIGURE 13.4 Exploded view of the MSA radiator assembly. (*Source*: JHU/APL.)

state is flaccid and must be stretched to provide the mechanical protection from debris impact. To ensure a taut connection, the CP1 is procured in a taut configuration, and then epoxied to one side of the window and then cured. Sandwiching the CP1 attach between the two windows, reinforces the connection. With the window assembly in place, the CP1 film is suspended several millimeters above the shutters, thus providing a barrier layer between the actual die and the environment.

Electrical conductivity of the film is achieved through application of a thin coating of indium tin oxide (ITO), a transparent electrical conductor. In sufficiently thin coatings ITO does not change the IR performance of the window. ITO coating serves to protect the CP1 from degradation in the presence of atomic oxygen. All the structural members of the MEMS shutter array radiator assembly were made of aluminum 6061 and finished with a clear anodize treatment, followed by a yellow irridite.

An exploded view of the MSA radiator assembly is shown in Figure 13.4. Additional information on the packaging of MEMS devices is found in Chapter 12 but clearly contamination, handling concerns, and functionality are the key ingredients to successful packaging scheme.

13.7 CONCLUSION

For space applications, MEMS devices are susceptible to environment-induced damage both on-ground and on-orbit. The potential damage may occur at any stage of the mission but they are especially prone to surface contamination prior to the prepackage phase.

The damage impact is alleviated by implementing prudent handling and contamination control practices. Facility for manufacturing and assembly must be maintained at adequate cleanliness conditions with proper procedures established. Personnel handling MEMS devices must be properly trained with special attention to preclude ESD damage to the devices. To achieve the best protection, MEMS devices must be isolated in a hermetic package or protected with covers whenever possible.

CCP delineates a comprehensive contamination control program for a mission. MEMS devices as an integral part of the mission must follow handling and contamination guidelines established in the CCP in order to meet mission requirements.

REFERENCES

1. C.H. Mastrangelo and G.S. Saloka, Dry-release method based on polymer columns for microstructure fabrication, *Proceedings of the 1993 IEEE Micro Electro Mechanical Systems — MEMS, February 7–10 1993, Fort Lauderdale, FL, USA*, IEEE, Piscataway, New Jersey, pp. 77–81 (1993).
2. G.T. Mulhern, D.S. Soane, and R.G. Howe, Supercritical carbon dioxide drying for microstructures, *Proceedings of the 7th International Conference on Solid-State Sensors and Actuators, Transducers '93, Yokohama, Japan*, pp. 296–299 (1993).

3. H. Watanabe, S. Ohnishi, I. Honma, H. Kitajima, H. Ono, R.J. Wilhelm, and A.J.L. Sophie, *Journal of the Electrochemical Society*, 142, 237–243 (1995).
4. S. Brown, C. Muhlstein, C. Abnet, and C. Chui, MEMS testing techniques for long-term stability, *Proceedings of the 1998 ASME International Mechanical Engineering Congress and Exposition, November 15–20 1998, Anaheim, CA, USA*, ASME, Fairfield, NJ, USA, p. 145 (1998).
5. R. Ramesham, R. Ghaffarian, and N.P. Kim, *Proceedings of SPIE — Reliability Issues of COTS MEMS for Aerospace Applications*, 3880, 83–88 (1999).
6. R.J. Markunas, New solution to an old problem: MEMS contamination, *A2C2 Contamination Control for Life Sciences and Microelectronics*, (February 2003).
7. P. Nesdore, Output: zip up your MEMS, *A2C2 Contamination Control for Life Sciences and Microelectronics*, (November 2002).
8. JEDEC Publication EIA 625, *EIA and JEDEC Standards and Engineering Publications* (1994).
9. MIL-STD-1686 (1992), Electrostatic Discharge Control Program for Protection of Electrical and Electronic Parts, *Assemblies and Equipment (Excluding Electrically Initiated Explosive Devices)*. Department of Defense, Washington, DC.
10. R.D.S. Blanton and N. Deb, Built-in self test of CMOS–MEMS accelerometers, *Proceedings International Test Conference, October 7–10 2002, Baltimore, MD, U.S.*, Institute of Electrical and Electronics Engineers, Inc., pp. 1075–1084 (2002).
11. N. Deb and R.D.S. Blanton, Analysis of failure sources in surface-micromachined MEMS, *Proceedings International Test Conference, Atlantic City, NJ, USA*, Institute of Electrical and Electronics Engineers, Inc., Piscataway, NJ, pp. 739–749 (2000).
12. N. Deb and R.D.S. Blanton, *Analog Integrated Circuits and Signal Processing*, 29, 151–158 (2001).
13. A. Kolpekwar, C. Kellen, and R.D.S. Blanton, MEMS fault model generation using CARAMEL, *Proceedings of the 1998 IEEE International Test Conference, October 18–21 1998, Washington, DC, USA*, IEEE, Piscataway, NJ, USA, pp. 557–566 (1998).
14. MSFC-SPEC-1238 (1986), *Thermal Vacuum Bakeout Specification for Contamination Sensitive Hardware*. George C. Marshall Space Flight Center, Madison, AL, USA.
15. MSFC-SPEC-1443 (1987), *Outgassing Test for Non-Metallic Materials Associated with Sensitive Optical Surfaces in a Space Environment*. George C. Marshall Space Flight Center, Madison, AL.
16. D. Farrar, W. Schneider, R. Osiander, J.L. Champion, A.G. Darrin, D. Douglas, and T.D. Swanson, Controlling variable emittance (MEMS) coatings for space applications, *8th Intersociety Conference on Thermal and Thermomechanical Phenomena in Electronic Systems, May 30–Jun 1 2002, San Diego, CA, USA*, Institute of Electrical and Electronics Engineers, Inc., pp. 1020–1024 (2002).

14 Material Selection for Applications of MEMS

Keith Rebello

CONTENTS

14.1 INTRODUCTION

Microelectromechanical systems (MEMS) were born out of the integrated circuit revolution of the 1960s, 1970s, and 1980s, and as such share much of the same fabrication technology and materials. While many of these materials are space qualified, and have been used in space electronics for decades, new issues must be accounted for when using them in sensors and actuators. When choosing which materials to use in a MEMS device, it is important to look at the device as a system of materials. The fabrication processes used to make the device, as well as their intended use will control the material selection. These selections will in turn affect the final performance of the MEMS device.

This chapter discusses some of the issues involved in selecting MEMS materials for space applications. Myriads materials available for MEMS fabrication prevent inclusion of all of them in this chapter; however, key materials and their properties are reviewed here. The reader is referred to the ever growing literature for materials that are not discussed.

14.2 SCALING LAWS

When dealing with objects on the microscale it is first useful to understand how physical phenomena here can differ from the macro world we are all accustomed to living in everyday. If we let l represent a linear dimension of an object, and then make the object 1000 times smaller, all of the linear dimensions decrease by a factor of 1000. We say this object has been scaled by $l/1000$. As the size of an object shrinks the surface area shrinks as a function of l^2 while volume decreases as l^3. So our object's surface area has decreased by $(1/1000)^2 = 1/1,000,000$ and volume has decreased by $(1/1000)^3 - 1/1,000,000,000$. Table 14.1 summarizes how physical phenomenon behaves and changes as dimensions scale. While these scaling laws are important when designing MEMS devices, they also play a role in material selection.

One of the important outcomes is a general increase in material strength. For example, single-crystal silicon whiskers[1] and SiC fibers[2] may be an order of magnitude stronger than their bulk counterparts.[3,4] The thin film materials commonly used in MEMS devices are typically stronger than their bulk values.[5] As size is further and further decreased down to the nanoscale, material properties approach their ideal values. Macroscale and microscale materials have defects and dislocations that can severely compromise their mechanical performance. Nanoscale materials are capable of becoming perfect with no defects or dislocations and consequently tend to have extraordinary properties. Carbon nanotubes for instance have a Young's modulus of 1.28 TPa and are capable of strains exceeding 15%.[6]

Since the surface area to volume ratio increases as dimensions shrink, surface effects become dominant. Van der Waals, surface tension, and frictional forces increase. Heat dissipation increases as thermal isolation becomes difficult, and cooling improves. Note that MEMS devices exposed to the vacuum of space have much lower heat dissipation due to the lack of convective heat transfer in space. The thin films typically used in MEMS devices behave differently from their bulk

TABLE 14.1
Scaling Laws[7,29]

Scaling Laws	Factor
Time	l^0
Diffusion	$l^{1/2}$
van der Waals force	$l^{1/4}$
Distance	l^1
Surface tension	l^1
Velocity	l^1
Area	l^2
Electrostatic force	l^2
Friction	l^2
Piezoelectricity	l^2
Thermal loss	l^2
Gravity	l^3
Magnetics	l^3
Mass	l^3
Power	l^3
Torque	l^3
Volume	l^3

properties. Due to scaling, the material defects and dislocations in their surfaces are no longer small with respect to the volume of the samples. Material properties tend to be specific to their individual deposition processes and material thicknesses. Therefore materials should be characterized before designing devices. Most testing and standards have been done for bulk materials and no standards currently exist for MEMS materials. Therefore it may not be possible to rely on published data for your individual process.

14.3   MATERIAL SELECTION

NASA space missions require strict adherence to reliability and quality standards. These requirements have been well defined for electronic components, but since the MEMS field is a relatively new technology for space applications, standards do not currently exist. MEMS devices are subject not only to typical electronic component failures and mechanical component failures; but, also as in its name, to those failures that deal with the electro-mechanical interaction between the two. While MEMS devices share many of the same materials as their integrated circuit (IC) cousins there are significant differences in function that can lead to different types of failure mechanisms. Spaceflight MEMS devices must not only be able to survive the same manufacturing, test, packaging, and storage environments that traditional MEMS devices do on Earth, but must also survive qualification, integration, and launch and operation in space. The space mission must be taken into account when selecting MEMS materials including mission duration, thermal swings, radiation, acceleration, vibration, shocks, and locations such as low-Earth orbit (LEO), geo-stationary orbit (GEO), deep space, or exploratory planetary environments.

14.4 MATERIAL FAILURES

This section describes some common MEMS material failure mechanisms, but the coverage is limited to the issues specific to the space environment. The reader is referred to the classic text books for material failure issues due to device fabrication and standard operation.[7,8]

14.4.1 STICTION

Stiction occurs when attractive surface forces cause MEMS components that touch to stick together. This adhesion can be caused by capillary forces, van der Waals forces, hydrogen bonding, or electrostatic forces. Stiction is one of the greatest reason for the failure of MEMS devices. MEMS structures are typically very smooth and polished causing two surfaces that touch to have many points or surface area in contact and thus generate large adhesive forces. Rough surfaces have less surface area in contact and thus lower adhesive forces. Some design solutions that can help mitigate surface interactions are:

- The addition of bumps or dimples to structures. These surfaces prevent the whole structure from making contact with the substrate, lessening adhesive forces.
- The use of self-assembled monolayer (SAM) coatings to reduce surface adhesion.[9] This surface modification uses chemicals to coat the surface (covalent bonding of monolayers) and change its surface properties. Silox-ane-based chemistries are used for coating silicon surfaces and thiol chemis-tries are used for gold.
- The deposit of diamond like carbon (DLC) films. Diamond is hydrophobic and will prevent capillary forces from causing stiction.[10]
- The use of flourinated polymer coatings.[11]
- The use of ammonium fluoride coatings.[12]
- The use of stiff materials for suspended structures.
- The roughening of contact surfaces.[13]
- The use of hermetically sealed packages with getters to prevent stiction due to humidity.[14]
- The use of leaky dielectrics in RF MEMS devices will prevent dielectric charging.[15]

14.4.2 DELAMINATION

Delamination occurs when a material interface loses its adhesive bond. It can be the result of fatigue, induced by the long-term cycling of structures with mismatched coefficients of thermal expansion or temperature-induced stresses. Delaminating materials can cause shorting, stiction, and mechanical impedance failures. Altering the mass and composition of the structures can affect the designed performance of the devices such as displacement and resonant frequencies.

14.4.3 FATIGUE

Fatigue is caused by the cyclic loading of a structure below the yield or fracture stress of the material. This can cause microcracks to form, which, over time can lead to localized plastic deformations, weakening, and ultimately failure of the material. Ductile materials like metals tend to exhibit fatigue more than brittle materials such as silicon.

14.4.4 WEAR

Wear is caused by the motion of one surface over another. This motion causes material to be removed from the surfaces. There are four types of wear: adhesion, abrasion, corrosion, and surface fatigue. Adhesive, abrasive, and surface fatigue are forms of wear that most moving MEMS devices in contact with another surface encounter. Hard materials such as silicon carbide or diamond are used to reduce this failure mechanism. Exploratory missions may cause MEMS devices to be exposed to chemical environments, which can cause chemical interactions at the device surfaces. The movement of the MEMS device can strip away the reaction products of surfaces leading to more corrosion.

14.5   ENVIRONMENTAL CONSIDERATIONS

Typical space applications expose devices to harsh environmental conditions. Radiation, extreme temperatures, pressures, shock, vibrations, thermal cycles, corrosive atmospheres, dust, and fluid environments are environmental considerations that should be addressed. Table 14.2 shows how some planetary conditions compare to the Earth's environment. Packaging may address some of these issues in part, but may not be sufficient to protect the MEMS devices completely. In addition some MEMS devices must be exposed to the environment in order to function, which may require specific materials to be used for device survival.

14.5.1   VIBRATION

Vibrations are typically low acceleration, long duration events. They have been shown not to be a large reliability concern in MEMS. Long-term vibrations can contribute to fatigue failures, however. For space applications, initial vibrations encountered at launch have been shown to be less than 13 g, as shown in Table 14.3. Dynamic shocks encountered in space flight and surface landings are more important considerations.

TABLE 14.2
Mission-Specific Environments[30]

	Mercury	Venus	Earth	Mars	Jupiter
Average temperature (°C)	350	465	15	−63	−144
Diurnal temperature range (°C)	−173⟶452	0	10⟶20	−133⟶27	
Solar irradiance (W/m^2)	9127	2660	1380	595	51
Surface pressure	10^{-9} mbar	95 bar	1013 mbar	6.1 bar	≫100 bar
Other considerations	Vacuum environment	H_2SO_4	H_2O	Oxidants, dust	Aerosols: NH_3 ice, H_2O ice, NH_4SH

TABLE 14.3
Launch Vibrations (All Entries in Grams)[31,32]

Vehicle	Axial Load (g)	Lateral Load (g)
T34D/IUS	±4.0	±5.0
Atlas-II	5.5	±1.2
Delta	6.0	3.0
H-II	±5.0	±1.0
Ariane ASR44L	4.5	±0.2
Shuttle	3.5	3.4
Pegasus	13	±6

14.5.2 SHOCK

Shock differs from vibration in that shock is a single mechanical impact event where mechanical energy is directly transferred into the device. MEMS devices will fail when the shock event exceeds a critical stress and causes a fracture or adhesion and delamination failures. Shock events can also cause stiction failures when the induced displacement exceeds the critical design displacement and causes the microstructure to touch the substrate or another microstructure. Most MEMS devices are capable of surviving high shocks, but failures often occur from the device packaging. Shearing off of the PC-board, package cracking, or wire bond shearing are typical failure mechanisms. Encapsulation potting can be used to help mitigate these effects. COTS accelerometers have been tested up to 120,000 g.[16–19]

14.5.3 TEMPERATURE

Space missions typically require that a MEMS device be exposed to extreme temperature changes. Internal stresses and many material properties are temperature-dependent. Unfortunately most MEMS material properties are taken at room temperature. The temperature range in which a device will operate properly will partially be determined by the coefficient of (linear) thermal expansion (CTE) or temperature coefficient of expansion (TCE). Typical values are shown in Table 14.4. MEMS devices with poorly matched coefficients of thermal expansion will be more sensitive to temperature fluctuations as mismatches will cause bending. Bending or curling can reduce a sensor's sensitivity and lessen the strength of electrostatic actuators. In addition, it can potentially cause stiction, delamination, or fatigue failures. Since future space missions anticipate temperatures in the range of −100 to 150°C, thermal changes are a growing concern to MEMS designers. Temperature not only affects the MEMS device, but also how the MEMS device is packaged. The whole MEMS system must be modeled.

High temperatures also can change the properties of organic materials. Polymers tend to outgas more at high temperatures. The structure of the materials may also change with elevated temperature. For example, Teflon samples were removed from the Hubble Space Telescope after astronauts noticed cracking. Upon analysis it was determined that excessive heating caused an increase in crystallinity, density, and embrittlement.[20]

14.5.4 ATOMIC OXYGEN

The degradation of spacecraft surfaces due to erosion by atomic oxygen (AO) was discovered during the early Shuttle flights. Surface erosion was seen on ram or forward-facing surfaces of several types of materials. AO is formed by solar ultraviolet (UV) radiation, dissociating oxygen molecules (O_2) into free oxygen atoms. Oxygen atoms are highly corrosive to organic materials. In addition, a spacecraft's orbital velocity of 7.8 km/sec (17,500 mph) causes the oxygen molecules to impact the spacecraft with energy of approximately 5 eV, which is high enough to react with many materials. The reaction can further be enhanced by solar UV radiation, which energizes molecular bonds and makes the reaction easier. The

TABLE 14.4
Thermal Coefficients of Expansion

Material	TCE Microns/°C
Si	2.6, 4.2
Parylene	0.35
SiO_2	7
SiN	0.3
Al	23.6
Au	14.2
Cu	16.6
Polyimide	6
SU-8	52
CMOS dielectric	0.4

amount of oxygen that impinges on the spacecraft is dependent upon attitude, altitude, exposure time, and solar activity.

AO does not always lead to erosion. It reacts with certain materials to form a stable oxide that in turn protects the surface from further corrosion. Silicon and aluminum will form SiO_2 and Al_2O_3, respectively. These coatings along with indium tin oxide (ITO) are often sputter coated onto other materials as protection from AO.

Table 14.5 shows the reaction efficiencies for various materials. Most metals do not show macroscopic effects from atomic oxygen. Silver and osmium react rapidly, however, and are generally considered unacceptable for use in uncoated applications. Ion bombardment effects from atomic oxygen can be neglected as the energies are two orders of magnitude lower than those in a conventional reactive ion-etching machine. Microscopic changes have been observed, however, and should be investigated further for devices in which surface properties are critical.

14.5.5 RADIATION

Radiation can damage MEMS devices by causing failure in:

- analog and digital electronic components of MEMS device
- the transduction mechanism of the actuator or sensor
- the mechanical structures
- optical properties (absorptance and refractive index)

**TABLE 14.5
Reaction Efficiencies of Materials with
Atomic Oxygen**

Material	R_c (m^3/atom $\times$ 10^{30})
Kapton	3
Mylar	3.4
Tedlar	3.2
Polyethylene	3.7
Teflon	<0.1
Carbon	1.2
Polystyrene	1.7
Polyimide	3.3
Platinum	0
SiO_2	0
Indium tin oxide	0
Al_2O_3	0
Copper	0.05

Although much work has been done on characterizing the effects of radiation on microelectronics, little has been done for MEMS devices. It is recommended that gamma, proton, and x-ray testing be done on MEMS devices to better understand the effects on devices destined for flight. Excitation and sensing voltages can be effected by dielectric charging and dielectric failures can be accelerated by radiation effects, and therefore electrostatic devices show the largest sensitivities to radiation.[21–24] Minimizing the use of dielectrics, employing radiation shielding, leaky dielectrics, and grounded conductive planes are mitigation strategies. The reader is referred to the chapter on radiation for further details, but major effects are summarized in Table 14.6.

14.5.6 PARTICLES

Particulates are fine particles that are prevalent in the atmosphere as well as in space. While particulates generally will not affect hermetically packed MEMS devices, those directly exposed to the space environment will need to be protected. On atmospheric missions dust will potentially clog moveable devices. Meteoroids and other orbital debris will be a concern for MEMS devices on the outside of spacecraft which are exposed to the space environment during orbit.

14.5.7 VACUUM

In vacuum, polymer materials tend to lose volume as their solvents outgas. All materials intended for spacecraft use must first pass the outgassing data as specified in NASA Reference Publication 1124, revised by the Jet Propulsion Laboratory (JPL) using an apparatus developed at Stanford Research Institute (SRI) that measures the mass loss in vacuum and collects the outgassed products. The original

TABLE 14.6
Radiation Effects[33]

Radiation Effect	Cause	Physical Impact
Single event upset (SEU)	High energy ions, protons	Formation of electron–hole pairs
Single event latch-up (SEL)	High energy ions, protons	Localized high current condition in semiconductor materials
Single event hard error (SHE)	High energy ions, neutrons, protons	Permanent localized charging of oxide
Single event burnout (SEB)	High energy ions, neutrons, protons	Increased parasitics
Single event gate rupture (SEGR)	High energy ions, neutrons, protons	Breakdown of oxide insulator
Lattice damage	High energy ions, neutrons, protons	Displacement of lattice atoms; minority carrier lifetime doping level effects
Total ionizing dose (TID)	Electrons, protons	Charge trapping, interface state growth at oxide–silicon interfaces

SRI report[25] contained data from June 1964 to August 1967 and provided a reference for choosing materials for use in spacecraft with low outgassing properties. The SRI apparatus was constructed at Goddard Space Flight Center (GSFC) in 1971, and the GSFC report[26] came up with two criteria: a maximum total mass loss (TML) of 1.0% and a maximum collected volatile condensable material (CVCM) of 0.10%. Eventually, an American Society for Testing and Materials (ASTM) Standard Test Method was developed and is identified as E 595-77/84/90.

The GSFC equipment is the SRI-described micro-CVCM apparatus, which condenses micro-quantities of volatile liquids to determine the amount of volatile condensable materials. The testing is done in vacuum, and critical dimensions of the apparatus are given in ASTM E 595-77/84/90 in order to produce similar results. The set up allows 12 samples to be tested at once, with each in its own separate chamber in a solid copper bar covered by a solid copper cover. The bar is heated to 398 K for 24 h, which forces all volatile materials out through a 6.3-mm diameter escape hole. Materials are collected on their own chromium-plated disks, 12.7 mm from the hole, which are kept at a constant 298 K. The volatile materials will collect on the disk if their condensation temperature is 298 K or higher. The TML is calculated by taking the percent mass loss of the sample after heating at 398 K, and the CVCM is determined by calculating the mass of condensable material on the collector as a percentage of the total initial mass of the sample. The amount of water vapor regained (WVR) is also sometimes calculated, based on the percentage of mass gained due to water readsorption or reabsorption in 24 h at 25°C when the sample is in 50% relative humidity.

In the reports, materials are grouped in three ways: Section A groups materials based on 18 of their probable uses (adhesives, paints, etc.); Section B groups all materials in one alphabetical list; and Section C groups them by use as in Section A, but only with those materials that pass TML less than 1% and CVCM less than 0.1%.

14.5.8 HUMIDITY

The strongest surface forces are caused by capillary condensation. As the relative humidity increases stiction failures will rise. This will affect MEMS devices exposed to the outside environment waiting for launch and devices exposed to the environment of planetary exploration missions. Designing with stiction mitigation strategies is recommended in these cases.

14.6 MATERIALS

14.6.1 SINGLE CRYSTAL SILICON

Single crystal silicon (Si) is the most widely used semiconductor material and is the most common MEMS substrate. It has a diamond (cubic) crystal structure and an electronic band gap of 1.1 eV. It can also be doped with impurities to alter its conductivity, at the expense of introducing defects into the crystal lattice. Single crystal silicon is a brittle material and instead of undergoing plastic deformation like metals it yields via catastrophic failure. This is an advantage for sensor applications, as silicon structures will only undergo elastic deformation allowing for high mechanical stability. The mechanical properties of silicon are anisotropic and dependent on crystal orientation.

Silicon has many useful properties for MEMS devices. On the microscale it is often likened to stainless steel on the macroscale. It has higher yield strength than stainless and has a density lower than aluminum. The hardness of Si is slightly better than stainless steel. Silicon's specific strength, defined as the ratio of yield strength to density, is significantly higher than for most common materials as shown in Table 14.9.

Single-crystal silicon cleaves, or forms cracks, along its crystal planes. The more continuous the surface and edges of a piece of silicon are the less likely it will crack under mechanical stress. Sharp corners and edges can serve as crack initiation points when the material is stressed. Smooth curves and highly polished surfaces resist cracking best. Etching processes can also damage or alter silicon surfaces, affecting the material's mechanical properties.

14.6.2 POLYSILICON

Polysilicon or polycrystalline silicon is also widely used in the semiconductor industry. It is a crystalline material, but instead of being made up of a single crystal like bulk silicon, it is made up of many small crystal grains. ICs use polysilicon for resistors, gates, emitters, and ohmic contacts. It is widely used as a structural material for surface micromachining, heaters, and piezoresistive sensors in MEMS devices. Like single-crystal silicon it can be doped to change its conductivity. Polysilicon is typically deposited by low-pressure chemical vapor deposition (LPCVD), plasma-enhanced chemical vapor deposition (PECVD), or sputtering. The structure of polysilicon and hence its material properties change with deposition process, dopants, and temperature. After deposition thermal annealing can be used to drive dopants, recrystallize the structure to reduce the grain size, and reduce stress. Since the material properties of polysilicon reported in the literature vary widely, it is best to characterize the material to be used in order to extract the proper properties for the design process.

14.6.3 SILICON NITRIDE

Silicon nitride is a dielectric material with good passivation properties. It is often used as the topmost barrier layer on ICs, and forms a good barrier for H_2O and ions. It is also used as a capacitor dielectric material, etch mask, and wear-resistant coating. Silicon nitride can be deposited by LPCVD, PECVD, physical sputtering, or reactive sputtering. Again the deposition process greatly influences the material properties of these films. LPCVD is a high-temperature process (700 to 800°C), which produces the best quality stoichiometric Si_xN_y films. By controlling the amount of silicon in the film, the refractive index and stress can be changed. Increasing the Si content in silicon nitride films lowers the tensile film stress, increases transparency, and improves HF etch resistance. PECVD systems can control stress by deposition frequencies, power, and pressure and etch resistance by impurities. Sputtering systems can also be tuned to control the stress of the film by altering the temperature, power, pressure, and gas flow rates of depositions.

14.6.4 SILICON DIOXIDE

Silicon dioxide is the native oxide of silicon and one of the most common MEMS materials. It is used as a dielectric insulator, etch mask, or as part of a mechanical structure. It can be thermally grown on silicon, or deposited by LPCVD, PECVD, or sputtering. Thermal oxides require high temperature (~1000°C), so low temperature oxide (LTO) LPCVD or PECVD processes are used to coat over metals. PECVD and sputtering processes can be tuned to control the stress of SiO_2-deposited films.

14.6.5 METALS

There are a wide number of metal materials which can be used in MEMS as electrical conductors, structural material, and low emissivity coatings. These materials can be deposited by evaporation, sputtering, CVD, laser deposition, or electroplating. Metals are ductile materials which will plastically deform when stressed past their yield strength. In this section we will cover some of the most commonly used metals.

Aluminum is one of the most commonly used IC conductor materials and therefore also one of the most common MEMS materials. It is either evaporated or sputtered to form surface micromachined structures. Its native oxide is Al_2O_3, which is a hard and chemically resistant material. Aluminum adheres well to silicon dioxide by forming Al_2O_3–SiO_2 bonds.

Gold is a soft material which is used as a conductor or emissivity coating. Gold along with platinum are fairly inert and do not oxidize. Being nonreactive they have a hard time adhering to materials. A thin layer (10–50 nm) of a reactive metal such as Ti or Cr is typically used between surfaces to promote adhesion. Silicon will diffuse into Au at temperatures of 100°C. To prevent this, a barrier layer is needed between the Si and Au interface. Typically a thin layer of SiO_2 will suffice to prevent Si diffusion.

There are a number of metals to avoid in space applications:

- Tin can form whiskers in vacuum environments and can undergo a low temperature (−40 to 60°C) transformation from a stable tetragonal structure to a crack-prone powdery diamond cubic structure.
- Silver can also form whiskers and easily corrodes in sulfur-rich environments. Silver is also susceptible to galvanic corrosion with other metals.
- Mercury, cadmium, zinc, magnesium, selenium, tellurium sublime in vacuum and can redeposit potentially causing shorting or damaging optics.

14.6.6 POLYCRYSTALLINE DIAMOND

Polycrystalline diamond films have a high hardness and high thermal conductivity. This makes them an attractive material for high wear and high-temperature environments. Diamond is naturally hydrophobic and structures made from diamond are inherently striction resistant. Diamond films are typically grown using a hot-filament CVD or microwave plasma system at ~900°C. It is a wide bandgap material with a bandgap of 5.5 eV. P-type regions can be formed by doping with boron. Diamond also has the highest stiffness amongst material, making it ideal for high Q resonators. It is typically etched using an O_2 ion source or reactive ion etch (RIE).

14.6.7 SILICON CARBIDE

Silicon carbide has many properties that make it well suited for MEMS applications, due to its chemical resistance and good mechanical properties. Its high-temperature resistance, radiation resistance, electronics capability, extreme hardness, and high stiffness make it a good choice for missions in harsh environments. SiC is much stiffer than Si and thus makes good resonant structures.

The properties of SiC can vary significantly depending upon how it is grown and processed. This is because SiC is polymorphic, and exists in many polytypes. The three most common crystal types are cubic, hexagonal, and rhombehedral. Of these cubic 3C-SiC, and hexagonal 4H-SiC and 6H-SiC are the most common. SiC does not have a defined melting point; however it breaks down at 2830°C where it decomposes into graphite and a silicon-rich melt. It is typically grown using APCVD or LPCVD processes at 1300°C. Polycrystalline SiC can be grown at temperatures as low as 500°C using APCVD, LPCVD, PECVD, or reactive sputtering processes. Silicon carbide is chemically resistant but can be etched electrochemically or with a plasma process.

Silicon carbide is a wide-bandgap semiconductor material with a bandgap of approximately 3 eV. The exact bandgap depends on the crystal structure. Silicon carbide also oxidizes readily above 600°C to form a native silicon dioxide. SiC is a better natural insulator than Si or GaAs, but can be doped with aluminum or boron to form p-type material or nitrogen or phosphorus to form n-type materials. SiC electronics have the potential to operate at temperatures of 400 to 600°C, which may make them attractive for future missions with high-temperature constraints such as Venus.

14.6.8 POLYMERS AND EPOXIES

Organic polymers can be deposited via evaporation, CVD, plasma deposition, spin on, and spray techniques. Nonreactive and nonoutgassing polymers are required for space missions. Polymers, epoxies, and polyimides must pass an acceptance criteria of <0.1% CVCM and <1.0% TML. Teflon, kapton, and mylar have been used extensively in space. Polyvinylchloride materials should be avoided due to outgassing in vacuum and temperature constraints.

14.6.9 SU-8

SU-8, an EPON® epoxy-based negative photoresist, originally developed at IBM. It is known for its high thickness films and is often used as a poor man's LIGA technique for electroplating metal MEMS. Thicknesses up to 2 mm can be achieved with aspect ratios greater than 20. Since the stiffness and strength are low and thermal coefficient of expansion is high it is primarily used as a mask. It has been used as a structural post in thermal switches scheduled to fly on Midstar 1 due to its 5× better thermal insulating performance over silicon dioxide and 100× improvement over silicon nitride.[27] Material properties of SU-8 are given in Table 14.7.

TABLE 14.7
Properties of SU-8[34,35]

Properties	SU-8
Young's modulus (GPa)	1.5 to 3.1
Yield strength (GPa)	0.03 to 0.05
Ultimate strength (GPa)	0.05 to 0.08
Thermal conductivity (W/mK)	0.2

14.6.10 CP1®

CP1 is a fluorinated polyimide that was originally developed at the NASA Langley Research Center as a thermal coating and later used for large thin-film concentrators on large space-based antennas. CP1 is licensed by SRS Technologies. It is space qualified and has a tested lifetime of 10 years. It is a transparent, low dielectric constant, UV radiation-resistant, and moisture-resistant material developed for high-temperature applications and is now being used to manufacture solar sails, large antennae, solar thermal propulsion systems, and flat film panels. CP1 is deposited by a spin on process and has been used to protect the MEMS thermal shutters that are to be flown on Space Technology 5 (ST5).[28] Material properties of CP1 are given in Table 14.8.

TABLE 14.8
Properties of CP1

Properties	CP1
CTE	51.2 ppm/°C
Dielectric constant (at 10 GHz)	2.4 to 2.5
Film density	1.434 g/cc
Glass transition temperature	263°C
Imide IR bands	1780, 1725, and 745 cm^{-1}
Inherent viscosity of polyamic acid (at 35°C)	1.2 to 1.6 dl/g
IR emittance. Hemispherical, 300 K (coated film: aluminum)	0.03
IR emittance. Hemispherical, 300 K (uncoated film)	0.194 (0.25 mil)
Polymer decomposition temperature	530°C
Refractive index	1.58
Solar absorbance. Full spectrum (coated film. aluminum)	0.106 (0.25 ml)
Solar absorbance. Full spectrum (uncoated film)	0.072 (0.25 ml)
Specific heat. C_p (at 25°C)	1.094 J/g°C
Tensile modules	315 ksi
Tensile strength	14.5 ksi
UV cut off (0.2 ml film)	320 nm

TABLE 14.9
Material Properties and Performance Indices[36,37]

Material	Density, ρ (kg/m³)	Young's Modulus, e (GPa)	Fracture Strength, σ (MPa)	Specific Stiffness, E/ρ (MN*m/kg)	Specific Strength, σ/ρ (MN*m/kg)	Strain Tolerance, $\sigma^{3/2}/E$ (vMPa)	Knoop Hardness (kg/mm²)	Thermal Conductivity (W/m/K)	Thermal Expansion (10^{-6}/K)
Silicon	2,330	129 to 187	4,000	72	1.7	1.5	850 to 1,100	150	2.35
Polysilicon	2,330	176	1,800	76	0.77	0.43	1,070 to 1,275	150	2.8
Silicon dioxide	2,200	73	1,000	36	0.45	0.43	820	1.38	0.55
Silicon nitride	3,300	304	1,000	92	0.3	0.1	3,486	19	0.8
Nickel	8,900	207	500	23	0.06	0.54	251	91	13.4
Aluminum	2,710	69	300	25	0.11	0.75	130	235	25
Aluminum oxide	3,970	393	2,000	99	0.5	0.228	2,100	25	8.1
Silicon carbide	3,300	430	2,000	130	0.303	0.208	2,480	490	3.3
Nanocrystalline diamond	3,510	967	5,030	295	0.28	0.31	7,500 to 8,500	1,200	1
Single-crystal diamond	3,500	1,035	53,000	296	15.14	11.79	9,000	2,000	1
Iron	7,800	196	12,600	25	1.62	7.22	400	80	12
Tungsten	19,300	410	4,000	21	0.21	0.62	485	178	4.5
Stainless steel	8,050	221	1,000	27	0.12	0.14	660	33	17.3
Quartz (Z-axis)	2,650	97	600	37	0.23	0.15	850	1.4	7.1, 13.2

14.7 CONCLUSION

Materials selection is an important consideration when designing and operating MEMS devices in the space environment. Material properties can greatly affect device performance. Table 14.9 shows performance indices for various materials. Specific stiffness is a good metric for high-frequency resonating structures. Specific strength is a good metric for pressure sensor and valves. Strain tolerance is a good metric for devices which need to stretch and bend. Table 14.9 also lists thermal and mechanical properties of various materials used in MEMS; however the reader is reminded that real world material properties can vary widely. They are useful as a starting point, but again the material properties of the MEMS materials will vary based on the fabrication processes used.

The following design features and materials should be avoided:

1. Large temperature coefficient of expansion mismatches, unless designed as a sense or actuation mechanism
2. Pure tin coatings, except that electrical or electronic device terminals and leads may be coated with a tin alloy containing not less than 3% lead only when necessary for solderability
3. Silver
4. Mercury and mercury compounds, cadmium compounds and alloys, zinc and zinc alloys, magnesium, selenium, tellurium and alloys, and silver which can sublime unless internal to hermetically sealed devices with leak rates less than 1×10^{-4} atm-cm/sec^2
5. Polyvinylchloride
6. Materials subject to reversion
7. Materials that evolve corrosive compounds
8. Materials that sublimate

REFERENCES

1. Voronin, V., et al., Silicon whiskers for mechanical sensors. *Sensors and Actuators, A: Physical; East–West Workshop on Microelectronic Sensors, May 7–9 1991*, 1992. 30(1–2): p. 27–33.
2. Yun, H.M. and J.A. DiCarlo, Comparison of the tensile, creep, and rupture strength properties of stoichiometric SiC fibers. *Ceramic Engineering and Science Proceedings; Proceedings of the 1999 23rd Annual Conference on Composites, Advanced Ceramics, Materials, and Structures*, 1999. 20(3): p. 259–272.
3. Ruoff, A.L., *On the Ultimate Yield Strength of Solids*. 1978. 49(1): p. 197–200.
4. Shackelford, J.F. and W. Alexander, *CRC Materials Science and Engineering Handbook*. 3rd ed. 2001. Boca Raton, FL: CRC Press. 1949 p.
5. Sharpe Jr., W.N., The MEMS handbook. In *The Mechanical Engineering Handbook Series*, Gad-el-Hak, M. Editor. 2002. Boca Raton, FL: CRC Press. p. 1v. (various pagings).
6. Falvo, M.R. and R. Superfine, Mechanics and friction at the nanometer scale. *Journal of Nanoparticle Research*, 2000. 2(3): p. 237–248.

7. Madou, M.J., *Fundamentals of Microfabrication*. 1997. Boca Raton, FL: CRC Press. 589 p. [22] of plates.
8. Gad-el-Hak, M., *The MEMS handbook. The Mechanical Engineering Handbook Series.* 2002. Boca Raton, FL: CRC Press. p. 1v. (various pagings).
9. Alley, R.L., et al., Effect of release-etch processing on surface microstructure stiction. In *Proceedings of the 5th IEEE Solid-State Sensor and Actuator Workshop, June 22–25 1992. Hilton Head Island, SC, USA*. 1992. Piscataway, NJ: IEEE.
10. Houston, M.R., et al., Diamond-like carbon films for silicon passivation in Micro Electro Mechanical devices. In *Proceedings of the 1995 MRS Meeting, April 17–20 1995*. San Francisco, CA, USA. 1995. Pittsburgh, PA: Materials Research Society.
11. Man, P.F., B.P. Gogoi, and C.H. Mastrangelo, Elimination of post-release adhesion in microstructures using conformal fluorocarbon coatings. *Journal of Micro Electro Mechanical Systems*, 1997. 6(1): 25–34.
12. Houston, M.R., R. Maboudian, and R.T. Howe, Ammonium fluoride anti-stiction treatments for polysilicon microstructures. In *Proceedings of the 1995 8th International Conference on Solid-State Sensors and Actuators and Eurosensors IX. Part 1 (of 2), June 25–29 1995*. Stockholm, Sweden. 1995. Piscataway, NJ: IEEE.
13. Komvopoulos, K. Surface texturing and chemical treatment methods for reducing high adhesion forces at micromachine interfaces. In *Proceedings of the 1998 Conference on Materials and Device Characterization in Micromachining, September 21–22 1998*. Santa Clara, CA, USA. 1998. Bellingham, WA: SPIE.
14. Shores, A.A. Effective moisture getter coating for hermetic packages. In *5th Annual International Sampe Electronics Conference, June 18–20 1991*. Los Angeles, CA, USA. 1991. Covina, CA: SAMPE.
15. Ehmke, J., et al., *Method and Apparatus for Switching High Frequency Signals*. 2002:/ US patent applications/0036304.
16. Peregino, P. and E. Bukowski, *Development and Evaluation of a Surface-Mount High-G Accelerometer*. 2004. Army Research Laboratory, Aberdeen Proving Ground. p. 1–44.
17. Ghaffarian, R., et al., *Thermal and Mechanical Reliability of Five COTS MEMS Accelerometers*. 2002. Pasadena, CA: NASA Jet Propulsion Laboratory. p. 1–7.
18. Sharma, A. and A. Teverovsky, *Evaluation of Thermo-Mechanical Stability of COTS Dual-Axis MEMS Accelerometers for Space Applications*. 2000. Greenbelt, MD: NASA GSFC Component Technologies and Radiation Effects (Code 562). p. 1–8.
19. Togami, T.C., W.E. Baker, and M.J. Forrestal. Split Hopkinson bar technique to evaluate the performance of accelerometers. In *Proceedings of the 1995 Joint ASME Applied Mechanics and Materials Summer Meeting, June 28–30 1995*. Los Angeles, CA, USA. 1995. New York, NY: ASME.
20. Dever, J., et al., Physical and thermal properties evaluated of teflon FEP retrieved from the hubble space telescope during three servicing missions, In *Research and Technology Report 2001*. 2001. Cleveland, OH: NASA Glenn.
21. McClure, S.S., et al., Radiation effects in micro-electromechanical systems (MEMS): RF relays. *IEEE Transactions on Nuclear Science*, 2002. 49 I(6): 3197–3202.
22. Knudson, A.R., et al., Effects of radiation on MEMS accelerometers. *IEEE Transactions on Nuclear Science; Proceedings of the 1996 IEEE Nuclear and Space Radiation Effects Conference, NSPEC*, 1996. 43(6 pt 1): p. 3122–3126.
23. Caffey, J.R. and P.E. Kladitis. The effects of ionizing radiation on micro electro mechanical systems (MEMS) actuators: electrostatic, electrothermal, and bimorph. In

17th IEEE International Conference on Micro Electro Mechanical Systems (MEMS): Maastricht MEMS 2004 Technical Digest, Jan 25–29 2004. Maastricht, Netherlands. 2004. Piscataway, NJ: Institute of Electrical and Electronics Engineers, Inc.

24. Edmonds, L.D., G.M. Swift, and C.I. Lee, Radiation response of a MEMS accelerometer: an electrostatic force. *IEEE Transactions on Nuclear Science; Proceedings of the 1998 IEEE Nuclear and Space Radiation Effects Conference, NSREC'98*, 1998. 45(6 pt 1): p. 2779–2788.

25. Muraca, R.F. and J.S. Whittick, *Polymers for Spacecraft Applications*. 1967. NASA 7–100.

26. Fisher, A., *A Compilation of Low Outgassing Polymeric Materials Normally Recommended for GSFC Cognizant Spacecraft*. 1971, NASA TM X-65705.

27. Beasley, M.A., et al., Design and packaging for a micro electro mechanical thermal switch radiator. In *ITherm 2004 — Ninth Intersociety Conference on Thermal and Thermomechanical Phenomena in Electronic Systems, June 1–4 2004*. Las Vegas, NV, United States, 2004. Piscataway, NJ: Institute of Electrical and Electronics Engineers, Inc.

28. Osiander, R., et al., Micro electro mechanical devices for satellite thermal control. *IEEE Sensors Journal; Microsensors and Microacuators: Technology and Applications*, 2004. 4(4): 525–531.

29. Trimmer, W.S.N., Microrobots and micromechanical systems. *Sensors and Actuators*, 1989. 19(3): 267–287.

30. Miller, L.M., MEMS for space applications. *Proceedings of SPIE — The International Society for Optical Engineering; Proceedings of the 1999 Design, Test, and Microfabrication of MEMS and MOEMS, March 30–Apr 1 1999*, 1999. 3680(I): 2–11.

31. Larson, W.J. and J.R. Wertz, Space mission analysis and design. In *Space Technology Library*. 1999, Torrance, CA: Microcosm; Dordrecht/Boston: Kluwer. 740.

32. Isakowitz, S.J., *International Reference Guide to Space Launch Systems*, 2nd ed. 1991. Washington, D.C.: American Institute of Aeronautics and Astronautics.

33. Muller, L., et al., Packaging and qualification of MEMS-based space systems. In *Proceedings of the 1995 9th Annual International Workshop on Micro Electro Mechanical Systems, February 11–15 1996*. San Diego, CA, USA. 1996. Piscataway, NJ: IEEE.

34. Chang, S., J. Warren, and F.P. Chiang, Mechanical testing of EPON SU-8 with SIEM. In *Proceedings of Microscale Systems: Mechanics and Measurements Symposium*. 2000. Orlando, FL: Society for Experimental Mechanics.

35. Beasley, M.A., et al., MEMS thermal switch for spacecraft thermal control. In *MEMS/MOEMS Components and Their Applications, January 26–27 2004*. 2004. San Jose, CA: The International Society for Optical Engineering.

36. Spearing, S.M., Materials issues in micro electro mechanical systems (MEMS). *Acta Materialia*, 2000. 48(1): 179–196.

37. Madou, M.J., *Fundamentals of Microfabrication*. 1997. Boca Raton, FL: CRC Press. pp. 373.

15 Reliability Practices for Design and Application of Space-Based MEMS

Robert Osiander and M. Ann Garrison Darrin

CONTENTS

15.1 INTRODUCTION TO RELIABILITY PRACTICES FOR MEMS

Reliability is the ability of a system or component to perform its required functions under stated conditions for a specified period of time.[1]

This chapter begins with the classification of failures for spacecraft components. They are generally categorized as:

(1) Failures caused by the space environment, such as damage to circuits by radiation
(2) Failures due to the inadequacy of some aspect of the design
(3) Failures due to the quality of the spacecraft or of parts used in the design or
(4) A predetermined set of "other" failures, which include operational errors[2]

The first two types of failures are emphasized in this chapter, while the latter two types are emphasized in the next chapter covering quality assurance. The relationship of microelectromechanical systems (MEMS)-known failure modes and the uniqueness of the space environment stresses are covered in detail in this section. As with any emerging technology field, the absence of historical data reduces the ability to depend on known techniques to assure the reliable insertion of new systems. Considering that the majority of MEMS devices are silicon-based, it is natural to look to the integrated circuit domain for the base of all quality assurance and reliability knowledge. For this reason both the traditional specifications derived from statistical approaches and the use of the physics of failure (POF) approach to reliability are discussed.

15.2   STATISTICALLY DERIVED QUALITY CONFORMANCE AND RELIABILITY SPECIFICATIONS

The impact of increasingly complex and dense integrated circuits upon the civilian and military space programs easily relate to this current problem. The emergence of the electronic, electrical, and electromechanical part programs for NASA during the 1960s and 1970s produced a dependence on the military specifications and standard programs that have continued to evolve over the years. Reliance on the military program is understandable due to its dominance, which minimized civilian space consumption of electronics. The Jet Propulsion Laboratory (JPL)[3] MEMS developers at NASA Goddard Space Flight Center (GSFC), and staff members of the Johns Hopkins University Applied Physics Laboratory (JHU/APL) have all used this approach. Where the adoption of microcircuit testing from highly used military specifications and standards is appropriate, this technique is endorsed. For example, when a hermetic microcircuit package is used for MEMS packaging, traditional packaging qualification is appropriate. Today, one sees this relationship more in consumer electronic consumption, which dwarfs military consumption. Anomalies from the military system may be found in the electrical, electronic, and electromechanical (EEE) parts program in the requirements for such tests as salt spray, which appears more appropriate for components in use on a Navy ship than for those in use in a NASA Aerospace program. NASA in turn supplements the military documents with its own requirements, adding another layer. The significance of these additional documents provides increased traceability, rigorous de-rating systems, and a forced usage of a restricted range of components.

There are no prescribed requirements relative to the quality assurance and reliability for aerospace applications for MEMS. The argument that these rigid and risk-avoidant approaches have led to overdesigned, expensive, low-technology systems, and stymied the use of new (and often better) components has some legitimacy. What worked extremely well with the emerging integrated circuit (IC) industry may be entirely inappropriate for the mature microcircuit market of today. Building the MEMS inspection and qualification plans around the current techniques sets the community up for the same problems seen in the EEE world today.

An additional quandary in turning to the "known world of integrated circuits" is that the program bases its strength in rigid piece-part testing. Technically, the MEMS-based device falls under the NASA definition of a part. However, in some cases, a MEMS-based device could fall under the NASA definition of an assembly, where a functional group of parts such as a hinge assembly and antenna feed of a deployment boom would be included. Given this cross-level and lack of a direct fit into standard NASA hardware nomenclature, the traditional test methodologies are not always a natural fit. The workhorse for EEE parts has been the QPL documents and MIL-STD-883: *General Test Methods for Microcircuits.*

The importance of beginning a rigorous test program at the lowest element possible and building a rugged program cannot be underestimated. However, building on the current test methods where direct fits are often missing requires tailoring at each step. In addition, in order to work with new technology and untried systems, a reliance on good process control must be built in. Plans for inspection, quality assurance, and specifications are provided as guidelines with the intent of tailoring and adding process control steps at each interface level.

The original reliability prediction handbook was MIL-HDBK-217, the Military Handbook for "Reliability Prediction of Electronic Equipment." MIL-HDBK-217 is published by the Department of Defense based on work done by the Reliability Analysis Center and Rome Laboratory at Griffiss AFB, NY.

The MIL-HDBK-217 handbook contains failure rate models for the various part types used in electronic systems, such as active microcircuits, semiconductors, and passive components such as resistors, capacitors, relays, switches, connectors, etc. These failure rate models are based on the best field data that could be obtained for a wide variety of parts and systems; this data is then analyzed and evaluated, with many simplifying assumptions thrown in, to create usable models. In the absence of a large utilization and knowledge base for MEMS, the use of MIL-STD-883 for test method and either MIL-PRF-38535 or MIL-H-38534 are reasonable interim steps. Each interface level is then available to be qualified along with the series of electrical, mechanical, and environmental tests meant to assure long life and final performance. These tests, where definitive, realize the reliability predictions of MIL-HDBK-217 and are driven by known activation energies of silicon-based microcircuits. Unfortunately, there is no equivalent to MIL-HDBK-217 for MEMS; however, using the documents mentioned before for guidelines is a reasonable approach.

15.3 PHYSICS OF FAILURE (POF) APPROACH

Military specifications and reliability work have historically been based on the MIL-HDBK-217, "Reliability Prediction of Electronic Equipment," approach. Transition from statistical-field failure-based models to POF-based models for reliability assessment has successfully been demonstrated for MEMS.[4-6] Although the POF approach is not a recent development, the Computer Aided Life Cycle Engineering (CALCE) Electronic Products and Systems Center has become the focal point for developing the knowledge base relative to microelectronics and packaging[7–9]. In comparing the two approaches, there are problems with using statistical field-failure models for the design, manufacture, and support of electronic equipment. The U.S. Army began a transition from MIL-HDBK-217 to a more scientific, POF approach to electronic equipment reliability. To facilitate the transition, an IEEE Reliability Program Standard is under development to incorporate physics of failure concepts into reliability programs.[10] The POF approach has been used quite successfully for decades in the design of mechanical, civil, and aerospace structures. This approach is almost mandatory for buildings and bridges because the sample size is usually one, affording little opportunity for testing the complete product or for reliability growth.[10,11] POF is an engineering-based approach to determining reliability. It uses modeling and simulation to eliminate failures early in the design process by addressing root-cause failure mechanisms in a computer-aided-engineering environment. The POF approach applies reliability models, built from exhaustive failure analysis and analytical modeling, to environments in which empirical models have long been the rule.[7,10] The central advantage of the POF in spacecraft systems is that it provides a foundation upon which to *predict* how a new design will behave under given conditions, an appealing feature for small spacecraft engineers. This approach involves the following:[12]

- Identifying potential failure mechanisms (chemical, electrical, physical, mechanical, structural, or thermal processes leading to failure); failure sites; and failure modes
- Identifying the appropriate failure models and their input parameters, including those associated with material characteristics, damage properties, relevant geometry at failure sites, manufacturing flaws and defects, and environmental and operating loads
- Determining the variability for each design parameter when possible
- Computing the effective reliability function
- Accepting the design, if the estimated time-dependent reliability function meets or exceeds the required value over the required time period.

A central feature of the POF approach is that reliability modeling, which is used for the detailed design of electronic equipment, is based on root-cause failure processes or mechanisms. These failure-mechanism models explicitly address the design parameters which have been found to influence hardware reliability strongly, including material properties, defects and electrical, chemical, thermal, and mechanical stresses. The goal is to keep the modeling in a particular application as simple as possible without losing the cause–effect relationships, which benefits corrective action. Research into physical failure mechanisms is subjected to scholarly peer review and published in the open literature. The failure mechanism models are validated through experimentation and replication by multiple researchers.[12]

An approach in the same vein emphasizing process monitoring and quality assurance methods has been applied to MEMS components. These methods include techniques to study:

(1) Process bias
(2) Material microstructure and mechanical properties
(3) Mechanical response of spring-supported structures, and
(4) Actuator performance

Characterization of the as-produced components and materials serves as the starting point for future studies of reliability of MEMS components and systems. Extensive process monitoring is performed at every step.[13]

In recent years, the POF approach has been used for new and emerging technologies such as multichip modules for insertion into space flight applications.[14] The POF approach has been applied to MEMS reliability by representatives from the French Space Agency, Centre National d'Etudes Spatiales[4,6] and at JPL, Caltech[15] among others.

15.4 MEMS FAILURE MECHANISMS

MEMS reliability and failure mechanisms concerns for the space environment include: material mismatches, fracture and fatigue, adhesion and stiction, friction and wear, electrostatic interference, radiation damage, and thermal effects.

15.4.1 MATERIAL INCOMPATIBILITIES

Process incompatibilities, materials issues, and fabrication limitations still present formidable challenges to any practical commercialization of most developmental microsystems.[16] Processing may induce thermomechanical stresses due to mismatch of the coefficients of thermal expansion (CTE) of the base material. Due to the unique structure and small scale of MEMS, residual stresses during the deposition processes can have a profound effect on the functionality of the structures. Typically, material properties of thin films used in surface micromachining are not controlled during deposition. The residual stress, for example, tends to vary significantly for different deposition methods. Currently, few techniques are available to measure the residual stress in MEMS devices. Differences in stress between the multiple metal and dielectric layers may cause vertical stress gradients and curl. Additionally, misalignment between layers may cause lateral stress gradients and curl. This curl that may also be induced through stresses in plating and postprocessing yields, an effect often termed "potato chipping."

At all steps, the concerns of material compatibilities will need to be addressed. During packaging of MEMS, stresses will be distributed within the die attach, with die and substrate contributing to the reliability or lack of reliability of the packaging structure. Numerous studies in the literature are available relative to the potential of decreasing or inducing stress during processing steps.[17–20]

Contamination from outgassing may bind to other materials present in the environment, leading to clogging or build up of material. Depending on the configuration, the device may become inoperable. Contamination binding with other materials or allowing a build up have been found to cause device failure when in crucial active areas.[21]

15.4.2 STICTION

With their small dimensions, MEMS structures are dominated by surface forces, especially the van der Waals force, that cause microscopic structures to stick together. Van der Waals forces bonding two clean surfaces together are a result of instantaneous dipole moments of atoms. If two flat parallel surfaces become separated by less than a characteristic distance of z_0, which is approximately 20 nm, the attractive pressure will be given by:

$$P_{vdW} = \frac{A}{6pd^2} \tag{15.1}$$

where
   A is Hamaker constant (1.6 eV for Si) and
   d the separation between the surfaces.

While this equation ignores the repulsive part of the surface forces and over-estimates the force of adhesion by at least a factor of two, it allows for an order of approximations for adhesive forces. Typical values of d are in the order of several angstroms.[2,22] As soon as a flexible structure comes close enough to another surface so that this force is stronger than the elastic force retracting the structure, the two surfaces will almost permanently stick together.

The probability of stiction occuring may be reduced with designs where surfaces that can contact other surfaces are minimized, for example, by using small dimples which hold the structures at a distance. Forces that can cause stiction in MEMS devices are capillary force and electrostatic force. Causes of stiction also include shock-induced stiction (mechanical overstress) and voltage overstress, which can both result in large areas in contact and allow stiction to occur.[21]

Processing techniques, such as critical drying after the release, may reduce the potential for stiction to occur as a result of the capillary forces. The ability to successfully release a MEMS device is a critical processing step of a MEMS device. Due to the inherent proximity of the moving structure and the surrounding surfaces, the final drying process on a surface micromachined polysilicon structure can lead to permanent stiction of the structure dependent upon the various drying techniques employed.[23] Stiction induced by capillary forces during the postrelease drying step of MEMS fabrication can substantially limit the functional yield of complex devices. Supercritical CO_2 drying provides a method to remove liquid from the device surface without creating a liquid or vapor interface, thereby mitigating stiction.[24] Fluoro- or hydrocarbon coatings can be used on MEMS surfaces after they are released to lower the surface interaction energy and prevent stiction during operation. These coatings provide a hydrophobic surface on which water cannot condense. Therefore, the most important stiction force by capillary condensation will not occur.[25]

15.4.3 CREEP

Reliability of components due to creep properties of materials is important to structural integrity. Reliability of MEMS devices will greatly be affected by creep of components that operate at high temperatures. The reliability will also suffer when MEMS components are made of materials, which creep at room temperature. Electrothermal microactuators, considered as the driver components for micromotors, are examples of structures prone to creep deformation upon actuation. Additionally, components made of polymers, such as polyimides, will undergo creep at room temperature.[26] Creep behavior of all materials exposed to thermal cycling, including solders and other attached materials should be reviewed.

15.4.4 FATIGUE

MEMS are often chosen for their long life and intrinsic strength. High cyclic fatigue failure results tend to be impressive. Results from a research team at Pennsylvania State University provide the most comprehensive, high-cycle, endurance data for designers of polysilicon micromechanical components available to date. These researchers evaluated the long-term durability properties of materials for MEMS. The stress-life cyclic fatigue behavior of a 2-μm thick polycrystalline silicon film was evaluated in laboratory air using an electrostatically actuated notched cantilever beam resonator. A total of 28 specimens were tested for failure under high-frequency (40 kHz) cyclic loads with lives ranging from about 8 sec to 34 days (3×10^5 to 1.2×10^{11} cycles) over fully reversed, sinusoidal stress amplitudes varying from 2.0 to 4.0 GPa. The thin-film polycrystalline silicon cantilever beams exhibited a time-delayed failure that was accompanied by a continuous increase in the compliance of the specimen. This apparent cyclic fatigue effect resulted in endurance strength at greater than 10^9 cycles, similar to 2 GPa, that is, roughly one-half of the (single cycle) fracture strength. Based on experimental and numerical results, the fatigue process is attributed to a novel mechanism involving the environmentally assisted cracking of the surface oxide film (termed reaction-layer fatigue).[27] In silicon, a fatigue-like phenomenon has been observed, but it occurs only at very high stress intensity levels, at which it is hardly a good idea to use brittle materials anyway. On the other hand, sudden fracture due to a short "overload" condition below the yield strength is likely to destroy brittle materials (containing small flaws), but not tough materials like metals, although the ultimate fracture strength of a metal components of a MEMS structure may well be lower than that of its brittle counterpart.[25] In accelerated life testing analysis, thermal cycling is commonly treated as a low-cycle fatigue problem, using the inverse power law relationship. Coffin and Manson suggested that the number of cycles-to-failure of a metal subjected to thermal cycling is given by:[28]

$$N = C/(\Delta T)^{\gamma} \tag{15.2}$$

where
   N is the number of cycles to failure
   C is a constant, characteristic of the metal
   γ is another constant, also characteristic of the metal
      and
   ΔT is the temperature range of the thermal cycle.

This model is basically the inverse power law relationship, where instead of the stress, V, the range ΔV is substituted to give

$$L(V) = 1/KV^{n} \tag{15.3}$$

where
   L represents a quantifiable life measure, such as mean life, characteristic life, median life, $B(x)$ life, etc.
   V represents the stress level
   K is one of the model parameters to be determined ($K > 0$)
      and
   n is another model parameter to be determined.

   This is an attempt to simplify the analysis of a time-varying stress test by using a constant stress model. It is a very commonly used methodology for thermal cycling and mechanical fatigue tests. However, by performing such a simplification, the following assumptions and shortcomings are inevitable. First the acceleration effects due to the stress rate of change are ignored. In other words, it is assumed that the failures are accelerated by the stress difference and not by how rapidly this difference occurs. Secondly, the acceleration effects due to stress relaxation and creep are ignored.

15.4.4.1   Fracture

Fracturing occurs when the load on the device is greater than the strength of the material. Clearly good design with proper margins or alternately, less brittle materials is the solution. In addition, debris can form, leading to additional failure modes. In the space environment applications, this is particularly a concern as conductive particles could induce numerous failures. Additional failure mechanisms such as radiation degradation and thermally induced reliability concerns are handled in other sections of this book.

15.5   ENVIRONMENTAL FACTORS AND DEVICE RELIABILITY

Environmental factors that strongly influence MEMS reliability are included in Table 15.1, which provides a checklist for typical environmental factors to be considered. Specific components may need to take extra factors into account or

TABLE 15.1
Environmental Factors Checklist (Typical)

Natural Occurring	Application Induced
Albedo, planetary IR	Acceleration
Electromagnetic radiation	Chemicals
Electrostatic discharge	Corona
Gravity, low	Electromagnetic, laser
Humidity, high	Electromagnetic radiation
Ionized gases	Electrostatic discharge
Magnetics, geo	Explosion
Meteoroids	Icing
Particulate levels, high	Magnetics
Pollution, air	Moisture
Pressure, high	Nuclear radiation
Pressure, low, vacuum	Particulate levels, high
Radiation, cosmic, soar	Shock, pyro, thermal
Temperature, high	Space debris
Temperature, low	Temperature, high, aero. heating, fire
	Temperature, low, aero. cooling
	Turbulence
	Vapor trails (plumes)
	Vibration, mechanical, microphonics
	Vibration, acoustic

may be able to ignore some other factors. Other natural environmental concerns are seen in a long duration balloon or unmanned aerial vehicles (UAV) type applications but primarily lower atmosphere and terrestrial could include: wind, rain, salt spray, sand and dust, sleet, snow, hail, lightning, ice, fog, clouds, freezing rain, frost, and fungus.

15.5.1 COMBINATIONS OF ENVIRONMENTALLY INDUCED STRESSES

Concurrent (combined) environments may be more detrimental to reliability than the effects of a single environment. In characterizing the design process, design or test criteria must consider both single and/or combined environments in anticipation of providing the hardware capability to withstand the hazards identified in the system profile. The synergistic effects of typical combined environments can be illustrated in a matrix relationship, which shows combinations where the total effect is more damaging than the cumulative effect of each environment acting independently. For example, an item may be exposed to a combination such as temperature, humidity, altitude, shock, and vibration while it is being transported. The acceptance to end-of-life history of an item must be examined for these effects. Table 15.2 provides reliability considerations for pairs of environmental factors.[29]

TABLE 15.2
Various Environmental Pairs

High Temperature and Humidity	High Temperature and Low Pressure	High Temperature and Solar Radiation
High temperature tends to increase the rate of moisture penetration. High temperatures increase the general deterioration effects of humidity. MEMS are particularly susceptible to deleterious effects of humidity.	Each of these environments depends on the other. For example, as pressure decreases, outgassing of constituents of materials increases; as temperature increases, outgassing increases. Hence, each tends to intensify the effects of the other.	This is a man-independent combination that causes increasing effects on organic materials.

High Temperature and Shock and Vibration	High Temperature and Acceleration	High Temperature and Explosive Atmosphere
Since both environments affect common material properties, they will intensify each other's effects. The degree to which the effect is intensified depends on the magnitude of each environment in combination. Plastics and polymers are more susceptible to this combination than metals, unless extremely high temperatures are involved.	This combination produces the same effect as high temperature and shock and vibration.	Temperature has minimal effect on the ignition of an explosive atmosphere but does affect the air–vapor ratio, which is an important consideration.

Low Temperature and Humidity	High Temperature and Ozone	High Temperature and Particulate
Relative humidity increases as temperature decreases, and lower temperature may induce moisture condensation. If the temperature is low enough, frost or ice may result.	Starting at about 300°F (150°C) temperature starts to reduce ozone. Above about 520°F (270°C), ozone cannot exist at pressures normally encountered.	The erosion rate of sand may be accelerated by high temperature. However, high temperature reduces sand and dust penetration.

Low Temperature and Solar Radiation	Low Temperature and Low Pressure	Low Temperature and Sand and Dust
Low temperature tends to reduce the effects of solar radiation and vice versa.	This combination can accelerate leakage through seals, etc.	Low temperature increases dust penetration.

TABLE 15.2
Various Environmental Pairs — Continued

Low Temperature and Shock and Vibration	Low Temperature and Acceleration	Low Temperature and Explosive Atmosphere
Low temperature tends to intensify the effects of shock and vibration. However, it is a consideration only at very low temperatures.	This combination produces the same effect as low temperature and shock and vibration.	Temperature has minimal effect on the ignition of an explosive atmosphere but does affect the air–vapor ratio, which is an important consideration.

Low Temperature and Ozone	Humidity and Low Pressure	Humidity and Particulate
Ozone effects are reduced at lower temperatures but ozone concentration increases with lower temperatures.	Humidity increases the effects of low pressure, particularly in relation to electronic or electrical equipment. However, primarily the temperature determines the actual effectiveness of this combination.	Sand and dust have a natural affinity for water and this combination increases deterioration.

Humidity and Vibration	Humidity and Shock and Acceleration	Humidity and Explosive Atmosphere
This combination tends to increase the rate of breakdown of electrical material.	The periods of shock and acceleration are considered too short for these environments to be affected by humidity.	Humidity has no effect on the ignition of an explosive atmosphere but a high humidity will reduce the pressure of an explosion.

Humidity and Ozone	Humidity and Solar Radiation	Low Pressure and Solar Radiation
Ozone meets with moisture to form hydrogen peroxide, which has a greater deteriorating effect on plastics and elastomers than the additive effects of moisture and ozone.	Humidity intensifies the deteriorating effects of solar radiation on organic materials.	This combination does not add to the overall effects.

Low Pressure and Particulate	Low Pressure and Vibration	Low Pressure and Shock or Acceleration
This combination only occurs in extreme storms during which small dust particles are carried to high altitudes.	This combination intensifies effects in all equipment categories but mostly with electronic and electrical equipment.	These combinations only become important at the hyperenvironment levels, in combination with high temperature.

Continued

TABLE 15.2
Various Environmental Pairs — Continued

Low Pressure and Explosive Atmosphere	Solar Radiation and Explosive Atmosphere	Solar Radiation and Particulate
At low pressures, an electrical discharge is easier to develop but the explosive atmosphere is harder to ignite.	This combination produces no added effects.	It is suspected that this combination will produce high temperatures.

Solar Radiation and Ozone	Solar Radiation and Vibration	Solar Radiation and Shock or Acceleration
This combination increases the rate of oxidation of materials.	Under vibration conditions, solar radiation deteriorates plastics, elastomers, oils, etc. at a higher rate.	These combinations produce no added effects.

Shock and Vibration	Vibration and Acceleration	Particulate and Vibration
This combination produces no added effects.	This combination produces increased effects when encountered with high temperatures and low pressure in the hyper-environmental ranges.	Vibration might possibly increase the wearing effects of sand and dust.

Each environmental factor that is present requires a determination of its impact on the operational and reliability characteristics of the materials and parts comprising the equipment being designed. Packaging techniques should be identified that afford the necessary protection against the degrading factors.

In the environmental stress identification process that precedes selection of environmental strength techniques, it is essential to consider stresses associated with all life intervals of the MEMS. This includes operational and maintenance environments as well as the preoperational environments, when stresses imposed on the parts during manufacturing assembly, inspection, testing, shipping, and installation may have significant impact on MEMS reliability. Stresses imposed during the preoperational phase are often overlooked; however, they may represent a particularly harsh environment that the MEMS must withstand. Often, the environments MEMS are exposed to during shipping and installation are more severe than those encountered during normal operating conditions. It is probable that some of the environmental strength features that are contained in a system design pertain to conditions that will be encountered in the preoperational phase rather than during actual operation. Environmental stresses affect parts in different ways and must also be taken into consideration during the design phase. Table 15.3 illustrates the principal effects of typical environments on MEMS.

TABLE 15.3
Environmental Effects and the Principal Failures Induced on MEMS Devices

Environment	Principal Effects	Typical Failures Induced
High temperature	Thermal aging	Insulation failure
	Oxidation	Alteration of electrical properties
	Structural change	Leaching of gold or other materials into silicon substrate
	Chemical reaction	
		Purple plague
		Kirkendahl voids
	Softening, melting, and sublimation	Structural failure
	Viscosity reduction or evaporation	Loss of lubrication properties
	Physical expansion	Structural failure
		Increased mechanical stress
		Increased wear on moving parts
Low temperature	Increased viscosity and solidification	Loss of lubrication properties
	Ice formation	Alteration of electrical properties
	Embrittlement	Loss of mechanical strength
		Cracking, failure
	Physical contraction	Structural failure
		Increased wear on moving parts
High relative humidity	Moisture absorption	Sealing, rupture of container
	Chemical reaction	
		Physical breakdown
	Corrosion	
		Loss of electrical strength
	Electrolysis	
		Loss of mechanical strength
		Interference with function
		Loss of electrical properties
		Increased conductivity of insulators
		Increased opportunity for failures due to stiction
Low relative humidity	Desiccation	Loss of mechanical strength
	Embrittlement	Structural collapse
	Granulation	Alteration of electrical properties, "dusting."
		Increased chance of ESD induced failures
High pressure	Compression	Structural collapse
		Penetration of sealing
		Interference with function
		Ruptures of fragile structures

Continued

TABLE 15.3
Environmental Effects and the Principal Failures Induced on MEMS Devices —
Continued

Environment	Principal Effects	Typical Failures Induced
Low pressure	Expansion	Fractures
		Explosive expansion
	Outgassing	Alteration of electrical properties
		Loss of mechanical strength
	Reduced dielectric strength of air	Insulation breakdown and arc-over
		Corona and ozone formation
Solar radiation	Actinic and physicochemical reactions	Surface deterioration
		Alteration of electrical properties
	Embrittlement	
		Discoloration of materials
		Ozone formation
Particulate	Abrasion	Increased wear
	Clogging	Interference with function
		Alteration of electrical properties
High air or gas pressure	Force application	Structural collapse
		Interference with function
		Loss of mechanical strength
	Deposition of materials	Mechanical interference and clogging
		Abrasion accelerated
	Heat loss (low velocity)	Accelerates low-temperature effects
	Heat gain (high velocity)	Accelerates high-temperature effects
Temperature shock	Mechanical stress	Structural collapse or weakening
		Seal damage
High-speed particles (nuclear irradiation)	Heating	Thermal aging
		Oxidation
	Transmutation and ionization	Alteration of chemical, physical, and electrical properties
		Production of gases and secondary particles
Zero gravity	Mechanical stress	Interruption of gravity-dependent functions
	Absence of convection cooling	Aggravation of high-temperature effects
Ozone	Chemical reactions	Rapid oxidation
	Crazing, cracking	Alteration of electrical properties
	Embrittlement	Loss of mechanical strength
	Granulation	Interference with function
	Reduced dielectric strength of air	Insulation breakdown and arc-over
Explosive decompression	Severe mechanical stress	Rupture and cracking
		Structural collapse

TABLE 15.3
Environmental Effects and the Principal Failures Induced on MEMS Devices —
Continued

Environment	Principal Effects	Typical Failures Induced
Dissociated gases	Chemical reactions	Alteration of physical and electrical properties
	Contamination	
	Reduced dielectric strength	Insulation breakdown and arc-over
Acceleration	Mechanical stress	Structural collapse
		Separation from substrate
Vibration	Mechanical stress	Loss of mechanical strength
		Interference with function
		Increased wear
	Fatigue	Structural collapse
Magnetic fields	Induced magnetization	Interference with function
		Alteration of electrical properties
		Induced heating

15.5.2 THERMAL EFFECTS

High temperatures impose a severe stress on most electronic items including MEMS devices, since it can cause catastrophic failure. High temperature also causes progressive deterioration of reliability due primarily to chemical degradation effects. The nature of MEMS design requires small sizes, often with high part densities. This generally requires a cooling system to provide a path of low thermal resistance from heat-producing elements to an ultimate heat sink of reasonably low temperature. Reliability improvement techniques for high-temperature stress include the use of heat dissipation devices, cooling systems, thermal insulation, and heat-withstanding materials.

Low temperatures experienced by MEMS can cause reliability problems. These problems usually are associated with mechanical system elements. They include mechanical stresses produced by differences in the coefficients of expansion (contraction) of metallic and nonmetallic materials, embrittlement of nonmetallic components, mechanical forces caused by freezing of entrapped moisture, stiffening of liquid constituents, etc. Typical examples include cracking, delaminations, binding of mechanical linkages, and excessive viscosity of lubricants. Reliability improvement techniques for low-temperature stress include the use of heating devices, thermal insulation, and cold-withstanding materials.

Additional stresses are produced when MEMS are exposed to sudden changes of temperature or rapidly changing thermal cycling conditions. These conditions generate large internal mechanical stresses in structural elements, particularly when dissimilar materials are involved. Effects of thermal shock-induced stresses include cracking of seams, delamination, loss of hermeticity, leakage of fill gases, separation of encapsulating materials from components and enclosure surface, leading to the creation of voids, and distortion of support members.

A thermal shock test may be specified to determine the integrity of solder joints since such a test creates large internal forces due to differential expansion effects. Such a test also has been found to be instrumental in creating segregation effects in solder alloys, leading to the formation of lead-rich zones, which are susceptible to cracking effects.

15.5.3 SHOCK AND VIBRATION

MEMS are often subjected to environmental shock and vibration during both normal use and testing. Such environments can cause physical damage when deflections cause mechanical stresses that exceed the allowable working stress of the constituent parts.

Natural frequencies of items comprising the MEMS are important parameters that must be considered in the design process since a resonant condition can be produced if a natural frequency is within the vibration frequency range. The resonance condition will greatly amplify subsystem deflection and may increase stresses beyond the safe limit.

The vibration environment can be particularly severe for electrical connectors, since it may cause relative motion between connector elements. In combination with other environmental stresses, this motion can produce fretting corrosion. This generates wear debris and causes large variation in contact resistance. Reliability improvement techniques for vibrational stress include the use of stiffening, control of resonance, and reduced freedom of movement.

15.5.4 HUMIDITY

Humidity can cause degradation of MEMS as discussed previously. Reliability improvement techniques for humidity and salt environments include use of hermetic sealing, moisture-resistant material, dehumidifiers, protective coatings or covers, and reduced use of dissimilar metals.

Deleterious effects may be exacerbated with high humidity; for example, crack growth has a dependence on moisture that is well documented.[30] Electrical performance may change as moisture enters gaps in a vapor form and condenses as water droplets, causing surface tension which may induce a piezoresistive stress effect.[31] Perhaps best known is the relationship of adhesion and friction of polycrystalline silicon MEMS.[32] This dependence is reduced, but not eliminated, when molecular coatings are applied to the surfaces. Antistiction coatings have the ability to penetrate into the intricate side wall and under-surface spaces in three dimensions. Thus, these coatings extend the operating life of MEMS devices by reducing stiction.[33]

15.5.5 RADIATION

Electromagnetic and nuclear radiation can disrupt performance levels and, in some cases, cause permanent damage to exposed devices. Therefore, it is important that such effects be considered in determining the environmental strength for electronic equipment that must achieve a specified reliability goal.

Electromagnetic radiation often produces interference and noise effects within electronic circuitry, which can impair system performance. Sources of these effects include corona or lightning discharges, sparking, and arcing phenomena. These may be associated with high-voltage transmission lines, ignition systems, brush type motors, and even the equipment itself. Generally, the reduction of interference effects requires incorporating filtering and shielding features or specifying less susceptible components and circuitry.

Nuclear radiation can cause permanent damage by alteration of the atomic or molecular structure of dielectric and semiconductor materials. High-energy radiation also can cause ionization effects that degrade the insulation levels of dielectric materials. The migration of nuclear radiation effects typically involves materials and parts possessing a higher degree of radiation resistance, and the incorporation of shielding and hardening techniques.

Each environmental factor experienced by an item during its life cycle requires consideration in the design process. This ensures that adequate environmental strength is incorporated into the design for reliability.

In conclusion, failure to perform a detailed life cycle environment profile can lead to overlooking environmental factors whose effect is critical to MEMS reliability. If these factors are not included in the environmental design criteria and test program, environment-induced failures may occur during space flight operations. Therefore, it is recommended that at the onset of the design process, researchers identify the operating conditions that will be encountered during the life of the equipment.

15.5.6 ELECTRICAL STRESSES

Civilian and military space missions are susceptible to corona and high breakdown voltage. Understanding the role and the potential degradation caused by these events is important for the MEMS designer. Historically, spacecrafts are vulnerable to corona when exposed to regimes of critical pressure during ground test and flight. NASA has encountered this problem many times. These coronal discharge problems have occurred many times in NASA history and can cause serious damage among craft components. Hardware susceptibility to corona-induced damage should be addressed in subsystem design and in test and operational procedures.

Ionizing portion of the atmosphere may subject a spacecraft to unequal flux of ions and electrons that can induce a charge. In low-earth orbit (LEO) a spacecraft travels through dense but low energy plasmas and the spacecrafts are negatively charged and may charge to thousands of volts. In geostationary orbit (GEO) there is a greater concern where biased surfaces, such as solar arrays, can affect the floating potential.[29] Particular attention must be paid to prevent arcing to MEMS devices if placed on the surface or skin of the satellite. Traditional approaches to assure that the surface of the satellite is conductive to bleed off charges can be used with MEMS devices on the surfaces with a conductive plating or coating depending on design and application.

Electrostatic discharge (ESD) or electrical overstress (EOS) occurs when a device is improperly handled. A human body routinely develops an electric potential in excess of 1000 V. Upon contact with an electronic device, this build-up will discharge, creating a large potential difference across the device. The effect is known to have catastrophic effects in circuits and could have similar effects on MEMS devices where ESD may cause attractions or shifts. While the deleterious effects of ESD on MEMS structures are just beginning to be published,[34,35] one should assume that certain electrostatically actuated devices will be susceptible to ESD damage.

15.6 CONCLUSION

For MEMS devices to be properly operated in space, materials and hardware reliability is essential. MEMS reliability can be achieved by applying conventional reliability practices for electronics while taking space environmental effects into consideration. For a space application, reliability practices are validated as requirements and reflected in a mission design and review cycle with key milestones such as preliminary design review (PDR) and critical design review (CDR) which are covered in the next chapter on quality assurance.

The performance of MEMS devices is strongly affected by environmental factors and the effect may vary according to specific MEMS applications. The environmental impact occurs during all mission phases. Generally, the impact is more severe under the preoperational environments than more benign operational and maintenance environments. Under the worst scenario, the synergy of environmental factors may cause detrimental effects and render the devices useless.

To warrant a successful mission, MEMS designers need to pay attention to prevent potential failure mechanisms in space. MEMS devices are susceptible to corona and high breakdown voltage. Lessons learned from numerous space mission failures provide important information for future MEMS design. During MEMS design and operation, additional effort is required to mitigate failure mechanisms related to materials and structures.

The study of POF, like any other scientific discipline, requires testing to validate hypotheses and gather data on failure mechanisms. A significant amount of research can be conducted on the ground, but some amount of space-based research is likely to be necessary. POF research could make extensive use of low-cost "time-in-space" facilities, such as Shuttle deployed free-flying spacecraft, balloon demonstrations and sounding rockets. Inexpensive long-duration missions might allow data to be gathered on actual performance in space, with components being returned to Earth for analysis.

REFERENCES

1. *IEEE Standard Computer Dictionary: A Compilation of IEEE Standard Computer Glossaries*. Institute of Electrical and Electronics Engineers, New York, NY, 1990.

2. Stark, B., Failure modes and mechanisms. *JPL Publication*, 99–1, 21–47, 1999.
3. Stark, B. and Kayali, S., Qualification testing protocols for MEMS. *JPL Publication*, 99–1, 209–234, 1999.
4. Schmitt, P., et al., Application of MEMS behavioral simulation to physics of failure (pof) modeling, *Microelectronics Reliability*, vol. 43, Elsevier Ltd, London, 2003, 1957–1962.
5. Lafontan, X., et al., MEMS physical analysis in order to complete experimental results return, *Proceedings of SPIE — 4019*, 2000, 236–243.
6. Pressecq, F., et al., CNES reliability approach for the qualification of MEMS for space, *Proceedings of SPIE — 4558*, 2001, 89–96.
7. Pecht, M. and Dasgupta, A., Physics-of-failure: an approach to reliable product development, International Integrated Reliability Workshop Final Report, IEEE, Piscataway, NJ, 1995, 1–4.
8. Pecht, M., Nash, F.R., and Lory, J.H., Understanding and solving the real reliability assurance problems, *Proceedings of the Annual Reliability and Maintainability Symposium*, Ed. IEEE, Piscataway, NJ, 1995, 159–161.
9. Zhang, Y., et al., Trends in component reliability and testing, *Semiconductor International*, 22 (10), 1999, 4.
10. Stadterman, T.J., et al., *Transition from Statistical-Field Failure Based Models to Physics-of-Failure Based Models for Reliability Assessment of Electronic Packages*, American Society of Mechanical Engineers, EEP10–2, ASME, New York, NY, 1995, 619–625.
11. Pecht, M. and Dasgupta, A., Physics-of-failure: an approach to reliable product development. *Journal of the Institute of Environmental Sciences*, 38 (5), 1995, 30–34.
12. Cushing, M.J., et al., Comparison of electronics–reliability assessment approaches. *IEEE Transactions on Reliability*, 42 (4), 1993, 542–546.
13. Last, H.R., Dudley, B., and Wood, R., MEMS reliability, process monitoring and quality assurance. *Proceedings of SPIE — 3880*, 1999, 140–147.
14. *Proceedings of the 1999 MEMS Reliability for Critical and Space Applications*, September 21–22 1999, 3880, 1999, 140–147.
15. Cushing, M.J. and Bauernschub, R., Physics-of-failure (pof) approach to addressing device reliability in accelerated testing of MCMS, *Proceedings of the IEEE, Los Alamitos, CA*, 1995, 14–25.
16. Man, K.F., MEMS reliability for space applications by elimination of potential failure modes through testing and analysis. *Proceedings of SPIE — 3880*,1999, 120–129.
17. Collins, S.D., Microsystems engineering, *Proceedings of SPIE — 4334*, 2001, 214–222.
18. Walwadkar, S.S., et al., Effect of die-attach adhesives on the stress evolution in MEMS packaging, *Proceedings of SPIE — 2003*, 847–852.
19. Starman Jr., L., et al., Stress measurement in MEMS devices, *2001 International Conference on Modeling and Simulation of Microsystems — MSM 2001*, Computational Publications, Cambridge, MA, 2001, 398–401.
20. Zhang, X., Chen, K.-S., and Spearing, S.M., Residual stress and fracture of thick dielectric films for power MEMS applications, *Proceedings of the IEEE Micro Electro Mechanical Systems (MEMS)*, 2002, 164–167.
21. Walwadkar, S.S., et al., Tailoring of stress development in MEMS packaging systems, *Materials Research Society Symposium — Proceedings* 741, 2002, 139–144.
22. Walraven, J.A., Failure mechanisms in MEMS, *IEEE International Test Conference (TC)*, Institute of Electrical and Electronics Engineers, Inc., Piscataway, NJ, 2003, 828–833.

23. Maboudian, R. and Howe, R.T., Critical review: adhesion in surface micromechanical structures. *Journal of Vacuum Science & Technology B: Microelectronics Processing and Phenomena*, 15 (1), 1, 1997.

24. Denton, H. and Davison, M., Processing variables for the reduction of stiction on MEMS devices. *Proceedings of SPIE* — 3874, 113–119, 1999.

25. *Proceedings of the 1999 Micromachining and Fabrication Process Technology V*, September 20–22, 1999, 3874, 1999, 113–119.

26. Allameh, S.M., An introduction to mechanical-properties-related issues in MEMS structures. *Journal of Materials Science*, 38 (20), 2003, 4115–4123.

27. Muhlstein, C.L., Brown, S.B., and Ritchie, R.O., High-cycle fatigue and durability of polycrystalline silicon thin films in ambient air. *Sensors and Actuators, A: Physical*, 94 (3), 2001, 177–188.

28. Nelson, W., *Accelerated Testing: Statistical Models, Test Plans and Data Analyses*. New York, NY, 1990.

29. Bonnie, F. and James, C., The natural space environment effects on spacecraft. In *NASA Reference Publication 1350*, 1994.

30. Brown, S.B., Van Arsdell, W., and Muhlstein, C.L., Materials reliability in MEMS devices, *IEEE Proceedings*, 1997, 591–593.

31. Chiou, J.A., Chen, S., and Jiao, J., Humidity-induced voltage shift on MEMS pressure sensors. *Journal of Electronic Packaging, Transactions of the ASME*, 125 (4), 2003, 470–474.

32. de Boer, M.P., et al., *Adhesion, Adhesion Hysteresis and Friction in MEMS under Controlled Humidity Ambients*, American Society of Mechanical Engineers, Materials Division MD84, ASME, Fairfield, NJ, 1998, 127–129.

33. Gunda, N., Jha, S.K., and Sastri, S.A., Anti-stiction coatings for MEMS devices, *Advanced Materials and Processes*, 162 (9), 2004, 27–28.

34. Walraven, J.A., et al., Electrostatic discharge/electrical overstress susceptibility in MEMS: a new failure mode, *Proceedings of SPIE* — *2000*, 30–39.

35. Walraven, J.A., et al., Failure analysis of electrothermal actuators subjected to electrical overstress (EOS) and electrostatic discharge (ESD), *Proceedings of the 30th International Symposium for Testing and Failure Analysis, ISTFA 2004*, 9639, 2004, 225–231.

16 Assurance Practices for Microelectromechanical Systems and Microstructures in Aerospace

M. Ann Garrison Darrin and Dawnielle Farrar

CONTENTS

16.1   INTRODUCTION

The objective of this chapter is to supplement the strong infrastructure in space
mission quality assurance with information for microelectromechanical systems
(MEMS) and microstructure-related space activity. The generic elements of any
good quality assurance plan apply to the use of microtechnologies in critical and
noncritical space flight applications. The quality assurance plan should be carried
out during the formulation phase of the project. Generic categories of the quality
assurance program include but are not limited to:

- Quality planning
- Design and development
- Change control
- Contractor surveillance
- Procurement
- Receiving, processing, fabricating, assembly, test, and inspection control
- Contamination control
- Metrology and calibration
- Handling, packaging, packing, and storage controls
- Quality records
- Quality audits
- Process improvement
- Reliability
- Safety
- Software quality

16.1.1   COMMERCIAL VS. SPACE ENVIRONMENT

The use of MEMS in space does not have the volume benefits of the commercial
world or the knowledge base seen in space-grade integrated circuits (IC). Commercial production of MEMS devices is a high-volume manufacturing activity where
reliability, efficient process, product characterization, and testing are well defined
from the very earliest development phase up. Elimination of process and design-
related failure mechanisms through statistical analysis and understanding of the
physics of failure yields defect-reduction programs. Successful commercial programs nurture high yield and profitable, yet reliable production lines.[1] In addition,
simulation tools, process-monitoring tools, and advanced characterization tools
are tailored to the product developed.[2–5] These tools and process monitors ensure a
reduction in the risk of processing errors, along with an integrated process or
product approach using quality systems and high-volume manufacturing. This is
critical to the production of high-quality products. Unfortunately, a key element
here is volume production, which is not common in most spacecraft applications.
None of the commercial lines in the United States (and perhaps the world) are
developed with the intent to produce space-grade MEMS, and most facilities

(excluding government laboratories such as Sandia National Laboratory) have been developed to produce MEMS solely for commercial and terrestrial applications. This chapter will emphasize the noncommercial high volume environment and assumes that production runs will be an iterative process using prototypes and small wafer runs. Therefore, the focus will be on custom and prototype activity.

16.1.2 TAILORING OF TEST PLANS

As a small volume, custom-type activity, test plans are expected to modify or supplement standard test plans. These tailoring activities should have the following attributes:

- It should be a standard methodology — not necessarily a standard test.
- It should be concurrent with other engineering activities — not a final pass or fail gate.
- It should be easily applicable to a given design — rather than being a standard test.
- It should be easily portable across processes — not requiring reinitialization of all steps taken to date.
- It should be quick and inexpensive — not requiring months of the design process and tens of thousands of dollars.
- It should be based on understanding of reliability — not the lack of it.
- It should be based on all data sources — not just a single qualification test.

An example of reliability testing that uses the above principles is product testing at Analog Devices, Inc. A series of mechanical tests confirm resistance to mechanical shock, stiction, and other MEMS-specific failure modes. These reliability tests can be applied at the technology, component, or system level,[3] but all fundamentally depend on the interactions of MEMS parts at their most basic level. The test conditions used in these reliability tests use MIL-STD-883 ("Test Methods for Microcircuits") as the base. MIL-STD-883 is a widely used and accepted document for prescribing test methodology. These MIL-STD-883 tests include:

- High-temperature operating life (HTOL at condition C)
- Temperature cycle (condition C)
- Thermal shock (condition C)
- High temperature storage (condition C)
- Mechanical stress sequence (group D, subgroup 4).

In addition, analog devices developed stress tests called "random drop" and "mechanical drop." Random drop is the random-orientation batch drop of packaged devices from a height of 1.2 m onto a marble surface. The drop is repeated about 10 times, and a basic functionality check is done between each drop. In the mechanical drop test, devices are dropped one by one from a height of 0.3 m onto a marble surface, first in the X-axis, then the Y-axis, and finally the Z-axis. An electrical screen is performed, and the same procedure repeated from a height of

1.2 m.[6] This work by Analog Devices, Inc. is an excellent example of the need to tailor test plans to achieve a reliable program. An understanding of the failure mechanisms specific to MEMS materials helps in developing and carrying out quality assurance tests for MEMS devices in space. Tests dealing with temperature, stiction, vibration, and shock will not be the same for all MEMS pieces, as their size, material properties, and fragility make their failures in these aspects unique to their experience in space. Chapter 15 discusses MEMS-specific failure modes in greater detail.

16.2   DESIGN PRACTICES FOR THE SPACE ENVIRONMENT

To ensure a reliability-oriented design, researchers should first determine the needed environmental resistance of the MEMS devices and its related subsystems. The initial requirement is to define the operating environment for the equipment. The Life Cycle Environment Profile (LCEP) is a tool used to define these requirements. In application, the use of de-rating and, in some cases, redundancy is also included to assure the reliability of the design.

16.2.1   LIFE CYCLE ENVIRONMENT PROFILE

The LCEP is the starting point in tailoring application-specific tests. This analysis is used in developing environmental design criteria consistent with the expected operating conditions, evaluate possible effects of change in environmental conditions, and provide traceability for the rationale applied in criteria selection for future use on the same program or other programs.

The LCEP is a forecast of events and associated environmental conditions that an item experiences from manufacturing to retirement. The life cycle includes the phases that an item will encounter such as: handling, shipping, or storage before use; disposition between missions (storage, standby, or transfer to and from repair sites); geographical locations of expected deployment; and platform environments. The environment or combination of environments the equipment will encounter at each phase is also determined. All deployment scenarios should be described as a baseline to identify the environments most likely to be associated with each life cycle phase.

To develop a life cycle profile, the expected events should be described for an item of equipment from final factory acceptance through terminal expenditure or removal from inventory. Then identify significant natural and induced environments or combination of environments for each expected shipping, storage, and logistic event (such as transportation, dormant storage, standby, bench handling, and ready modes, etc.). Finally, describe environmental and stress conditions (in narrative and statistical form) to which equipment will be subjected during the life cycle. Data may be derived by calculation, laboratory tests, or operational measurements, and estimated data should be replaced with actual values as determined. The profile should show the number of measurements used to obtain the average value of these stresses and design achievements as well as their variability (expressed as standard deviation). Given the dependence of MEMS reliability on the operating conditions encountered during the life cycle, it is important that such conditions be identified accurately at the beginning of the design process.

16.2.2 DE-RATING AND REDUNDANCY

One method to develop reliable systems is the use of redundancy. Civilian and military project engineers design systems and electronic circuits with redundancy so that if one system fails, the second or even third system will operate in its place. Use of redundancy in critical electronic systems can cover for unexpected or unpredictable failure mechanisms during the required mission lifetime. There are different levels of redundancy that are used on spacecraft. The geostationary operational environmental satellites (GOES) each have two parallel systems to operate their instruments. The Earth Observing System (EOS) can require redundancy down to individual electronic parts.

In determining redundancy requirements, a design engineer considers past experience, the additional costs, the additional weight, the additional space required, the particular project's requirements, and especially the criticality of each function. Failure modes and effects analysis (FMEA) are performed in the design phase of a spacecraft to determine the criticality of a function. Other analyses such as stress analyses, worst-case analyses, and trend analyses assess the reliability and criticality of a system. Statistical analyses determine how many redundant systems will meet the reliability requirements of the project. The space station program specifies requirements for the criticality of particular functions. For Space Station Manned Base (SSMB) functions for crew survival, two redundant systems are required. For SSMB functions for station survival, a single redundant system is required.

Another method used to develop a reliable system is to de-rate parts for their respective applications. Although de-rating programs are not available for MEMS devices, the same principle of operating well within a parts margin is applied. The approach NASA takes to de-rating is to run all electrical, electronic, and electromechanical (EEE) parts well within their respective safe operating areas (SOA). The SOA of a part depends on its design and performance ability. Each part type is derated to the guidelines found in MIL-STD-975 or in accordance with the individual program de-rating requirements (e.g., SSP 30312, *EEE Parts Derating and End of Life Guidelines*).[7] In general, parts de-ratings reduce the factors that limit the SOA of a part to increase reliability and device longevity. These include temperature, voltage, current, cycles, and power consumption. Space flight parts have specified operating areas between −55 and 125°C. By de-rating the operating temperature of a specific component, the failure rate may reduce by a factor of five for active devices. Certain part types will have an extended operating life when de-rated in terms of power consumption. In addition, de-rating minimizes the impact of aging affects such as the drift of electrical parameters. Although the term de-rating applies to microelectronics and not to MEMS, operating within reduced margins is prudent and should be required on all space programs. The

SOA in terms of temperature, voltage, current, cycles, and power consumption definitions apply for each device.

16.3 SCREENING, QUALIFICATION, AND PROCESS CONTROLS

16.3.1 DESIGN THROUGH FABRICATION

The selection of the specific tools for the MEMS designer will be driven by compatibility with the foundry selection. The designer will select the appropriate foundry and follow the tool guidelines of that entity. Designing MEMS devices requires a strong link between design and process engineers. Establishing systematic design principles through a common computer-aided design (CAD) framework facilitates the design. MEMS design for manufacturing (DFM) techniques focus on process and design qualification through systematic parametric modeling and testing, from initial development of specifications to manufacturing. The overall result is a MEMS product design framework that incorporates a top-down design methodology with parametric reusable libraries of MEMS, IC, and other relevant system components. The framework should be capable of allowing one to design within a specific process (via a process design kit) that enables virtual manufacturing.[8] The MEMS designers must be able to design MEMS devices within the process limitations for a working and high yielding chip. Means are required to inform MEMS designer of those limitations. Design rules must also communicate the process limitations to those responsible for developing layout verification and layout design tools. The design rules will ensure the greatest possibility of successful fabrication and a specific foundry. Design rules define the minimum feature sizes and spaces for all levels and minimum overlap and spacing between relevant levels. The minimum line widths and spaces are mandatory rules. Mandatory rules are given to ensure that all layouts will remain compatible with the foundries lithographic process tolerances.

Failure mechanisms in the product may arise in the case of design rule violations. Violation of minimum line and space rules could potentially result in missing, undersized, oversized, or fused features. MEMS design rules must become increasingly more specific to reflect the changes in expertise of the people using the rules.[9] Process control monitors are used to verify control of parameters during the fabrication process. A verification system must be specified and in place to verify the ability to meet required performance in final application. The procedures to accept or reject criteria for the screens should be certified by the qualifying activity (QA). The manufacturer, through the technical review board (TRB), should identify which tests are applicable to guarantee the quality and reliability of the associated MEMS fabrication technique or end product (e.g., wafer or die level product, packaged product, etc.). The manufacturer may elect to eliminate or modify a screen based on supporting data that indicates that for the specific technology, the change is justified. If such a change is implemented, the producer is still responsible for providing a product that meets all the performance, quality, and reliability requirements. Devices that fail any screening test shall be identified, separated, or removed.

16.3.2 ASSEMBLY AND PACKAGING QUALIFICATION/SCREENING REQUIREMENTS

Particular attention must be paid to devices after delivery and release as they are in their most unprotected and vulnerable state. Therefore, an entire chapter (Chapter 13) of this book deals with "Handling and Contamination Control." The handling and storage procedures must be in place before receipt of any microsystem. Use only facilities with a strong background in microelectronic packaging for space flight hardware to perform assembly, and packaging activity. Using known steps and tests from the military specification world is useful.

16.3.2.1 MIL-PRF-38535 Integrated Circuits (Microcircuits) Manufacturing, General Specification

MIL-PRF-38535 specification establishes the general performance requirements for IC or microcircuits and the quality and reliability assurance requirements, which must be met for their acquisition. The intent of this specification is to allow the device manufacturer the flexibility to implement best commercial practices to the maximum extent possible while still providing product that meets military performance needs. Detailed requirements, specific characteristics of microcircuits, and other provisions that are sensitive to the particular use intended will be specified in the device specification. Quality assurance requirements outlined in MIL-PRF-38535 are for all microcircuits built on a manufacturing line, which is controlled through a manufacturer's quality management (QM) program and has been certified and qualified in accordance with requirements herein. Several levels of product assurance including radiation hardness assurance (RHA) are provided for in this specification. MIL-PRF-38535 is often used in connection with MIL-STD-883 microcircuit test methods.

16.3.2.2 MIL-STD-883 Test Method Standard, Microcircuits

This standard establishes uniform methods, controls, and procedures for testing microelectronic devices suitable for use within military and aerospace electronic systems including basic environmental tests. These tests determine resistance to deleterious effects of natural elements and conditions surrounding military and space operations. The standard covers other controls and constraints necessary for a uniform level of quality and reliability suitable to the intended applications of those devices. For this standard, the term "devices" includes such items as monolithic, multichip, film and hybrid microcircuits, microcircuit arrays, and the elements that form circuits and arrays. This standard applies only to microelectronic devices. However, MEMS devices in microcircuit packages may test in accordance with MIL-STD-883. Figure 16.1 provides a suggested test and inspection flow derived from MIL-PRF-38535 and microcircuit test methods MIL-STD-883 test methods for microelectronics.

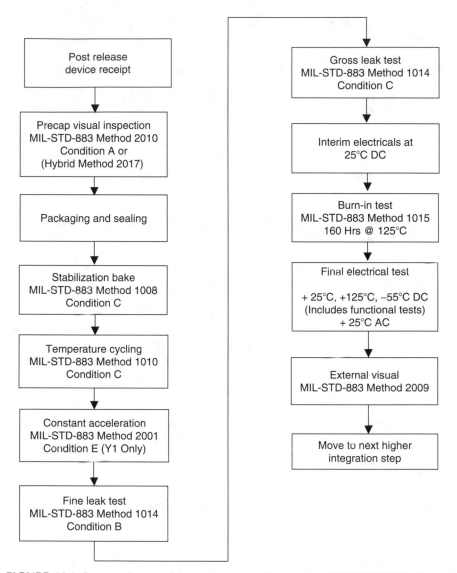

FIGURE 16.1 Suggested test and inspection steps derived from MIL-PRF-38535 General Specification IC (Microcircuits) Manufacturing and MIL-STD-883 Test Methods for Micro Electronics.

A generic screening procedure derived from the military specifications is shown in Table 16.1 and has been adapted from the microcircuit specifications MIL-PRF-38535F. This screening procedure then can be mission tailored to meet the environmental constraints required.

An example of a mission specific tailoring of the test plan is shown in Table 16.2. Tests may be added or deleted depending on the MEMS technology involved.

TABLE 16.1
Screening Procedure for Hermetic MEMS Adapted from MIL-PRF-38535

Screen	MIL-STD-883, Test Method (TM) and Condition
1. Electrostatic discharge (ESD) sensitivity	TM 3015 (initial qualification only)
2. Wafer acceptance	TRB plan
3. Internal visual	TM 2010, test condition A. Internal visual inspection shall be performed to the requirements of TM 2010 of MIL-STD-883, condition A. Devices awaiting preseal inspection, or other accepted, unsealed devices awaiting further processing shall be stored in a dry, inert, controlled environment until sealed.
4. Temperature cycling	TM 1010, test condition C, 50 cycles minimum
5. Constant acceleration	TM 2001, test condition E (minimum) Y1 orientation only. All devices shall be subjected to constant acceleration, except as modified in accordance with 4.2, in the Y1 axis only, in accordance with TM 2001 of MIL-STD-883, condition E (minimum). Devices which are contained in packages that have an inner seal or cavity perimeter of 2 in. or more in total length, or have a package mass of 5 g or more, may be tested by replacing condition E with condition D in TM 2001 of MIL-STD-883. For packages that cannot tolerate the stress level of condition D, the manufacturer must have data to justify a reduction in the stress level. The reduced stress level shall be specified in the manufacturers QM plan. The minimum stress level allowed in this case is condition A.
6. Serialization	In accordance with device specification
7. Interim (pre burn-in) electrical parameters	In accordance with device specification
8. Burn-in test	TM 1015, 160 h at $+125°C$ minimum Burn-in. Burn-in shall be performed on all packaged devices, at or above their maximum rated operating temperature (for devices to be delivered as wafer or die, burn-in of packaged samples from the lot shall be performed to a quantity accept level of $10(0)$). For devices whose maximum operating temperature is stated in terms of ambient temperature (T_A), table I of TM 1015 of MIL-STD-883 applies. For devices whose maximum operating temperature is stated in terms of case temperature (T_C), and where the ambient temperature would cause the junction temperature (T_J) to exceed $+175°C$, the ambient operating temperature may be reduced during burn-in from $+125°C$ to a value that will demonstrate a T_J between $+175°C$ and $+200°C$ and T_C equal to or greater than $+125°C$ without changing the test duration.
9. Interim (post burn-in) electrical parameters	In accordance with device specification
10. Percent Defective Allowable (PDA) calculation	5 percent, all lots

Continued

TABLE 16.1

Screening Procedure for Hermetic MEMS Adapted from MIL-PRF-38535 — Continued

Screen	MIL-STD-883, Test Method (TM) and Condition

11. Final electrical test
   a) Static test at +25°C,
      maximum and minimum
      rated operating temperature
   b) Dynamic or functional
      tests at +25°C, maximum and
      minimum rated operating
      temperature
   c) Switching tests at +25°C,
      maximum and minimum rated
      operating temperature
	In accordance with device specification

12. Seal
   a) Fine
   b) Gross
	TM 1014 Seal (fine and gross leak) testing. Fine and gross leak seal tests shall be performed, as specified between temperature cycling and final electrical testing after all shearing and forming operations on the terminals.

13. External visual	TM 2009

For critical space applications, burn-in times may be extended especially for qualification. Other tests that may be required and are found in MIL-STD-883 include destructive physical analysis (die related), residual gas analysis (package related), and radiation tests.

16.3.3 PACKAGING AND HANDLING

Packaging is sometimes an overlooked detail, but in fact, is one of the most difficult and expensive aspects of MEMS. MEMS devices contain exposed moving parts that can be made nonfunctional or unreliable by the presence of liquid, vapor, gases, particles, or other contaminants. Unlike a standard integrated circuit, it is not possible to clean a MEMS device once it has been released. For this reason, the MEMS wafers must be singulated (cut up into individual die) and assembled before they are released if possible. After the die release, they must be protected from particulates and contamination. Dust from machines or people making contact with active areas or regions can impede movement of a MEMS device, or affect the electrostatic fields that govern its motion.

Package cleanliness acceptable for a standard integrated circuit is a reliability concern for a MEMS device, again because particles and contamination that do not affect operation of an IC interact with the microelectromechanical device. The package environment, including such issues as outgassing of die attach, presence of particles, moisture levels, chemical interactions with antistiction coatings, assembly temperature, and other issues all must be evaluated and addressed in the quality and

TABLE 16.2
MEMS Sample Test Plan

Test Item	Qualification	Acceptance
Bond strength Die shear	Test method 2023 100% NDBP	Test method 2023 100% NDBP
High-temperature storage	150°C	150°C
Low-temperature storage	−55°C	−55°C
Burn-in	100 h total on-time	100 h total on-time
Thermal cycle or vacuum	Maximum or minimum design $\pm10°C$; four cycles 1×10^{-5} Torr	Maximum or minimum design $\pm15°C$; six cycles thermal cycle
Random vibration level duration	Flight (limit) level + 3 dB flight duration/axis; three axes	Flight (limit) level flight duration/axis[1]; three axes
Sinusoidal vibration level duration sweep rate	1.25 × flight (limit) level flight duration/axis; three axes 4 oct/min	Not required
Temperature cycle	−55 to +80°C	−55 to + 60°C
Mechanical shock analysis	1.4 × flight (limit) level	Not required
Structural loads test analysis	1.25 × flight (limit) loads 1.4 × flight (limit) loads	Not required
Thermal shock	Permission requirements	Permission requirements
Acoustics level duration	Flight (limit) level + 3 dB flight duration	Not required
EMI/EMC	Mission dependent (refer to ST5-495-007 for details on type and levels of testing required)	Mission dependent (refer to ST5-495-007 for details on type and levels of testing required)
Conducted emissions conducted susceptibility radiated emissions Radiated susceptibility		
Magnetics	Mission dependent (refer to ST5-495-007 for details on type and levels of testing required)	Mission dependent (refer to ST5-495-007 for details on type and levels of testing required)

reliability of a MEMS device. Because this is so critical, it is important to package the MEMS devices in a controlled, particle-free environment. Every step from die preparation to package seal must be performed in a class 100 cleanroom environment until the device is safely sealed in a hermetic package. Cleanroom techniques normally reserved for wafer fabrication must be extended for use in probing, die prep, and assembly. Thus, the packaging of the MEMS device is as challenging as building the MEMS die itself. Customers who purchase a raw unpackaged die from a

MEMS vendor and package the device themselves are more than likely underestimating the difficulty of the quality and reliability challenges involved.

MEMS reliability focuses on mechanical failure modes rather than electrical ones. One major failure mechanism is stiction, or the tendency of two silicon surfaces to stick to each other. Another concern is the release process and any postprocesses where contaminants and moisture may be present.

16.4 REVIEWS

Engineering design reviews and fabrication feasibility reviews should be held on every program considering the use of MEMS devices. These reviews may be held often and should include peer reviewers. For fabrication feasibility reviews, the team should be interdisciplinary and cover every area that will have impact on the design or build. The first major formal review of the detailed design including MEMS devices will be at the preliminary design review (PDR),[10] which nominally will cover the subsystem or the system, or the MEMS device(s). Areas of particular concern to the MEMS provider and user for the PDR are listed below. Since both the PDR and the critical design review (CDR) may be at a larger subsystems and systems level, additional guidance is given in this chapter specific to the incorporation of MEMS in designs for space programs.

The PDR is the first major review of the detailed design and is normally held prior to the preparation of formal design drawings, yet after the concept feasibility has been demonstrated in hardware. A PDR is held when the design is advanced sufficiently to begin some breadboard testing and/or fabrication of design models. Detail designs are not expected at this time, but system engineering, resource allocations, and design analyses are required to demonstrate compliance with requirements. The identification of single point failure modes needs to be assessed as well as critical design areas that may be life-limiting.

A PDR should cover the following items with the assurance that MEMS specific information be included in the highlighted sections:

- Science and technical objectives, requirements, general specifications
- Closure of actions from previous review or changes since the last review
- Performance requirements
- Error budget determination
- Weight, power, data rate, commands, EMI/EMC
- Interface requirements
- Mechanical or structural design, analyses, and life tests
- Electrical, thermal, optical, or radiometric design and analyses
- Software requirements and design
- Ground support equipment design
- System performance budgets
- Design verification, test flow and calibration or test plans
- Mission and ground system operations
- Launch vehicle interfaces and drivers

- Parts selection, qualification, and failure mode and effects analysis (FMEA) plans
- Contamination requirements and control plan
- Quality control, reliability, and redundancy
- Materials and processes
- Acronyms and abbreviations
- Safety hazards identified for flight, range, ground hardware, and operations
- Orbital debris assessment

The completion of the PDR and the closure of any actions generated by the review become the basis for the start of the detailed drafting and design effort and the purchase of parts, materials, and equipment needed.

The CDR is held near the completion of an engineering model, if applicable, or the end of the breadboard development stage. This should be prior to any design freezing and before any significant fabrication activity begins. The CDR presents a final detailed design using substantially completed drawings, analyses, and breadboard or engineering model evaluation testing to show that the design will meet the final performance and interface specifications and the required design objectives. MEMS selection, de-rating criteria, screening results, calculated reliability, and the results of a FMEA are to be presented. The CDR should include all of the items specified for a PDR, updated to the final present stage of development process, in addition to the following items:

- Evolution and heritage of the final design
- Combined optical, thermal, and mechanical budgets or total system performance
- Closure of actions from the previous review
- Interface control documents
- Final implementation plans including: engineering models, prototypes, flight units, and spares
- Engineering model or breadboard test results and design margins
- Completed design analyses
- Qualification and environmental test plans and test flow
- Launch vehicle interfaces
- Ground operations
- Progress and status and control methods for all safety hazards identified at, but not limited to, the PDR
- Reliability analyses results: FMEA, worst-case analysis, fracture control
- Plans for shipping containers, environmental control, and mode of transportation
- Problem areas and open items
- Schedules

The minimum requirements for submittal and approval by the program would include:

- De-rating system (allowing safe margins within a well-defined SOA [2x's where possible operating margins])
- Material identification and utilization logs (MIUL)
- Stress screening, qualification, and acceptance testing requirements
- FMEA
- Life cycle environment profile (defined more fully in the following section)

In addition, required documentation for submission should include all of the appropriate traceability records.

16.5 ENVIRONMENTAL TEST

At the assembly, subsystem, and system level, temperature cycling and vibration are often used in combination with a vacuum as the predominant test screens. Other screens that are used at these levels of assembly include low- and high-temperature burn-in, power cycling, shock, and electrical screening. Stress screening can dramatically benefit the system at various hardware assembly levels. Part and component screening can remove defects in a system prior to higher assembly level testing. At the subsystem level, screens can remove an additional percent of the remaining defects before system testing. It is important to identify defects at the lowest possible level of assembly in order to have the greatest impact on cost savings and timeliness. Parts screening and qualification test requirements must (1) remove defects at the earliest possible time in a system build cycle and (2) assure the parts will be able to fully perform in the required environment. For a good overview of component-level screening requirements, see the General Environmental Verifications Specification for STS and ELV Payloads, Subsystems, and Components (GEVS-SE document)[11] or MIL-STD-1540 Test Requirements for Vehicles.

16.5.1 Sample Environmental Component Test Requirements

An example of a GEVS-SE based set of test requirements for a specific mission is shown below. The "Component Test Requirements And Guidelines" ST5-495-007 used on the New Millennium Program (NMP) Space Technology 5 Project (ST5) is GEVS-SE based. Some of the examples illustrate a tailoring of the GEVS-SE based on mission specific requirements. The ST5 project is a mission utilizing a three-spacecraft constellation, where integrated on each vehicle is a suite of science validation instruments and new spacecraft technologies. Through flight validation of these technologies, ST5 will reduce the risks for future development of nano-satellites and constellation missions. One of the technologies to be demonstrated is the variable emissivity (Vari E) experiment using a MEMS-based approach. The following defines the verification and test process for all Vari E devices. Designed for a technology demonstration, this MEMS-based experiment is defined as a nonmission critical component. Critical components are components that have a *direct* effect on the spacecraft health and safety. Nonflight critical components are components that do not have a direct effect on the spacecraft health and safety. Both critical and nonflight critical components examples are shown herein. In addition, this sample test plan recognizes two levels of testing: protoflight and acceptance. Protoflight testing is performed on the protoflight unit or first flight unit if no protoflight unit is identified. Acceptance testing is performed on all follow-on flight units.

Performance verification (by test, analysis, or a combination of the two) for all components is performed at both ambient conditions and applicable test levels to determine the limits at which flight component can operate as intended. The test levels and exposure durations in this document were selected in view of the fact that ST5 is a short-term (3 months) demonstration mission.

Environmental tests will include functional and performance testing in as close to a space flight launch through on orbit environment as possible on Earth. Often operational tests will be divided into limited or comprehensive performance testing. As a minimum comprehensive performance testing is in the configuration to be used in application. The testing will require shock, vibroacoustic, structural tests, thermal cycling, and thermal vacuum, EMI and EMC, magnetic and Burn-In tests (operational life). Table 16.3 lists the ambient test conditions required to support the environmental testing.

16.5.1.1   Test Tolerances

Unless otherwise specified, the maximum allowable tolerances on component test conditions and measurements shall be as specified as in Table 16.4.

16.5.1.2   Test Documentation

Formal, controlled documentation is required for all critical component testing and recommended for nonflight critical component testing. Documentation of critical component testing shall be available for review by the project upon request. At the least, the critical component documentation shall include:

- Test procedure
- Test configuration
- Test records and logs
- Test data and reports

TABLE 16.3
Ambient Test Conditions

Condition	Environment
Temperature	$24 \pm 8°C$
Humidity (operating)	40 to 70%
Humidity (nonoperating)	20 to 80%
Pressure	12 to 15 psi

TABLE 16.4
Sample Test Tolerances

Condition	Tolerance
Temperature	$\pm$ 3 °C
Pressure	
Above 1 Torr	$\pm$ 10%
0.01 Torr to 1 Torr	$\pm$ 25%
Below 0.001 Torr	$\pm$ 80%
Relative Humidity	$\pm$ 5%
Electromagnetic Compatibility	
Voltage magnitude	$\pm$ 5% of the peak value
Current magnitude	$\pm$ 5% of the peak value
RF amplitudes	$\pm$ 2 dB
Frequency	$\pm$ 2 %
Distance	$\pm$ 5% of specified distance or $\pm$ 2.5 cm, whichever is greater
Magnetic Properties	
Mapping distance measurement	$\pm$ 1 cm
Displacement of assembly center of gravity (cg) from rotation axis	$\pm$ 5 cm
Vertical displacement of single probe centerline from cg of assembly	$\pm$ 5 cm
Mapping turntable angular displacement	$\pm$ 3 degrees
Magnetic field strength	$\pm$ 0.1 nT
Repeatability of magnetic measurements (short term)	$\pm$ 5% or $\pm$ 2 nT, whichever is greater
Demagnetizing and Magnetizing Field Level:	$\pm$ 5% of nominal
Mass Properties	
Weight	$\pm$ 0.2%
Center of gravity	$\pm$ 0.15 cm ($\pm$ 0.06 in.)
Moments of inertia:	$\pm$ 1.5%
Acoustic Vibration–Sound Pressure Levels	
1/3-Octave band	$\pm$ 3.0 dB
Overall SPL	$\pm$ 1.5 dB
Sinusoidal Vibration	
Amplitude	$\pm$ 10%
Frequency (Hz)	$\pm$ 2%
Random Vibration Acceleration	
RMS level power spectral density (G^2/Hz)	10%
20 to 500 Hz (25 Hz or narrower)	$\pm$ 1.5 dB
500 to 2000 Hz (50 Hz or narrower)	$\pm$ 3.0 dB
Random overall GRMS:	$\pm$ 1.5 dB
Shock Response (Q $=$ 10)	
1/6 Octave band center frequency amplitude (G's)	$\pm$ 6 dB with 30% of the response spectrum center frequency amplitude greater than nominal test specifications
0-2000 Hz $>$2000 Hz	$+$ 10 dB/-6dB
Load	
Static and steady-state (acceleration)	$\pm$ 5%

16.5.1.3 Test Methodology

Table 16.5 identifies the critical component protoflight and acceptance test levels.

1. Flight duration or minimum of 1 min.
2. Test not required for protoflight if addressed by random vibration levels.
3. The sweep direction should be evaluated and chosen to minimize the risk of damage to the hardware. If a sine sweep is used to satisfy the loads or other requirements, rather than to simulate an oscillatory mission environment, a faster sweep rate may be considered, for example, 6 to 8 oct/min to reduce the potential for over-stress.
4. Acoustic sensitive components only.
5. Component thermal acceptance tests shall be performed in vacuum if the component contains high junction temperature devices or high voltages (>100 V) or vacuum-sensitive parts such as moving mechanical assemblies, hermetic or O-ring sealed parts whose operation depends on the seal integrity, or is sensitive to outgassing.

TABLE 16.5
Component Test Levels

	Acceptance	Protoflight
Random vibration level duration	Flight (limit) level flight duration/axis three axes	Flight (limit) level + 3 dB flight duration/axis; three axes
Sinusoidal vibration level duration sweep rate	Not required	1.25 × flight (limit) level flight duration/axis; three axes 4 oct/min
Acoustics level duration	Not required	Flight (limit) level + 3 dB flight duration
Structural loads test analysis	Not required	1.25 × flight (limit) loads 1.4 × flight (limit) loads
Mechanical shock analysis	Not required	1.4 × flight (limit) level
Thermal cycle or vacuum	Maximum or minimum design ±15°C; six cycles thermal cycle	Maximum or minimum design ±10°C; four cycles 1 × 10^{-5} Torr
Pressure pressure vessel pressure profile	Proof test (1.5 × MEOP) leakage test (meop) not required	Proof test (1.5 × MEOP) Leakage Test (MEOP) 1.12 × predicted pressure rate
Burn-in	100 h total on-time	100 h total on-time
EMI/EMC conducted emissions conducted susceptibility radiated emissions radiated susceptibility	As specified in test plan	As specified in test plan
Magnetics	As specified in test plan	As specified in test plan

16.5.1.4   Protoflight Testing

All newly designed critical components undergo protoflight level environmental stresses and functional tests prior to integration with the spacecraft. In general, the critical component providers perform these tests as part of their process of delivering flight articles. The protoflight test as seen in Table 16.6 sequence may include the following:

These tests may be tailored for a particular component. The exact test sequence run on each component shall be documented.

16.5.1.5   Acceptance Testing

Acceptance tests as defined in this document shall be performed on all previously qualified critical components. A sample acceptance test is shown in Table 16.7.

Acceptance tests may be tailored for a particular component. The exact test sequence run on each component shall be documented. Performance testing may also need to be modified to assure that the MEMS components receive thorough testing as they will be applied. For ST5, the use of both comprehensive and limited performance testing was used to assure adequacy of the test regime without requiring excessive testing.

TABLE 16.6
Protoflight Testing

Comprehensive performance test (CPT)
EMI/EMC
Magnetics
Random vibration
Limited performance test (LPT)
Sinusoidal vibration (if required)
Limited performance test (LPT)
Acoustics (if required)
Limited performance test (LPT)
Structural loads
Limited performance test (LPT)
Thermal cycle or vacuum (CPT or LPT testing performed
   at extremes, as approved by the project)
Proof and leak (if required)
Limited performance test (LPT)
Burn-in
Comprehensive performance test (CPT)
Mass properties
Alignments (if required)
Deployments (if required)

TABLE 16.7
Acceptance Testing

EMI/EMC
Comprehensive performance test (CPT)
Random vibration
Limited performance test (LPT)
Thermal cycle or vacuum (CPT or LPT testing performed
  at extremes, as approved by the project)
Burn-in
Comprehensive performance test (CPT)
Mass properties
Alignments (if required)

16.5.1.6   Comprehensive Performance Testing

A comprehensive performance test (CPT) shall be conducted on each component. When environmental testing is performed, additional CPT shall be conducted during the hot and cold extremes of the temperature or thermal-vacuum test for both maximum and minimum input voltage, and at the conclusion of the environmental test sequence, as well as at other times as prescribed. The CPT shall be a detailed demonstration that the hardware and software meet their performance requirements within allowable tolerances. The test shall demonstrate operation of all redundant circuitry (if applicable) and satisfactory performance in all operational modes within practical limits of cost, schedule, and environmental simulation capabilities. The initial CPT shall serve as a baseline against which the results of all later CPTs can be readily compared. The test shall demonstrate that, when provided with appropriate inputs, internal performance is satisfactory and outputs are within acceptable limits.

16.5.1.7   Limited Performance Testing

Limited performance tests (LPT) shall be performed before, during, and after environmental tests, as appropriate, in order to demonstrate that functional capability has not been degraded by the tests. The limited tests are also used in cases where comprehensive performance testing is not warranted or not practicable. LPTs shall demonstrate that the performance of selected hardware and software functions is within acceptable limits. Specific times when LPTs will be performed shall be as prescribed.

16.6   FINAL INTEGRATION

Following successful environmental test, the MEMS incorporated in systems, subsystems, or instruments will move into integration and test (I&T) with the host vehicle. Mission integration and test practices and procedures will be the over arching guidance for all activity in the I&T phase. The I&T phase may provide potentially detrimental handling, storage, and test conditions. Caution should be exercised to assure protection from moisture and contaminants of MEMS devices. The use of red tag items (covers and protective devices) that will be removed prior to flight is encouraged. The provider must be prepared for excessively long storage periods that have been caused by drawn-out flight schedule delays. Although I&T is performed in a controlled environment, conditions during test may change dramatically and storage conditions will be less controlled. Storage due to standdown time of launch vehicles has traditionally varied from 4 months to close to 3 years.[12]

The qualification test of a spacecraft is a lengthy and demanding process. Besides proving the design, the entirety is demonstrated for the first time. The qualification test sequence normally matches the expected flight sequence: vibration, shock, and thermal vacuum. We may also configure the spacecraft to match the operational sequence by folding the solar array and deployables during the vibration and test and deploying or removing them during thermal vacuum.[13]

Any susceptibility of the MEMS during these tests should be identified and planned for early. The design may need to have aliveness test points rather than test MEMS out of a vacuum. The whole life cycle must be planned early to prevent problems encountered late in the build.

16.7 CONCLUSION

The lack of historical data and well-defined test methodologies for the emerging MEMS in space presents a problem for the flight assurance manager, quality engineer, and program manager among others. The well-defined military and aerospace microcircuit world forms the basis for assurance requirements for microelectromechanical devices. This microcircuit base, with its well-defined specifications and standards, is supplemented with MEMS-specific testing along with the end item application testing as close to a relevant environment as possible. This chapter provides a guideline for the user rather than a prescription; that is, each individual application will need tailored assurance requirements to meet the needs associated with each unique situation.

REFERENCES

1. Hartzell, A. and Woodilla, D., MEMS reliability, characterization and test, Presented at Reliability, Testing, and Characterization of MEMS/MOEMS, San Francisco, CA, October 22–24, 2001.
2. Pfeifer, T. et al., Quality control and process observation for the micro assembly process, *Measurement: Journal of the International Measurement Confederation*, 30(1), 1, 2001.
3. Ermolov, V. et al., MEMS for mobile communications, *Circuits Assembly*, 13(7), 46, 2002.

4. Neul, R., Modeling and simulation for MEMS design, industrial requirements, Presented at International Conference on Modeling and Simulation of Microsystems — MSM 2002, Cambridge, MA, April 21–25 2002, 2002.
5. Kolpekwar, A. and Blanton, R.D., Development of a MEMS testing methodology, Presented at IEEE International Test Conference, Washington, D.C., November 3–5 1997.
6. Delak, K.M. et al., Analysis of manufacturing scale MEMS reliability testing, in *Proceedings of SPIE MEMS Reliability for Critical and Space Applications*, 3880, 165, September 21–22, 1999.
7. Grumman Corporation, Electrical, electronic, and electromechanical parts de-rating and end of life guidelines, Appendix J. SSP 30312, Grumman Space Station Support, August 1990.
8. Schropfer, G. et al., Designing manufacturable MEMS in CMOS compatible processes — Methodology and case studies, Presented at MEM, MOEMS, and Micromachining, Strasbourg, France, April 29–30, 2004.
9. Juneidi, Z. et al., MEMS synthesis and optimization, Presented at Design, Test, Integration, and Packaging of MEMS/MOEMS, April 25–27, 2001.
10. NASA Goddard Space Flight Center, Materials Code 301.
11. NASA Goddard Space Flight Center, GEVS-SE General environmental verification specification for STS &: ELV payloads, subsystems, and components, January 1990.
12. Loftus, J.P., and Teixeria, C., Launch Systems, in *Space Mission Analysis and Design*, Larson, W.J., and Wertz, J., Eds, Microcosm, Inc. and Kluwer Academic, Torrance, CA, 1992.
13. Reeves, E., Spacecraft manufacture and test, in *Space Mission Analysis and Design*, Larson, W.J., and Wertz, J., Eds, Microcosm, Inc. and Kluwer Academic, Torrance, CA, 1992.

Index

382   Index